BARBARIANS AND BROTHERS

BARBARIANS AND BROTHERS

Anglo-American Warfare, 1500–1865

WAYNE E. LEE

OXFORD
UNIVERSITY PRESS

OXFORD
UNIVERSITY PRESS

Oxford University Press is a department of the University of Oxford.
It furthers the University's objective of excellence in research, scholarship,
and education by publishing worldwide.

Oxford   New York
Auckland   Cape Town   Dar es Salaam   Hong Kong   Karachi
Kuala Lumpur   Madrid   Melbourne   Mexico City   Nairobi
New Delhi   Shanghai   Taipei   Toronto

With offices in
Argentina   Austria   Brazil   Chile   Czech Republic   France   Greece
Guatemala   Hungary   Italy   Japan   Poland   Portugal   Singapore
South Korea   Switzerland   Thailand   Turkey   Ukraine   Vietnam

Oxford is a registered trade mark of Oxford University Press
in the UK and certain other countries.

Published in the United States of America by
Oxford University Press
198 Madison Avenue, New York, NY 10016

© Oxford University Press 2011

First issued as an Oxford University Press paperback, 2014.

Library of Congress Cataloging-in-Publication Data
Lee, Wayne E., 1965–
Barbarians and brothers: Anglo-American warfare, 1500–1865 / Wayne E. Lee.
p.   cm.
Includes bibliographical references and index.
ISBN 978-0-19-973791-8 (hardcover); 978-0-19-937645-2 (paperback)
1. Great Britain—History, Military—1485–1603. 2. Great Britain—History, Military—1603–1714.
3. United States—History, Military—To 1900. 4. War and society—Great Britain—History.
5. War and society—United States—History. 6. War (Philosophy) 7. Escalation (Military science)
8. Limited war. 9. Violence. I. Title.
DA66.L44   2011
355.020941´0903—dc22
2010029156

1 3 5 7 9 8 6 4 2

Printed in the United States of America
on acid-free paper

Contents

Part I

BARBARIANS AND SUBJECTS: THE PERFECT STORM OF WARTIME
VIOLENCE IN SIXTEENTH-CENTURY IRELAND

Part II

COMPETING FOR THE PEOPLE: CODES, MILITARY CULTURE, AND
CLUBMEN IN THE ENGLISH CIVIL WAR

Part III

Part IV

Acknowledgments

A BOOK STARTS OUT as a solitary effort. By the end, however, almost without noticing, the pile of debts mounts to astonishing heights. In trying to acknowledge those debts I am certain of two things. One is that those mentioned here deserve more than mere mention. The other is that some deserving individuals will go unmentioned through my own failure of memory. I hope both groups will forgive me. I can also confidently state that any errors that have crept in here are my fault, and none of theirs.

Quite a number of people have provided close readings of chapters, or portions of chapters, but none has been as assiduous and as helpful as my wife, Rhonda, whose gentle touch with my so-called style has made all the difference. Only the readers will be more grateful than I am. David Edwards, Rory Rapple, and Vincent Cary have aided me in coming to grips with the complex history of Ireland. Adam McKeown pointed me to some unexpected sources for English soldiers in Ireland, and then provided a marvelous reading of much of the manuscript. Charles Carlton, Cynthia Herrup, Barbara Donagan, Philippe Rosenberg, Clifford Rogers, Stephen Morillo, Everett Wheeler, and Jake Selwood all read much or all of Part II of this book, and saved me from any number of errors. It is only fair to say that Barbara Donagan's work is more than usually foundational to that part of this book, and I am grateful to be able to stand on her scholarship. The Triangle Early American History Seminar, especially the participants John Sweet, Kathleen Duval, Peter Wood, Lil Fenn, Holly Brewer, Trevor Burnard, and Cynthia Radding, provided a wonderful forum for discussing and improving Part 3. It has been a rare privilege to work with so many others experts in Native American studies. Earlier versions of Part III were commented on by Ian Steele, Fred Anderson, Karen Kupperman, and Stephen Carney. Furthermore, my work in Native American history has been greatly aided by my "side career" as an archaeologist. I have worked in the field for many years now with anthropologists who specialize in tribal societies, and I have learned much from Michael Galaty and Bill Parkinson, and more recently from consulting with Kevin McBride and Jason Mancini at the Mashantucket Pequot Research Center. Much of the argument in Part 3 was originally published in the *Journal of Military History* in 2007, and I am grateful for the journal's permission to reuse that material and for the exposure my argument received in that forum. Many of my old colleagues in the American Revolution field aided me in Part 4, especially Rick Herrera, Caroline Cox, Holly Mayer, Daniel Krebs, and Ira Gruber. Sections of Part 4 were published in the winter of 2006 in *Army History*, and that too has proven a useful sounding board. Mark Grimsley and Joe Glatthaar have been my unfailing tutors on the American Civil War.

Research like this depends on libraries. Even as ever more material appears online, the work of librarians and archivists remains indispensable. The staffs at Trinity College Dublin, the National Library of Ireland, the British Library, the Folger Shakespeare Library in Washington, DC, the David Library of the American Revolution, the New York Public Library, and the Newberry Library in Chicago, have all done yeoman's work in supporting my research. My home library at the University of North Carolina actually moved a section of the library around to make my life easier, while Margaret Brill at Duke University's Perkins Library helped me even while the staff was in the process of moving around their whole library.

My colleagues at the University of North Carolina and at Duke have been model collaborators and participants in the larger project. I'm particularly grateful for UNC's Institute for Arts and Humanities, which provided a semester's leave from teaching and ran a marvelous interdisciplinary seminar on writing that helped me focus my ideas. Karen Hagemann and Brett Whalen were particularly helpful in that process. Alex Roland and Richard Kohn have continued to push me intellectually, and Alex has repeatedly forced me to ask the hard questions about this project. The Medieval and Early Modern Studies Program at UNC provided funding for travel to England and Ireland, and my department granted me a semester's leave to do that research. Without that help I would still be writing. Erin Dinning-Hartlage, David Perry (whom I owe bourbon), Heather Rypkema, Tracy K'Meyer, Dirk Bönker, and the rest of the Durham and Louisville Thursday-night club have all been fantastic sources of encouragement, wisdom, and support.

Thank you all.

Notes on Style

Quotes from original sources have been printed here as found, with a very few exceptions explained in the notes. I have silently converted vv to w, u to v, and i to j to reduce confusion, so "iustice" becomes "justice"; similarly y = the, and superscripts have been brought down. When working from an edited version of a manuscript, I have reproduced the editor's spelling without change.

Dates from before 1752, when the English calendar was modernized, are old style, but the year is taken to begin in January and not March.

For those unfamiliar with English titles, I should explain the difference between a title and the family name. For example, the earl of Ormonde is called "Ormonde" in the documents. His birth name was Tom Butler. His brothers, Piers and Edmund, also Butlers, are the junior members of his generation, and so are not referred to as "Ormonde," while he is only rarely referred to as a "Butler." Three different earls of Essex appear in this book, two of them in the same chapter, so for clarity I will occasionally fall back on using their birth names (Walter and Robert Devereux).

BARBARIANS AND BROTHERS

Introduction

IN 1602 AN ENGLISH ARMY stormed the castle of Dunboy in southwest Ireland, rounded up about eighty Gaelic Irish survivors, and hanged them in the courtyard the next day.

In 1644 another English army, this time fighting a civil war at home, bombarded an enemy garrison in the castle of Sudeley in western England. After an initial offer to surrender was discussed and rejected, the alternate discourse of the cannon resumed. The garrison soon capitulated to a promise of mercy, and those inside were spared—probably to be recruited by the besiegers.

In 1675 a group of besieged Susquehannock Indians in Maryland were unable even to begin a discussion with their besiegers. When they sent out five representatives to parley with the attacking English army, the English simply killed them.

In 1780 an American army sat besieged by the British in Charleston, South Carolina. Seeing no hope of relief, they entered into days of discussions over the terms of their surrender. As at Sudeley, the discussion alternated between written demands and the resumption of combat. Eventually the two sides produced a detailed surrender agreement, specifying the treatment of various categories of defenders. The American army marched out and laid down their arms; the regular troops of which then went in one direction, and the militia in another.[1]

Behind these bare outlines of restraint and atrocity lie complex processes and beliefs about war making. A simple explanation for the killing in one circumstance and the restraint in another might focus on the fact that the Irish and the Indians were considered "barbarians," while the other examples come from wars between "brothers." But does this explanation suffice? And is that all that can be learned from these comparisons? Did the participants see such killings as "atrocities"? Were these exchanges a normal part of war, or were they exceptional?

Barbarians and Brothers aims to answer these questions and more. It is a book about restraint and atrocity, about the many ways societies seek to limit war's destructive power, and about the choices and systems that unleash it. But it is also a book about the nature of war. Its fundamental argument is that war is defined by *both* violence *and* restraint, consciously and unconsciously, materially and mentally. Patterns of force and individual choices in war reflect more than the abandonment of the shackles of peace in favor of violence unconstrained by any law save necessity. War is violence, but it is violence perpetrated by humans with the intent to communicate with other humans. War is intended to convey specific messages to an enemy; only rarely in history has that message been merely "die." As an act of communication, it has its own structures, patterns, and internally consistent logic, a "grammar," in which violent acts carry meaning and convey intention. It is true, as the nineteenth-century theorist Carl von Clausewitz suggested, that the intensity and violence of war escalates as each side seeks victory, but its practitioners also struggle to keep their actions within bounds that fit their understanding of war. For Clausewitz the main limit on war's escalation was political—the assessment of ends versus means, gains versus likely costs.[2] Clausewitz was too subtle a thinker to ignore other kinds of restraints, but they were not his focus. Truly understanding war, however, demands an examination of restraint. This is not an attempt to downplay atrocity, or to redeem war from the ignominy it usually deserves. Rather this book seeks to understand both restraint *and* atrocity through a holistic examination of societies' thoughts and assumptions about war—how they plan to win specific wars, how they commit resources to fight a war, and how they seek to enforce social norms on those chosen to fight.

To illustrate these processes, *Barbarians and Brothers* examines five of the most significant sets of conflicts in the founding of the English colonies and the American republic. The nature of English warfare in Ireland in the sixteenth century helped lay the groundwork for English assumptions about North America and for some military methods employed in the colonies. Similarly, the English experience of civil war in the mid-seventeenth century imprinted profound messages in the minds of American colonists about the evils of standing armies and the need to restrain soldiers' behavior. Despite that lesson, the many Anglo-American wars with Indians often were waged with terrifying violence. The Americans then struggled during the American Revolution to reconcile these two different trends of restraint and devastating violence and produced three distinct ways of war: one against the British, one against Indians, and a third, middle way for the partisan war between rebels and Loyalists. After the revolution, Americans convinced themselves of the virtuousness of their conduct compared to the rapacity of their enemies, and they entered the American Civil War expecting to wage another virtuously restrained war. Instead, the intensity of popular emotion, combined with the capabilities of the industrial era, convinced Union generals that this war required a return to strategies of devastation previously reserved for Indians and Irishmen—although with much greater control over the level of violence.

No single explanation of the nature of violence exists for these five cases, but there are connections, trends, and parallels among them. Of particular importance was that in all these conflicts Englishmen and Americans found themselves fighting either people they defined as "barbarians" or their own compatriots—their brothers. Both circumstances placed special burdens on their understanding of war and of appropriate conduct in war.

America in the early twenty-first century confronts a similar problem. After 150 years of conceiving of enemies as states, the nation now fights a "war on terror." In such a war terrorists are defined as barbarians, and many Americans live in fear that they may lurk among us disguised as brothers. The "normal" understanding of war has been upended, and we struggle to define appropriate conduct within the new order. Our fears led us to condone or permit the abuse of prisoners at Abu Ghraib and domestic surveillance, waterboarding and the erosion of constitutional protections. Similarly, in their struggles with barbarians and brothers from 1500 to 1865, Englishmen and Americans learned to demand protections for themselves while simultaneously writing other people out of the nation. These struggles produced good and evil, new beliefs in the necessity of restraint and liberty, and a new willingness to exclude and destroy. They have much to tell us about ourselves.

Restraint and "Frightfulness"

In all the cases discussed in this book, the participants started with assumptions about restraint. Combatants and noncombatants alike expected certain sufferings; they could not deny that blood would be shed, but they also imagined that there would be limits on behavior and on war's destructiveness. Combatants then wrestled with those limits, trying to find ways to fit their notions of restraint into their military and logistical calculations. Even in wars with people they called "barbarians," the hope of ultimately incorporating their opponents into their own society served as an initial restraint on violence. They voiced a kind of logic that suggested, "Better that the corrupt native elite goes down hard that we may save and civilize more in the long run." Not surprisingly, indigenous people labeled barbarians by their enemies generally proved reluctant to become subjects, and their leaders militantly and stridently resisted being supplanted.

When restraints fail, or are deliberately cast aside, violence in war escalates quantitatively or qualitatively, or both. Quantitative escalation brings the commitment of greater resources, the expansion of destruction, and often an enlargement of the conflict in space and time. Qualitative escalation refers to the way each side might adopt practices that they normally find disturbing. Common examples include mutilation, torture, and killing women, children, or the defenseless. This kind of qualitative escalation is culturally specific. Some societies regularly practiced behaviors that are now deemed atrocious, but which for them were normal. For example, if a Native American killed an enemy and then proceeded to scalp him, he did so within his traditional way of war. The act of scalping was not a qualitative escalation of violence; it was "normal." If a European witnessed that act, however, and, outraged, in turn scalped an Indian, that did mark an escalation—at least to fellow Europeans.[3] Similarly, Englishmen in Ireland, although long accustomed to judicial beheading, professed horror at the Irish custom of beheading enemies in battle. The English, in their own qualitative escalation, rapidly adopted the practice.[4]

When violence escalates it becomes more "frightful." That choice of words may sound strange, but it is intended to convey the combination of quantitative and qualitative escalation. War does not become merely more "destructive"; the very nature and quality of that destruction changes. In many cases, "frightful" also conveys the warring society's own view of events, as a war breaks through the boundaries of what a society considers normal behavior.

Understanding how a war becomes frightful requires first understanding the restraints placed on it. For the most part, historians have approached this issue through the relatively narrow lens of the so-called laws of war—the codified traditions that most Americans think of vaguely as the Geneva Conventions. That limited definition usually leads to two types of analyses. One tracks the religious or legalistic development of the codes as an exercise in intellectual history, usually with few references to the actual practice of warfare. The other takes a more "utilitarian" approach and dismisses the codes by suggesting that such shallow legalisms collapsed in the face of military necessity. It is all too easy to assume that military men past and present have dealt only in calculations of advantage. In fact, history shows that they and the societies around them struggled with the meanings and consequences of violence unleashed in war. A third, more recent tradition in the history of violence in war emphasizes ethnic or racial demonization as the root of atrocity. None of these interpretations is entirely wrong, but I believe restraint and frightfulness can better be explained through four categories of analysis: capacity, control, calculation, and culture.

Capacity, Control, Calculation, and Culture

One must begin with a state's *capacity* to mobilize force and that force's overall capacity to destroy, a subject all too often left unexamined. For most of history the ability of an army to destroy was defined primarily by its ability to burn, which was modified only by time and army size. Prior to the industrial production of explosives and their delivery systems, devastation was limited to the application of fire or the person-to-person infliction of violence. The most fundamental limit on frightfulness, therefore, has been the demographic and financial capacity of a society to produce larger armies and fight longer campaigns. Assuming a similar intent to destroy, an army of twelve thousand cannot commit as much violence as an army of a hundred thousand, unless its smaller size allows it to remain in the field much longer.[5] It is in this sense that the political and industrial revolutions of the late eighteenth century dramatically increased the capacity of a state to inflict damage.

The ability to raise an army was not the same thing as the ability to sustain and *control* it. Whatever strategic or moral reasons might exist to limit the frightfulness of an army, if the society lacked the institutional means to control it, including the ability to feed and pay it, then the soldiers in that army inevitably looked out for themselves. For the period covered in this book, state expenditures on war and the military accounted for most of their budgets. One historian has suggested that they consumed an average of 40 percent of the peacetime budget, and 80–90 percent of the wartime budget, of early modern European states. Given that those states were at war more than half of the time between 1500 and 1750, this suggests that they were operating on the very edge of their ability to provide for their armies.[6] Soldiers who were not provided for became a threat to anyone in their path.

Control is also defined as societal oversight that enforces the maintenance of normal social values. In part because early states failed to efficiently supply or pay their men, a separate, libertine military (or "soldiers'") culture developed, whose values diverged from mainstream society. Officers, representing society's elite, were expected to maintain control over the soldiers. All too often, however, the officers shared in the divergent military culture, or simply lacked the institutional tools to control their soldiers. Over time, the increasing weight given to collective synchronized discipline as a means to military success enhanced the desire of the state to control all aspects of their soldiers' behavior. The soldiers' culture of libertinism and plunder gave way to a culture of discipline backed by a bureaucracy of control. Then, in yet another turnabout, the culture of discipline slipped—as did the control of soldiers' violence off the battlefield—in the face of the political revolutions of the eighteenth century, the impassioned commitment of citizens to their state, and the narrowing social gap between officers and soldiers. Discipline nevertheless continued to be seen as a key element of battlefield success.

Societies at different times, and in different technological and political contexts, have had different perceptions of necessity. Desiring to "win," political leaders and military commanders turned to a strategic theory, or to a rough-and-ready *calculation* of how to do so, and shaped their decisions about where to go and what to destroy on that basis. This is not a claim for rationality in the sense of there being only one objectively rational choice. But these men were engaged in a conscious calculation of the material and moral factors within a specific vision of success as they perceived them.

Because it is an act based on perception, calculation is deeply influenced by culture, nevertheless, the focus here is on calculation as a select leadership's conscious balancing of a specific vision of victory against the limits of material reality. That process is very different from the diffuse, countless decisions of small-unit leaders or soldiers whose aggregate pattern of behavior represented a broader military culture. Because leaders typically arrived at their decisions consciously, the sources, often generated by the generals and their critics, usually clearly explain why certain military choices were made. One can analyze those choices for their effects on frightfulness. The effect is not always escalation: there are occasions when calculation might mandate restraint. For example, an army might try to win an opponent's "hearts and minds," or decide not to kill those farmers whose produce fed it.

Finally, there is the subtle but pervasive effect of *cultural values* and beliefs about war, which includes, but is not limited to, the particular set of values usually defined as the laws of war. Here "culture" refers to the patterns of meanings and beliefs expressed in symbols and actions, by which people communicate, perpetuate, and develop their knowledge about and attitudes toward life.[7] Most think of culture as pertaining to an entire society and the "patterns" involved being those which regulate and structure almost all aspects of life. However, there are subcultures within a society, especially in organizations that have an extended life and that acculturate new members. Historian Isabel Hull has suggested that to examine this kind of "organizational culture" requires seeking out the organization's basic assumptions, some of which may even "remain hidden from the actors and often contradict their stated beliefs." To discover this "constellation of basic assumptions . . . one must begin by examining [not only] the patterns in their practices . . . but also the group's language . . . , myths, explanations of events, standard operating procedures, and doctrines."[8] Thus a culture, whether at the societal or organizational level, holds beliefs, knowledge, and assumptions, not always explicitly stated, which are nevertheless transmitted from one generation to the next. Individuals continue to make choices, and by doing so they contribute to a body of precedent for future members. And all individuals retain the ability to improvise, to deviate, even if slightly, from past behaviors, especially when faced with changes in the material conditions of their lives. In this way culture evolves, as the participants in it continuously revise their patterns of belief and behavior.[9]

Culture is normally thought of as a restraint on violence, but that is not always the case. Military subcultures developed their own patterns of behavior that could amplify violence rather than restrain it. The common cultural insistence on retaliation, either out of passion or justice, could have similar effects. A desire for retaliation might lead to violence that takes no account of calculations of victory.

Furthermore, analyzing cultural values of war encompasses the levels or types of violence authorized by a society. Societies generally authorize killing armed enemies in wartime. Some scholars have argued that such authorization is designed to overcome any natural or socialized resistance to killing.[10] Accepting that most humans experience some degree of reluctance to kill, I argue that they do so while also feeling a sense of release or license provided either by intense personal fear or social certainty. By fear I mean the visceral, the instinctive act of self-defense, or even the soldier who fears punishment for not obeying an order to do something he otherwise might not. In a sense fear is merely another kind of certainty: kill or be killed. Social certainty is broader; it is the belief that one's killing accords with accepted community goals. License has been given and the conscience can remain clear.

Being "at war" provides the baseline social certainty (or authorization) to kill, but other questions arise. Once freed to kill by being at war, *how* is one expected to kill? And what limits exist on who and when one kills? Getting into soldiers' heads, understanding what they were afraid of and what values licensed them to kill and destroy, reveals how the landscape of violence within war transcended commanders' choices or calculations of necessity. Junior leaders and individual soldiers had (and have) great power to determine the frightfulness of war. The conditions that created those situations can be blamed on those at higher levels of command, but the individual soldiers' cultural predilections, combined with their experiences in combat and on campaign, shaped their individual choices and determined what happened to their victims. To understand those choices in their countless variations, *Barbarians and Brothers* explores the social background, composition, and experience of armies, and then asks deeper questions about what those men believed and how their experiences and beliefs shaped their actions.

Barbarians and Brothers

Capacity, control, calculation, and culture intersect in different ways, and their particular intersection determines the level of frightfulness within a conflict. This is not a story of decline or of progress over 365 years, but rather an examination of the conjunctions of the four factors within shifting contexts. Each individual conflict in history has been the result of a different conjunction, but it is also possible to identify patterns and developments over time.

The wars between brothers and against barbarians represent unique and particularly challenging conjunctions of all of these factors. One of the ironies of English and American history is that both societies regularly confronted the problem of trying to redefine a "normal" vision of warfare between European states into one appropriate for either wars with barbarians or among themselves. Deepening the irony, Englishmen and Americans imagined their society as relatively open and inclusive. English liberties available to all subjects of the crown became American freedoms guaranteed to all citizens in the state. In fact, in their wars with barbarians and brothers, Englishmen and Americans repeatedly struggled with the problem of defining who could be imagined as a subject or a citizen. Their answers to that question proved crucial to determining wartime violence.

For "barbarian" wars, part of the problem was making one's grammar of violence comprehensible to peoples who did not share the same language and logic of war. In wars with brothers, the options seemed confined to a binary opposition: showing terror or extending mercy to traitors who might yet be reincorporated into the nation. Both situations differed markedly from fighting a foreign enemy from a roughly similar cultural milieu. For most of this period, a European enemy was unlikely to entirely destroy and occupy another European polity (although it did happen on occasion, especially in eastern Europe). Violence against such an enemy could be tuned to necessity or expedience; it could escalate, but was normally bounded. The destruction of armies, capture of territory, or taking of cities generated new relative conditions of power that the competing sides acknowledged, debated, and then adjusted their relationships accordingly, but this violence did not lead to the wholesale destruction of relations. When fighting presumed barbarians, Englishmen and Americans rhetorically claimed that their enemies might one day be included as subjects or citizens, but such inclusion usually demanded that the "barbarian" side entirely remake the nature of their society. Such an all-encompassing goal for war created conditions in which frightfulness could escalate exponentially.

Brothers' wars held similar potential for escalation. As attacks upon normal internal social relations, they opened up an emotive sense of betrayal. But brothers' wars also involved worries about future relationships; violence left unrestrained would make reconciliation difficult. The obvious model for restraining the conduct of war was that used in international war—whatever its imperfections—because the meanings and purposes of violence in that kind of war would be intelligible to both sides. But other, less restrained models were also available. Where international martial traditions respected certain kinds of relationships between combatants, domestic civil law used the exigencies of war and words like "treason" to justify widespread, nominally "judicial" executions and wholesale confiscations of property. In those circumstances the goal of war again became more comprehensive and opened up the potential for escalation. The American Revolution is a good example of how both models existed side by side. The American leadership chose to fight their British "brothers" via the international model of conflict. Dealing with their Loyalist neighbors, however, they frequently applied the label of traitor and enacted policies through civil law that greatly escalated the violence of their war against them.

Englishmen and Americans of this era found themselves confronting these problems of restraint and frightfulness, inclusion and destruction, on a nearly constant basis. *Barbarians and Brothers* examines representative conflicts, keeping capacity, control, calculation, and culture always in mind, but looking at each from a slightly different emphasis and perspective.

Part I covers a century of conflict between the Irish and the English from roughly 1500 to 1603, providing a long look at what happens when "barbarian" and "brother" coexist in the same person, while also closely examining the role of soldiers' fears and officers' notions of chivalry. Part II examines just one war, the "first" English Civil War, from 1642 to 1646, with a special concern for the development of international codes of conduct in war. It also examines the emergence of two restraints on armies: a demand for collective discipline that clashed with a libertine soldiers' culture; and the ability of an aroused political public to limit military capacity, especially when they responded violently to the presence of armies. Part III returns to the long view, but adopts a Native North American perspective on war. It finds that Native Americans had highly developed restraints on war, and then demonstrates how that system failed to synchronize with the fundamentally different visions of restraint and war held by their English and American enemies. Part IV, on the American Revolution, compares two individual operations conducted by the same army and at times even the same regiments, but against entirely different enemies, yielding very different results. Finally, the conclusion briefly examines the American Civil War in the context of the massive shift in capacity created by the industrial and political revolutions of the late eighteenth century. Each part is broken into two chapters centered on a specific narrative, chosen for its exposition of important issues. Frequently the narrative involves a march rather than a battle or a siege. The story of an army on the march includes all the issues most closely related to violence: supply, time, fear, civilians, strategic choices, and eventually battle or siege. It is in the movement of an army and the choices of its commanders that we can see most clearly both restraint and escalation.

Barbarians and Brothers is also intended to reveal more broadly the nature and development of war in the early modern era. It is a reminder that only rarely is war truly unrestrained, fought without let or hindrance. In some ways, that fact is the simple result of capacity. Rarely indeed are all the resources of a society committed to destruction. Furthermore, no single atrocity defines the violence of an entire war, any more than a single negotiated surrender or orderly exchange of prisoners does. There is always to-ing and fro-ing, restraint and horror, discipline and mutiny, honor and cowardice. Studying armies on the march allows us to see both the banal and the grotesque, which in war are often not far apart.

It is simply impossible to try to cover the totality of all of these wars in a book of readable length, so the case studies are designed to limit the variables. Part II, for example, covers only the *English* Civil War, and for the most part only the so-called first one, ending early in 1646. This allows a sharper focus on the war between brothers, excluding the escalatory effect of the concurrent wars with those considered barbarians in Ireland and Scotland. Part IV uses General John Sullivan's 1779 campaign against the Iroquois as an example of white-Indian violence primarily because there were virtually no militia units present on that campaign. This confines the analysis to the supply, control, and culture of the Continental army alone.

In addition to being built around two specific narratives, each part shifts focus back and forth between high politics and the lowly soldier. The politics and the generals cannot be ignored, but the experiences, expectations, and choices of junior officers and soldiers were critical to defining the nature of wartime violence. Their choices were profoundly affected by the strategic vision of their superiors, but they had their own cards to play. They could occasionally force their commanders' hands, and at local and personal levels they had a surprising freedom of action. Narratives are the best way to convey those feelings and choices. Academic history rarely does justice to the blood, sweat, fear, and voided bowels of war's violence. On the other hand, mere stories rarely convey the complexity of the situations in which humans find themselves willing to kill or forced to die. What follows combines story and analysis, respecting the sources and conveying what was, in the end, a human experience.

Part 1

BARBARIANS AND SUBJECTS

The Perfect Storm of Wartime Violence
in Sixteenth-Century Ireland

By the sixteenth century the English had been in Ireland for hundreds of years. Their power and control had waxed and waned, but in the sixteenth century first Henry VIII, and then his children, sought to impose their will on an island that had never fully been conquered. New waves of Englishmen arrived in Ireland riding on the authority of their monarch but also carrying their own ambitions for gain and glory. The Gaelic Irish and the long-established Anglo-Irish inhabitants resisted this attempt to reorder their lives. Resistance made them rebels; cultural prejudice made them barbarians, and they soon found themselves on the receiving end of a frightful combination of greed, prejudice, and legal traditions justifying virtually any level of violence. The vitality of Irish resistance was as important to the story as English legal traditions. The ruggedness of their country and their elusive style of warfare dramatically escalated English commanders' calculation of the violence necessary for ultimate victory. English monarchs long held to a vision of incorporating the Irish as subjects, but frustration and the aggrandizing hopes of English "colonists" generated a spiral of violence that finally culminated in a new and devastating conquest from 1594 to 1603.

1

Sir Henry Sidney and the Mutiny at Clonmel, 1569

But to enterprise the whole extirpation and total destruction of all the Irishmen of the land, it would be a marvelous expensive charge, and great difficulty; considering both the lack of inhabitants, and the great hardness and misery these Irishmen can endure. . . . [And] we have not heard or read in any chronicle, that at such conquests the whole inhabitants of the land have been utterly extirped and banished. Wherefore we think the easiest way and least charge were to take such as have not heinously offended to a reasonable submission and to pros-ecute the principals with all rigor and extremity.

Lord Deputy and Council of Ireland to Henry VIII, 1540

LORD DEPUTY HENRY SIDNEY would probably have preferred for heads to roll. Decapitating his enemies was nothing new to him or to other English-men fighting in Ireland, but in this case beheading was not an option: the enemy was fear inside his own camp. Thirty years later another lord deputy would be less hesitant. In 1599, the earl of Essex revived the ancient Roman practice of decimation and executed one man in ten of a company that had broken and run from the Irish.[1] But here, in 1569, in Sidney's camp at Clon-mel in the province of Munster, Sidney's troops were prepared to run away even before the battle. Lurking enemies and scattered leaflets pronouncing their doom had undermined the soldiers' courage, and that, combined with a few idle days in camp, had left them listless, scared, and ready to go home. He had to do something, so he called for wine—a lot of wine—and now he stood in the camp's market square summoning the soldiers to hear him. Clearly too few in number for the threat they faced, they gathered round, expecting to hear that the campaign was over and they were going home.[2] Within days these same men would storm castles, execute the defenders, toss the dead over the walls, and proceed to deliberately devastate much of the countryside. How did this transformation in Sidney's men take place, and what does it say about the nature of war in sixteenth-century Ireland?

The usual explanations for the long history of English-Irish violence begin with English prejudice, readily found in their sustained rhetorical attacks on the irremediable barbarity of the Catholic Gaelic Irish and the equally degenerate (and Catholic) Anglo-Irish descendants of the twelfth-century English (really Norman) conquerors.[3] Further, the argument goes, since much of this ethnic demonization accompanied attempts to plant Englishmen on Irish soil, the wars and their extremities can be explained as a kind of colonial process of land theft, justified by religious and ethnic prejudice, that served the greed of a few elites (often disinherited second sons) pursuing vast new estates. Irish (and Anglo-Irish) resistance only intensified the level of violence in an escalatory cycle.

One can hardly doubt the role of prejudice and avarice in generating violence and even atrocity, but they are insufficient to explain the full complexity of the military and political culture of the era. A veritable laundry list of additional factors must be considered, starting with how contemporaries defined victory and calculated how to achieve it. But this was more than a purely "military" calculation. English monarchs, their advisors, their Irish deputies, and even the military men (or "captains") in Ireland struggled with the meaning of Irish subjecthood and its claim on royal consideration and protection. Subjects merited protection, but rebellious subjects might well merit severe punishment. On the other hand, pardon and mercy were at times offered rapidly and quickly. Traditional methods of war in Ireland, both English and Irish, relied on devastation, but it was often narrowly targeted at overthrowing an elite lord, without necessarily intending genocidal famine. Within the military culture of the day, sieges ending in assault might be expected to result in massive atrocities, but such killings and plunderings were seen as legitimate, and therefore not atrocities at all. The entire system of military mobilization in Ireland depended to a large extent on the serious oppression, if not outright devastation, of the peasantry. Control over common English soldiers frequently broke down. The troops sometimes lived in conditions of extreme fear and paranoia and could be expected to lash out when they had the chance. Irish methods of war, both their tactics and their customary practices, did not help those fears. Both sides, for example, were invested in a military culture that took the heads of enemies, but they interpreted the process in different ways at different times. In some circumstances it could be seen as a violation and a provocation, in others as a legitimate part of the due process of law.

All of these factors operated to produce what was an undeniably heinous brand of warfare, one that included true atrocities even by the looser standards of the sixteenth century. The sixteenth century saw the beginnings of some of the most severe religious warfare in European history, but it was also an era in which many of the basic codes of military behavior were becoming increasingly formalized and internationally shared. Ireland, nevertheless, fell victim to an unfortunate combination of English rationales and reasons for wartime atrocity. The English imagined it a country peopled by "savage" subjects frequently in rebellion, fiercely resistant to Protestantism, who, from tradition and from calculation, fought in a style that today would be called "guerrilla" warfare. Rebellions in Ireland escalated in number and length over the course of the century, feeding a profoundly ethnocentric English sense of grievance. Some English administrators and officers, aware of the lies behind calling the Irish "savage," offered more restrained strategic alternatives. Some commanders recognized the possibility of winning local hearts and minds and therefore sought to limit the depredations of their soldiers. One high commander in Ireland left an almost achingly painful statement of regret and guilt about the violence inflicted by both sides.[4] But such individuals struggled against a tempest of angry, or greed-inspired, or simply frustrated denunciations of the perfidy and incivility of the "mere" Irish. Finally, any complete explanation must also account for the fact that this was far more than a mere war, even five, ten, or thirty years long. This was a grinding, century-long contest for power and authority.

The English in Ireland: An Overview, 1169–1603

In fact, the English and the Irish had been struggling for a great deal longer than a century.[5] The "English" first appeared in Ireland in 1169 as Normans. They rapidly made themselves the political masters of most of the island, and in time imported some of their own peasants and dominated the population of many of the towns. The countryside, however, remained predominantly Gaelic. Over the ensuing centuries the Norman elite accommodated themselves to Gaelic political and social customs, continuing to rule, but increasingly ruling in ways that resembled their Gaelic neighbors.[6] Furthermore, many of these "Anglo-Irish" elite gradually shed their allegiance to the English crown. By 1500 royal influence in Ireland was confined to a relatively small enclave around Dublin called the Pale, and to a lesser extent the larger Anglo-Irish earldoms of Kildare, Desmond, and Ormond in the south and west. Even in those locations, English royal commands usually ran through the Anglo-Irish earl of Kildare, whose legal authority may have derived from his appointment as the King's deputy, but whose actual power rested on his manipulation of a complex network of kinship and alliances including both Gaels and Anglo-Irish. Despite this rolling back of Norman (later English) authority, the medieval conquest nevertheless left two key legacies. One was the existence of an Anglo-Irish elite, many of whom maintained their identity as Englishmen, who were willing to acknowledge loyalty to the English crown, but who also sought to maintain control over their domains with a minimum of royal interference. Second, a precedent of conquest had been established in English minds. Henry Sidney understood the value of such a precedent, and he made a point of restoring the tomb (at Christchurch in Dublin) of Richard Strongbow, the first Norman conqueror of Ireland. English kings, queens, councilors, writers, and even soldiers regularly pointed to the Norman conquest as having established the legal right of the English to rule the Irish. Irish resistance to being "proper" subjects only proved their barbarity.

Henry VIII, king of England from 1509, began his reign with the traditional assumption that the Irish were subjects to be incorporated into the realm, to their benefit and the crown's. One advisor suggested that the key problem was not the wildness of the Irish peasantry but instead the poverty they had been subjected to through the exactions of their lords. He reminded Henry that "as the common folk go, so goes the king," and that the king should "rendre accompte [to God] of his folke." Henry seemed to take the advice to heart, ordering his lieutenant in Ireland in 1520 to treat the captain and heads "aswell of the Inglishery, as Irishery, to come in to you, as our obeisaunt subgiettes." Crucially, this ideological position also lumped all those who resisted into the category of rebels.[7] Henry's take on the Irish could be seen as enlightened (all king's subjects were to be treated the same), while potentially paving the way to the more extreme forms of violence considered legitimate for rebels.

Henry flirted with more direct rule, bypassing the earl of Kildare and appointing deputies directly from England. That proved expensive and awkward, and Henry soon restored the deputyship to Kildare, but he also tried to control him more firmly. Henry's tighter leash produced a full-fledged rebellion in 1534, led by the earl's son, "Silken" Thomas Fitzgerald. What was in some ways a relatively pedestrian noble challenge to royal authority quickly

SIXTEENTH-CENTURY IRELAND. Adapted from Cyril Falls, *Elizabeth's Irish Wars* (London: Methuen, 1950). Map by Justin Morrill, the-m-factory.com.

got caught up in the issues of the Reformation. Fitzgerald assumed the mantle of leading a Catholic revolt, branded Henry a heretic, and drew heavily on Gaelic as well as Anglo-Irish resources.[8] Sir William Skeffington, acting for Henry in Ireland, attacked Fitzgerald in Maynooth Castle, battered down the walls, and executed the survivors.[9] With the end of the rebellion in Ireland after fourteen months, and the end of nearly contemporaneous ones in the peripheral areas of England, Henry moved to further centralize his control and end his reliance on major nobles.[10]

Nevertheless, transforming the "'sundry sorts' of people" of Ireland into "one class only, the king's subjects, all of whom would be anglicised" in their ways of war, language, inheritance and dress, remained central to Henry VIII's vision for tightening his lordship over the island.[11] In part he was left with little choice—his destruction of the Anglo-Irish Fitzgeralds after the revolt left him without a suitable magnate to form the hard core of royal control in Ireland. In one sense Henry needed first to expand his control in Ireland in order to more fully centralize it, and that expansive vision was inclusive of the Gaels. Increasingly violent alternatives were considered, but at the suggestion of "extirpation" in 1540 Henry's advisors demurred. They feared the probable "mervailous sumptious" (excessive) expense, but they also noted the lack of any legal precedent for such measures. They had neither heard nor read of any precedent "that at suche conquests the hole inhabitauntes of the lande have bene utterly extirped and banisshed." They recommended instead to accept the submission of all but the most egregious offenders.[12] Eventually the policy of accepting submissions evolved into the "surrender and regrant," in which Gaelic chiefs surrendered their lands and Irish titles to Henry VIII, now declared the King of Ireland, who returned them via a feudal grant. Under surrender and regrant, a spate of Gaelic chiefs became English barons, and three of the largest chiefs were ennobled as earls (Thomond, Clanricard, and Tyrone, with more to follow).[13] True subjects in law, Gaelic chiefs holding English titles could sit in the Irish Parliament in Dublin and participate in court life in England.[14]

Here, then, lay a central paradox that set a crucial tone for much of what would follow in the sixteenth century: the Irish were subjects who, when rebels, should be allowed to submit. If they refused to submit, however, they merited no mercy and could and should be "extirped." The severity of this attitude toward rebels increased during the sixteenth century, but the idea had deep roots. European legal tradition going back to Roman law established that only a sovereign prince held the right to wage war. Therefore any war waged against the prince by his subjects constituted treason. Attempts to enforce that belief consistently only became possible as centralized monarchies emerged, but Roman legal tradition clearly allowed kings to treat defeated opponents in civil wars as "bandits rather than legitimate political opponents." As the power of the English monarchy became more centralized in the late middle ages, this assumption became the basis for the English treason law, which initially defined treason simply as levying war against the king. The Tudors both expanded the definition of treason to include a variety of other affronts to the crown and extended it into Ireland. Elizabeth's conflict with the Catholic Church further heightened the state's sensitivity toward treason—publicly claiming that Elizabeth was not the rightful monarch came to constitute high treason.[15]

Closely tied to the expansion of the definition of treason was a parallel enlargement of the state's tools for punishing it, specifically a shift in the meaning and use of "martial law." Originally referring to the hasty trials necessary under wartime conditions, and usually applied to one's own soldiers, by the mid-sixteenth century it was increasingly used as a tool for suppressing disorder more generally.[16] Its use was particularly marked in the case of large-scale rebellion, although usually the state quickly shifted back to the ordinary course of law. If the king was "at war," often defined by having his standard displayed, martial law required no further legal proclamation. Offenders taken in the course of military operations could be swiftly executed. When not "at war," as, for example, in the comparatively placid aftermath of a rebellion, a martial-law commission was required, and after about 1550 those commissions became more and more common, both in England and in Ireland.[17] The problem, in terms of violence, was that martial law was still law. It rendered extreme methods legal, and desperate or greedy administrators in Ireland proved all too willing to invoke it.

In the wake of a rebellion, an administrator armed with martial law could execute a range of rebels with little ceremony and less evidence. In practice in England, the Tudors carefully calculated their response to rebellions. They asserted their unquestioned claim to power and eliminated the leadership of their defeated opponents, but they also assuaged important political players and displayed appropriately royal mercy. Justice in the early modern era was always a balance between terror and mercy, designed as much to impress the public as it was to punish the offenders. Even civil-law trials for treason, which unlike martial law could also result in confiscating property from offenders of a high social status, produced a surprisingly small number of actual executions. Rebellions usually fell into this pattern of balanced terror and pardon, but there were exceptions. If the rebellions had been particularly threatening to the crown itself, the response could be merciless.[18] Nevertheless, as the century wore on, the English suppression of rebellions grew only slightly more severe in England. In Ireland it escalated rapidly, resting on the legitimacy of the theoretical "no limits" warfare for rebels, and without the restraints that kept such actions in check in England.[19]

Despite this escalating preference for the terror of martial law against the Irish rebels, the pardon remained a key component of English policy; the numbers of participants pardoned usually defied tallying. Justice was about terror *and* mercy. Even in Ireland there would be frequent, rapid, and substantial numbers of pardons offered to rebels.[20] Therein lay one horn of the dilemma: a rebel was not a foreign enemy; he was a subject, for whom a pardon *should* be the norm.

The other horn was that the Gaelic Irish did not see it that way. The Irish had their own notions of sovereignty and subjecthood; their legal traditions defining the meaning and obligations of sovereignty differed from English ones. The Irish lords, unsurprisingly, tended to accept the benefits that came with their English titles while rejecting the intention to reform their political and social system. In one sense, those who accepted surrender and regrant found themselves straddling an unstraddleable fence. It committed them to a mode of behavior that fit English expectations but compromised their standing within the Irish political system where the realities of local power played out. The struggle between the pretense of centralized authority and local realities of power almost inevitably generated conflict.

As for the Anglo-Irish lords, even those who still retained a sense of themselves as English subjects and who conducted themselves more or less according to English law, had nevertheless long been free of serious royal interference in their affairs. And they, like the earls of Kildare discussed earlier, had accommodated themselves to Gaelic legal and social systems. They chafed under the centralizing state. They had their own separate vision of how to reform the Gaelic population. And they sought to protect their personal estates and retain an older, more autonomous form of subjecthood, all without rejecting an English identity. The Reformation greatly complicated those plans.

This political redefinition of the subject in the context of state centralization provided a kind of cultural and moral platform of legitimacy for much of the violence that followed. Henry's split from the Catholic Church, which provoked active hostility from his subjects in Ireland, further bolstered the moral legitimacy that each side claimed to justify their violence: both sides proclaimed themselves the representatives of God's true church, and as such they claimed divine authorization for their rebellions or their suppression of rebellions.

Greatly worsening the situation was the process of colonization, or, in the term of the time, plantation. This process has been extensively studied, with some historians suggesting that the English were inspired by the dramatic Spanish successes in the New World to seek outlets for what they imagined as a restless, burgeoning population, while also bolstering their position in a suddenly competitive Atlantic basin. Crucial to this vision was the growing belief that local control in Ireland required the importation of English settlers, who, in the natural course of things, would displace the local population. Early versions of colonization focused on claiming land either "vacated" by the dissolution of Catholic institutional land ownership in Ireland or the land of Scots whose status as hostile invaders merited their displacement. There were several problems even within this limited vision of plantation. Just changing the landlord did not "vacate" the land; the Gaelic tenants remained and resented being pushed off. Perhaps worse, English adventurers, fired by visions of gaining enormous personal estates, found it all too easy to twist reports of rebellion and the locals' Catholicism to their advantage.

It was in the adventurers' and administrators' personal motivations that the problems of plantation and martial law intersected, with enormous consequences. A commission of martial law granted officials great power to define who was a rebel meriting execution. Some English administrators in Ireland saw such power as a means for personal gain, and even the less corrupt imagined it as a core component of successful control of the countryside. Although execution at martial law did not technically convey land rights to the crown, attainder (removal of civil rights by act of Parliament) frequently followed retroactively, even years later. Furthermore, the complexity of competing English and Gaelic systems of land tenure allowed for corrupt English claims on "vacant" land, as well as corruption in the attainder process itself. The largest sixteenth-century plantation was on land legally confiscated from the attainted Desmond rebels in Munster in the 1580s. The legal tangles involved in distinguishing between those rebels pardoned versus those attainted versus those executed by martial law allowed ample scope for corrupt seizure of land by local officials, but the processes of law, even corrupt law, also resulted in a patchwork of small colonized parcels rather than one new, large, anglicized territorial unit. In this case it was primarily Anglo-Irish land being confiscated. Their seemingly more secure status as subjects demanded the creation of a rhetoric condemning them as even more barbarous and treacherous than the native Irish and thus worthy candidates for losing their land and position. Furthermore, as with the earliest plantations, changing the landlord did not automatically change the actual tenant on the ground. Dealings with the latter often turned violent, as did the Gaelic and Anglo-Irish reaction to the corrupt taking of land. Plantation, rather than pacifying, tended to aggravate.[21]

The contest over the nature of subjecthood under the Tudors, amplified by the radical nature of plantation, was superimposed on the normal frictions that existed between monarchs and regional magnates in the sixteenth century.[22] To these problems were added the more or less continual violence in the marchlands between the Gaels and English, as well as the endemic wars and raids among Gaelic lords.[23] The end result was a long series of rebellions that increased the English sense of grievance at disobedient subjects and led English administrators to entertain extreme solutions.[24] The whole process gained greater virulence with the Reformation and the threatened intervention of the Spanish, an intervention that sought the extinction of the Protestant monarchy in England, not just in Ireland.

The last rebellion at the very end of the century, the so-called Nine Years War (1594–1603) culminated all of these trends, weaving them into a conflict that became the quintessence of violence.[25] In brief, Red Hugh O'Donnell, and then Hugh O'Neill, both Gaelic lords in Ulster, the latter holding the English title of the earl of Tyrone, increasingly shifted away from their nominal loyalty to England. Ulster had long remained the region least subject to English control, and O'Neill not only reformed his military system in response to the English challenge, he also laid the foundations for an island-wide anti-English coalition aided by direct Spanish intervention. The war commenced in a fitful series of attacks and truces in 1594, and then saw major Irish victories at Clontibret (1595) and Yellow Ford (1598). This pair of disasters for the English (along with others) led to a vast expansion of English forces in Ireland, an expansion whose necessity seemed proved by the Spanish landing at Kinsale in 1601. By that time Elizabeth had found in Lord Mountjoy a commander possessing the requisite energy, understanding of the situation, and good fortune. The last attribute proved especially crucial in preventing the Spanish army from joining with O'Neill's; Mountjoy captured the former and destroyed the latter. With the Spanish threat seemingly checked, although still feared, Mountjoy progressively reduced Ulster, ably assisted by George Carew's operations in Munster. Both men relied on extensive campaigns of devastation, sustained by a network of forts and suffused with an element of terror in the treatment of those Gaels who risked surrender. In 1603 O'Neill was forced to submit, but on terms of life, liberty, and a pardon that eventually restored him to his lordship.

The seeming decisiveness of 1603 notwithstanding, the century as a whole had seen confusion and disagreement about exactly how to treat the Irish and the Anglo-Irish. That confusion in turn generated shifting state policies.[26] One deputy might emphasize reinvigorating surrender and regrant with its associated policy of including the Gaels in the administration of government. Another deputy, for reasons of personal interest, outrage over a particular rebellion, or a sincere belief in the necessity of hard measures, might instead pursue more radical policies of forcing the Gaelic lords out, exporting the mercenary "swordsmen" off the island, and supplanting Irish tenants with English ones. Royal policy was only slightly less variable than the policies of the deputies.[27] When it comes to violence on the ground, however, policy is merely an intention. The choices made at all the many levels between a royal proclamation of policy and the application (or not) of a torch to a farmhouse shaped the overall qualitative and quantitative nature of violence. Policy also depended on governmental capacity to execute it, and on the nature of the military and the administrative system of the time. In an effort to get closer to these complex interactions, let us now return to Sidney and his men at Clonmel.

Sidney, 1569

Sidney had a long history of dealing with Ireland and its complexities.[28] A vigorous and active man, he had been the deputy pro tem twice and had served a term as the actual lord deputy, the highest English official in Ireland, only the year before. He had arrived for that first tour as lord deputy with a detailed plan to reform the government and increase its stability, income, and Englishness. Sidney was one of a long parade of English lord deputies in the sixteenth century, each of whom sought to bring some kind of order to Ireland's four provinces. Every effort at greater English control, however, seemed to spawn another rebellion. Sidney was more energetic than most, but ruling Ireland seemed an intractable problem, and each of the provinces presented its own special challenges. In Ulster the Gaelic lord Shane O'Neill proved the most troublesome, and Sidney soon found himself warring rather than reforming. At least Ulster was inhabited almost entirely by the native Irish (with a heavy admixture of Scots in the northeast) and so one usually at least knew who the enemy was. Such clarity notwithstanding, Shane proved elusive, and when a frustrated Sidney returned to England, he was not well received at court. (To add injury to insult, once there he endured passing a kidney stone the size of a nutmeg seed.) Returning as lord deputy in October 1568, he found Ulster still a problem, but Shane, at least, had died.

The province of Munster should have been easier to handle. Although it was surrounded by and intermingled with independent Gaelic Irish chieftains, the major powers in Munster were the Anglo-Irish earls of Desmond and Ormond (the Fitzgeralds and Butlers, respectively). Munster also had a liberal sprinkling of heavily English towns, such as Cork and Waterford. Sidney was all the more aggravated, therefore, when the peace in Munster came under simultaneous, although mostly uncoordinated, assault in 1569 from the Butler and Fitzgerald families. As was usually the case in the Irish wars, an assortment of other Anglo-Irish and Gaelic families threw in their lot with the current crop of rebels.[29]

Sidney held a jaundiced view of both families. The head of the Butler family, the earl of Ormond, although temporarily in England and not directly implicated in his brothers' revolt, had stubbornly resisted Sidney's plans to reform the province.[30] Worse, the earl of Desmond, originally a supporter of Sidney, had persistently broken the peace by waging private wars against Ormond, leading to his arrest in 1567. In the spring of that same year, Sidney had toured Munster and was shocked by what he found and heard. Reporting to Queen Elizabeth, he lamented seeing "the bones and Sculles of your ded subjectes, who partelie by murder, partelie by famyn, have died in the feldes . . . [something] hardelie anny christian with drie eies coulde beholde." Worse yet, some poor women's rescuers came just too late to save them but soon enough to detect the continued stirring of their unborn children. The murderer being known to be one of Desmond's men, Sidney was then shocked to learn that Desmond had "lodged and banckett[ed]" with him soon after the murders.[31]

Despite his shock, Sidney could not immediately respond. Shane's continued rebellion kept Sidney occupied in Ulster until he had had to leave for England. Upon his return, reports kept filtering in of almost unheard-of atrocities in Munster as the Butlers and the Desmond Fitzgeralds inched toward open rebellion.[32] The previous fall a Gaelic lord, recently ennobled by Elizabeth as the earl of Clancare, attacked a neighboring Anglo-Irish lord, and in the way more or less traditional to Irish warfare, he "preyed" on his land, taking fifteen hundred cattle while burning seven thousand sheep and pigs and all the wheat of the fall's harvest. Less traditionally, however, he reportedly burned "a great nombre" of men, women, and children.[33] The same month a Gaelic chief complained to the English government in Dublin that the Anglo-Irish Butlers had preyed upon his country (again, more or less traditional practice), but in addition to their thieving and burning they also corralled "all the pore women yonge and olde maryed and unmaried" into two churches, and there "ravished" them "contynually daye and night" for two days.[34] There was surprisingly little reaction to this complaint. The government was in transition, with Sidney en route from England to take up office again, but soon the separate offenses of the Butlers and Fitzmaurice became impossible to ignore. Major rebel attacks began in June, but it was not until late July that Sidney finally moved, and by then a rebel army threatened the key town of Kilkenny. In the ensuing few months, restraints seemed to fall away, culminating in the Butlers' August 15 attack on Enniscorthy during the annual Lady Day fair. There the Butlers' forces killed numbers of Anglo-Irish merchants and committed "horrible rapes of young maidens and wifes before their parents and husbandes faces."[35] One of Sidney's captains reported the attack, pleaded for gunners to help him defend a nearby house, and promised to stake his life, or, in his words, to "adventure his carcas[s]," in holding it.[36] His soldiers, who presumably had heard of the rebels killing the entire garrison of Ballyknockane in early July, were probably less sanguine.[37]

Sidney's soldiers had a lot to worry about. James Fitzmaurice, leading the Desmond end of the revolt, was making contact with Spain and proclaiming the rebellion a "holy war." He denounced Elizabeth as a "pretensed Queen" whom the pope would shortly oust. The pope obliged by excommunicating Elizabeth in April 1570. Fitzmaurice called for Catholic unity in Ireland, claiming that once that was achieved, "there is no power in this realm able to withstand our forces." Fitzmaurice, an Anglo-Irish noble, still asserted his Englishness, claiming not to fight "against the crown of England, but only against the usurper thereof."[38]

Furthermore, the soldiers had to wonder: Who was the enemy? And what kind of enemy were they? Traitors, rebels, legitimate combatants, or irredeemable savages? The fighting in Ulster at least had the virtue of clarity in the Gaelic versus English contest, but here in Munster the rebels were Anglo-Irish lords, men of English descent who had been living and ruling in Ireland for over three hundred years. They were joined in revolt by major Gaelic chiefs, one of whom recently had been made an English earl. The rebels seemed to be attacking other Anglo-Irishmen as well as more recent English arrivals. The enemy armies were composed of Anglo-Irish horsemen, Scottish-descended "galloglass"—ax-bearing infantrymen, and the lightly equipped, mobile, but often despised Irish "kerne" infantrymen. But so was the government army. Sidney had a personal bodyguard that included fifty galloglass captained by one Callough McTurlough, and there were 225 kerne in the permanent pay of the royal army, not to mention the service of Gaelic Irish chiefs who remained loyal to the queen and were even now bringing their own kerne into Sidney's camp. Further increasing the confusion, many "English" companies by the 1560s contained any number of Gaelic Irish recruits; one accounting in 1564 found 17 percent of the companies to be Irish. The soldiers also surely knew of Sidney's anger at the "insolency" of Waterford, a city they had marched to help protect, whose city fathers refused to send even a few soldiers for a mere three days' service.[39] One can hardly blame Sidney's soldiers for wondering who they could trust or how to categorize their foes.

Worse, even without its still imprisoned earl, the Desmond lordship was no mean opponent. A near contemporary intelligence summary reported the Desmond kin as having at their disposal 120 knights, four hundred horse, eight "battles" of galloglass mercenaries, one battle of crossbowmen, and three hundred "gunners" or shot. And if they could tap the local lords of the earldom, they could raise as many as 5,600 Irish kerne.[40] The Butlers' numbers are harder to determine, but Edward Butler had reportedly attacked in 1568 with six hundred gunners and kerne, one hundred galloglass, sixty horsemen, and three hundred "slaves, knaves and boys."[41] In the most significant act of cooperation between the Butlers, the Desmond Fitzgeralds, and their major Gaelic ally, Donal MacCarthy—he was now rejecting his English title of earl of Clancare—put together 3,700 men (including a remarkable four hundred pike and four hundred shot) to lay siege to Kilkenny.[42] In contrast, the mobile English army in Ireland in March 1569 consisted (on paper) of 231 foot, 449 horsemen, and 225 kerne with a couple hundred men scattered in tiny garrisons or serving the royal artillery.[43]

Nevertheless, trusting as usual in the power of the heavier English horse to offset their weaker numbers, Sidney immediately dispatched three companies under Peter Carew, Nicholas Malby, and William Collier, the former two to guard the boundaries of the English Pale while Collier went to reinforce and hold the town of Kilkenny. Meanwhile the threatened towns and communities in Munster sought their own solutions, engaging in the time-honored practice of buying off the rebels, and even exchanging hostages with them in a devil's bargain: the rebels promising not to plunder if the town supplied their wants.[44] Sidney hurried to the scene from Dublin with a scant six hundred men.

It is difficult to know much for certain about those six hundred men, but they were likely hard-luck, hard-bitten, or both. Sidney once expressed amazement that people in Ireland could wish for "soldiers nothinge insolente, nothinge sensuall, nothinge gredie, no quarrelers? So wishe I," he exclaimed, "but I scarse hope for it."[45] For the most part England recruited its soldiers for Ireland from the West Country, especially Wales and the Welsh marches, but also from the turbulent Scottish border region. Only the demands of the

Nine Years War at the end of the century would expand recruitment across the rest of southern England. The process usually involved delivering quotas to the counties whose officers then hurriedly raised men. Individual villages and towns, asked for their quota, cast about for the most expendable of men, men who likely believed, as ran the proverb in Chester, that "ytt weare better bee hanged att home then dye lyke dogges there" in Ireland.[46] A record of soldiers embarking for Ireland in 1596 recorded the despair of one John Lewys, who had run away, later to be "found in a bushe and he cut his owne throte and so dead." The Irish service had a bad reputation, and even men accustomed to war feared the unpredictable crossing of the channel and the even less predictable war on the other side. Ireland's grim reputation produced correspondingly grim recruits for its wars. The commissioners gathering recruits in England remarked, "There was never beheld such strange creatures brought to any muster"; "They are most of them either lame, diseased boys, or common rogues. Few of them have any clothes; small weak starved bodies, taken up in fairs, markets and highways, to supply the place of better men kept at home."[47] Most English companies, composed of "poor old ploughmen and rogues," came as reluctant, probably conscripted individuals.[48] Veteran companies arriving from other theaters of war in Scotland or on the continent were noted as exceptions.[49] There were individual veterans too. In a later war, for example, a Lieutenant Vickers appears, who came to Ireland around 1574 as a private soldier. He was already experienced in the wars on the continent, and he stayed in service in Ireland for at least twenty-five years, rising into the officer ranks, only to find himself vilified by some Englishmen for marrying an Irish woman and thus "Ally[ing] himself with the kinrad of Traytors" and not to be pitied for thus nourishing the "snake" in his own bosom.[50]

The men were organized into companies of foot and horse, led by captains, often with a "petty" captain or two as assistants, and a standard bearer ("lieutenants" sometimes appear as the commanders of smaller detachments or stood in for absent captains). Most companies, or "bands," were 40 to 150 men strong. The English horsemen fought as heavy cavalry, armed with a lance they could couch under their arm and seated in a heavy, stable saddle. Horsemen on campaign invariably had two or three horses each, one for fighting and the others for marching, and they were attended by the infamous "horseboys," mounted servants capable of wide-ranging depredations on civilians. In 1569 English footmen were still in the middle of a transition from archers to "shot," and the pike was just beginning to play a major role in English armies in Ireland. In the 1560s, therefore, most English foot companies had a mix of shot (men with a harquebus) and archers, and a few companies with pikes.[51] In time more and more of those companies combined pike and shot, and the archers faded away. The pikemen fought in tight squares, and in theory the shot stayed close to them in battle, fearing the charge of horsemen. In Ireland, however, the light, stirrupless Gaelic horse was rarely perceived as very dangerous. The shot, therefore, were often "loose," and moved about the battlefield rapidly and irregularly.

To add to the variety, the English army on campaign was fleshed out with galloglass mercenaries and Irish kerne, hired directly by the government or in the service of local lords allied to the English. The galloglass were long-established septs or clans of mercenaries originally from Scotland who fought as heavily armored, ax- and longsword-wielding infantry. The shift to pikemen was leaving them behind, but in midcentury they still were considered the core of a Gaelic army and useful auxiliaries to the English.[52] A "kerne" was both a soldier with a particular armament and a class of mercenary. In Irish armies the majority of the kerne were simply land-poor Irish freeholders who could not afford armor and heavy weaponry. When their lord demanded their service in a "rising out," they came equipped as "a kinde of footeman, sleightly armed with a sworde, a targett [shield] of woode, or a bow and sheafe of arrows with barbed heades, or else 3 dartes, which they cast with a wonderfull facillity."[53] Some men made such contract military service a profession, serving the English or wealthier Gaelic or Anglo-Irish lords who retained them as standing forces and quartered them among the peasantry.[54] Later in the century, innovative Irish lords rearmed the permanent soldiers as pike or shot, and English records began to refer to these permanent soldiers as *buannadha*, or bonaghts, a term derived from the Gaelic name for the tax in kind that Irish lords were using to pay and feed them.[55]

This chaotic cavalcade of horsemen, harquebusmen, archers, pikemen, kerne, and galloglass shambled out into Munster, falling ill along the way, wearing out their shoes, rotting off their feet, and struggling through bogs. As they pushed through ambushes by small Irish bands, they absorbed rumors of atrocities and the alarming report of rebels hanging the surrendered English soldiers from a small garrison at Tracton in early June.[56] Many of Sidney's men recently had campaigned against Shane O'Neill in Ulster, where the Queen had directed Sidney to behave as though at "oppen warrs with hym."[57] At least that made him a clear enemy. What do you do with rebels who might submit to pardon at any moment? Indeed, although their commander had just marched them through Kilkenny, where not long before thousands of rebels had besieged the city, now he was announcing the

Defeat of Sir Henry Harrington, 1599. Detail from a contemporary drawing by or for Captain Mountague, commander under Sir Henry Harrington. Here Irish "rebbels" emerge from the bog to attack the "loose shott" of the English. A pike square can be seen entering from the left. John T. Gilbert, *Facsimiles of National Manuscripts of Ireland* (London: HMSO, 1882), Part IV.1, pl. 26 "Defeat of Sir Henry Harrington and His Forces near Wicklow, A.D. 1599." From the copy in the Rare Book Collection, The University of North Carolina at Chapel Hill.

promise of pardons to all who would "desist from rebellion." This despite reports that the Butlers had set up straw dummies dressed in the clothes of dead Englishmen as targets for their kerne.

On August 2 the army arrived outside the walls of Clonmel and went into camp. In camp or on the march the soldiers were accustomed to a daily ration of alcohol, usually anywhere from a quart to a gallon of beer daily.[58] In camp they expected a little more. They were even known to become obstreperous when rations ran short or if the quality did not measure up. Captain Nicholas Dawtrey's troops once threatened to stuff their bread supplier ("vittler") into one of his own ovens. Dawtrey quelled the mutiny by giving the vittler a lesson in the proper brewing of beer and baking of bread.[59] The troops also took advantage of not marching or being shot at to engage in earthier pleasures. Captain John Harington was startled to see some Irish soldiers "withouten clothes on their backes or foode in their bellies, . . . under hedges withe marvelous ill favourede wenches."[60] Sidney finally had the burgesses of Clonmel post a guard at the city gates to stop unauthorized visits by his soldiers. The urban proximity of more alcohol and sex notwithstanding, Clonmel proved an unhappy camp. Adding to rumor, the rebel enemy had the audacity to throw leaflets into the camp, promising battle and death as soon as the English moved again—or even before. All this created, as Sidney recorded somewhat blandly, "an impression of fear in my men's hearts as it was most strange for me to behold." In this case their fear translated into action. The "private soldiers came to their officers, the officers to their captains, the captains to my counsellors, and the counsellors to me, and nothing in their mouths but 'home! home! home! or else we are all undone.'"

Thus the scene returns to Sidney standing in the camp, waiting to speak to his soldiers. No fool, Sidney knew something had to be done. English armies in Ireland had mutinied before and would again. Soldiers had even dictated march routes and strategies to their commanders. Walter Ralegh later opined that English soldiers proved willing to do things in pursuit of Spanish gold in the new world that in Ireland would have had them turning "their Peeces and Pikes against their Commanders, contesting that they had beene brought without reasone to the Butcherie and slaughter."[61] Although the social divide between the soldiers and Sir Henry Sidney gaped wide indeed—the office of the Queen's lord deputy in Ireland was symbolized by a sword, signifying the power of life and death—the administrative distance was quite short. Sidney commanded his men in person in the field, and between him and his soldiers there were really only one or two administrative links: captains and petty captains. Sidney adeptly mixed the high ceremony of his position with a personal appeal. He began quietly by ordering more clothes for the ragged men, opening his own purse to do so. He then sounded the assembly with trumpet and drum, and the men gathered in the camp marketplace. The soldiers clustered around, eagerly hoping for word that they were returning home, and made more eager by the unusual present of a tun (roughly 250 gallons) of wine now being served out at Sidney's order. He fired their courage with words and stoked it with alcohol, and by the end of his speech (of which we sadly know nothing), their "cowardish coldness was turned into a martial-like heat," and they were crying "Upon them! upon them! upon them! Lead you, and we will follow to the land's end, or die by the way." Sidney reported a long night of the sounds of weapons being sharpened.

Such enthusiasm had consequences. En route to Clonmel, Sidney had come across one of Edmund Butler's small castles. The garrison seemed at first defiant, firing their outer buildings and waiting for Sidney to begin preparations for an assault. They quickly changed their minds, and Sidney quietly accepted their surrender. There were rules to this game, and Sidney and his soldiers appear to have observed them, but after Clonmel Sidney's newly inspirited soldiers, when confronted by other defiant castles, added their own frightful touches. As a general rule in the sixteenth century, and for the next two hundred years or so, a besieged garrison was expected to hold out at least until an assaultable breach had been made in the walls. Much of contemporary fortification design sought merely to delay the making of such a breach in the hopes of relief or that the besiegers would run out of food. On the other hand, the longer that process lasted, and the greater the privations suffered by the thousands or tens of thousands of men camped in the increasingly filthy cesspit of trenches outside the walls, the more likely further resistance would render the soldiers uncontrollable once they assaulted the city in, as the saying went, "hot blood." Consciences on this subject were clear. In Shakespeare's immortal depiction, from the nearly contemporaneous *Henry V*, Henry threatens the besieged Harfleur, warning the citizens that a successful assault had consequences:

> The gates of mercy shall be all shut up,
> And the flesh'd soldier, rough and hard of heart,
> In liberty of bloody hand shall range
> *With conscience wide as hell*, mowing like grass
> Your fresh-fair virgins and your flowering infants[62]

Officers and soldiers understood this system, as did Shakespeare's audience. To "cry havoc" was to signal that the battle was over and the plunder and destruction could begin.[63] It was part of the military culture of the day, and it had become a virtual rule that officers would not try to control the soldiers in those moments. They were thought to have earned their day or two, or at most three (for really big cities) to kill, plunder, pillage, and rape. The defenders, after all, were felt to have brought it on themselves by not surrendering in time, and the soldiers expected the pillage as part of their pay.[64]

After Clonmel, Sidney and his men marched into the country of the "White Knight" (an ally of James Fitzmaurice) and there found one of the castles held against them. The defenders, wary of past massacres by the English and uncertain of their status as rebels or soldiers, chose to hold out.[65]

Sidney lacked the kind of artillery usually needed to batter down the walls, so the task fell to the soldiers. They assaulted through the relatively insignificant courtyard wall (or "bawne"), dodged the rocks thrown down at them, and pressed up against the walls and the wooden portcullis, which they broke and set on fire. Now truly desperate, the defenders set their own castle on fire and hid in a deep vault, hoping the fire would keep the English out. Sidney pulled his soldiers back and allowed the embers to cool all that night and the next day. The fire merely delayed what was now inevitable. In the version of this story that Sidney wrote years later, he claimed the defenders believed that help was on the way and therefore refused his renewed offer for them to surrender. In any event, the English renewed their assault the next day, fighting hand to hand, stair by stair, and room to room. They finally put "the whole warde . . . to the sword" and threw them "over the topps of the castell to the terror of all over." This seems extreme, given that there were only *eight* surviving defenders, and the English had taken no casualties.[66] Perhaps disappointment at all the plunder going up in smoke spurred the soldiers' anger. Perhaps they were frustrated at so much effort expended to defeat a mere eight men. Their luck improved at the next castle. This one also appeared to bid defiance, but after a minimum of preparations for an assault, the garrison slipped out at night through an unguarded bog. Not only were the soldiers allowed free plunder, but one of their captains, Jasper Horsey, was rewarded with the post of seneschal of the region, a position that lent itself to corruption and profiteering.

As with most campaigns in Ireland, there was more to the process than marching and laying siege. A commonly calculated strategy throughout late medieval and early modern Europe, although potentially applied to different purposes and to greater and lesser extents (about which more later), was to devastate the countryside along the way.[67] Both sides employed this tool consistently. Sidney's descriptions of the devastation that he wrought are quite nonchalant: speaking of the White Knight's country, he said only, "I therefore passed in effect through all his countrey burning all the corne that was gathered and spoiling the rest, I rased one of his castells[,] burnt and spoiled all his other houses."

The pressure of devastation was intended to undermine a particular rebel's ability to fight, to prove his inability to defend his territory, and to encourage him to submit. It was not intended to kill the local population. In essence, the tactic of devastation in most sixteenth-century contexts was logistical and emotional, not demographic. This limited intention affected the qualitative nature of the violence. Civilians suffered or were killed in the now invisible microscale of soldiers' choices, but the local logistical and symbolic intention did not demand widespread killing. For the Butlers, this kind of pressure produced rapid results. Never as strong as the Desmond rebels, and pushed by their still-loyal brother, the earl of Ormond, who had returned from England, the Butlers submitted relatively quickly.[68] This was part of Sidney's calculation: to force a rebel to submit and his household and allies with him. Naturally enough, therefore, Sidney accepted their submission and extended them a pardon, but, oddly, found himself rebuked by the Queen for humiliating them a bit too much. The Butlers were new to rebellion, she wrote, while the Desmonds and McCarty More had been "disobedyent subjects and of the nature of rebels about the space of these xii monethes and more."[69]

Sidney made his way back to Dublin, but the Desmonds and their Gaelic adherents persisted in rebellion in the far west of Munster and in Connaught. Sidney left in charge Humphrey Gilbert, formerly a captain, now knighted in the field and promoted to colonel of the army. In this role Gilbert became infamous in Ireland, lionized in contemporary pamphlets for his policy of terrorizing the countryside and remembered by historians for his later role in colonizing voyages to the new world.[70] Gilbert had ambitions for estates in Ireland as well as in a whole new hemisphere. In short, Gilbert was a colonizer, and colonizers calculated by different rules.

Gilbert's atrocities in Ireland, although well documented and often retold, were far from unique. Although the state of the earlier records makes it difficult to compare one part of the century to another, it seems likely that the situation probably worsened after about 1569. And there is little doubt that the Nine Years War of 1594 to 1603, which completed the Tudor conquest of the island, was particularly filled with horrors, but the whole century was rife with deliberate devastation and the killing of women, children, and prisoners.[71] A full explanation must look deeper into the role of colonization, but also at the nature of military calculation and the military system and the military culture with which the English tried to conquer and administer Ireland, especially in the cataclysm of the Nine Years War at the end of the century. For these purposes there is no better example than the experiences of Walter and Robert Devereux, the first and second earls of Essex.

2

The Earls of Essex, 1575 and 1599

I will pass over usual matters such as burning, killing and spoiling wherewith I am troubled and even in conscience pinched almost every hour of the day for God I fear will demand at my hands the shameful effusion of innocent blood whereat I may shake the scabbard with the sign of a sword but have not a sword to draw. And whereon when I should look, I am driven with shame enough, either to look beside or wink and with a wounded soul pass over horrible injuries. . . . There lies some secret misery in this universal [Irish] rebellious disposition which God for his mercy's sake grant not only to be revealed but also provided for.

Lord Deputy Fitzwilliam, 1572

WALTER DEVEREUX, THE FIRST EARL OF ESSEX, was deeply involved in one of the processes that most undermined restraint in the Irish wars and the consideration of the Gaels as subjects: colonization. Essex took up Henry Sidney's suggestion to bring English settlers into Ireland and establish a private-enterprise colony.[1] He chose the rocky, fractured northeast coast of Ireland, almost within sight of Scotland, where three peoples, not just two, competed for authority and control. The Scots had long provided mercenaries for the Irish wars, a role that was expanding in new directions late in the sixteenth century in the form of large bodies of "redshanks"—men who fought bare-legged. Although redshanks had periodically come to Ireland at the behest of Gaelic lords in need of troops, their now long-established connections with resident Scottish families and new marriage ties with Irish chiefs provided an opportunity for the potential widening of Scottish power and presence in Ireland.[2] Elizabeth viewed them as foreign interlopers in her realm, and where the Irish might have a claim on her protection, the Scots in Ireland had none.

Although the Scots in Ulster were deeply intertwined by blood and geography with the Gaelic Irish, Essex arrived in August 1573 with virtual carte blanche to "expel the said Scotte out of those countreys" in favor of English colonists. He also carried orders to treat the native Irish well, and not to harm "any person that is knowne to be our good subject." He in turn promised not to "imbrue my hands with more blood than the necessity of the cause requireth."[3] Elizabeth's instructions and Essex's response are a perfect example of the weaknesses in the ideology of "subject" protection in Ireland. Subjecthood was supposed to confer protection under the law, but on-the-ground commanders distant from London found it easy to focus on the qualifier "good" and then proceed according to their own ambitions. Later, they could explain that the victims were not, in fact, *good* subjects, and point to the "necessity of the cause." Essex's project hit snag after snag, from weather to local opposition to the departure of many of his men back for England.

Growing desperate, Essex identified the Scottish settlement on the isle of Rathlin as a threat to the waterborne supplies intended for his movements deeper into Ulster, and in July 1575 he dispatched Captain John Norris to attack the island.[4] Norris's men forced their way ashore and pursued the Scots to a castle. Bringing two cannon into play, they eventually broke the walls down into an assailable breach. As the defenders continued to resist, the English (on the third day of fighting) made their first assault but were forced back. The English launched their assault anew, and again it failed. It did, however, convince the Scottish commander to call for a parley. Norris refused to guarantee their lives and goods, offering only to protect the commander and his family, with the rest to "stand upon the courtesy of the soldiers." Left with little choice, the Scots accepted the terms and hoped for the best. From the English soldiers' point of view, however, two failed (and costly) assaults upon a breach added up to a right of revenge, and the men "made request, or rather pressed, to have the killing of them." The soldiers slew some five to six hundred men, women, and children in the castle or in caves around the bay.[5]

On Rathlin the various elements of wartime violence came together in a perfect storm. Scots migrants and mercenaries, outside the protection of subjecthood, standing in the way of colonization, were freely massacred under a harsh code of war, in this case demanded by angry and fearful soldiers. For Essex the violence did not stop there. In the context of a colonizing effort and his hope to clear the land of its inhabitants, it proved a short ideological step from the massacre of a "foreign" invading force (women and children included) to using any sign of rebellion as an excuse to place even Gaelic Irish subjects outside the protection of law. Traitors, after all, were easy to tar, provided one's interest was not true pacification but expropriation. Essex was a private-enterprise colonist; his interest lay in land and security. Seven months previous to the massacre at Rathlin, Essex had grown frustrated with the intransigence of Brian McPhelim O'Neill. Initially submissive, Brian had shifted to cooperating with the Scots and with other Irish rebels. Essex highlighted this "breach of their faiths" as giving him "just cause to govern . . . in the most severe manner, which I could not without evil opinions have offered if their revolt had not been manifest."[6] Notwithstanding Brian McPhelim O'Neill's English knighthood, Essex defined "severe" to include taking Brian and his wife prisoner while they shared a Christmas feast under his protection, killing some two hundred of Brian's followers, and then executing Brian and his wife.[7] In the hands of a colonist pursuing his own interest, the crown's open definition of subject proved all too malleable. Brian O'Neill might have an English knighthood, but his Gaelic identity rendered him suspect, and any accommodation he made to the realities of local power dynamics involving other Gaels (or Scots) made it easy to label him a rebel who forfeited the protection of law.

Agents of colonization like Walter Devereux and Humphrey Gilbert naturally drifted toward the extreme solutions appropriate to their extreme goals. In general, however, the process of colonization is not the focus here. Some horrific violence was associated with early colonization efforts, but it nonetheless appears that the worst of the Nine Years War, which was in turn the worst of the Irish wars, became frightful for reasons unrelated to colonization.[8] The war of devastation waged by George Carew and Lord Mountjoy in 1601–3 was not primarily about expropriation. The issue was their vision of how to win the war, in part informed by their easy identification of the enemy as rebels. They did not have to ignore Hugh O'Neill's English title; by the time they began managing the war he had formally rejected it. Another key element was the practical calculations made by experienced captains in Ireland and their recommendations about how to proceed. Some of their visions were, or became, quite extreme; others less so. But with a few exceptions growing out of self-interest, their calculations about how to win were based on an assessment of the material factors present in Ireland, and on what "winning" meant to them. Those calculations also were affected by cultural and customary notions of martial honor and a proper way of war, as well as by colonial ambitions. There are issues of fear and prejudice to consider as well. Recall Essex's explanation of the final massacre at Rathlin in 1575: "The soldiers being moved and much stirred with the loss of their fellows who were slain and desirous of revenge made request or rather pressed to have the killing of them."[9] Without shifting the blame from Essex and his senior commanders (who acceded, after all, to the soldiers' demands, and not in a hot-blooded context), this explanation suggests the need to examine more deeply the experiences and choices of the common soldiers, whose fears could play a key role in the escalation of frightfulness. But both captains and soldiers were led in war by the highest-ranking aristocrats in English society, and so one must consider their values too. There is no better example than Walter Devereux's son, Robert, the second earl of Essex.

The Second Earl of Essex, 1599

Robert Devereux remains a fascinating and colorful figure, whose tortured and eventually fatal relationship with Elizabeth reads like fiction. Among other things, Essex was a believer—he believed in the rightness and justice of England and English laws. He believed in the acquisition of personal glory through martial success. He believed in living the code of chivalry.[10] Above all, he believed in himself. Essex was the kind of man who challenged the enemy commander at the siege of Rouen in 1591 to a duel, declaring, "The king's cause is more just than the League's, I am better than you, and my mistress is more beautiful." His erstwhile opponent disagreed with the first two points, claimed the last point irrelevant, and declined to duel.[11] At the failed siege of Lisbon in 1589, a disappointed Essex spurred his horse to the gates of the city, drove a lance into the door, called out for anyone inside willing to duel with him, and then wheeled about to lead his troops away.[12]

To depict Essex this way is not to detract from the value of his experience. Essex fought in a variety of campaigns on the continent, usually in the company of officers with Irish experience, some of whom would soon join him in Ireland. He was a hard man who had seen hard things. In 1589 Essex was a latecomer to an English expedition that had just raided Corunna. There drunken English soldiers had cut the throats of five hundred Spanish prisoners after putting many other inhabitants "to the sword in furie." Later, during the siege of Corunna's upper town, when someone in the Spanish garrison rashly fired on the English drummer beating a parley, the worried Spanish, fearful of a reprisal, quickly hung the offender over the wall. On that same expedition, Essex indulged his penchant for duels by challenging an enemy commander to duel, either singly or in a group of six, eight, or ten.[13]

Essex's sense of drama might seem to undermine his value as an exemplary figure, but he and his father loom large in the history of Ireland, and in some ways Robert exemplified key characteristics of the English military aristocracy. Robert's role in Ireland began at the end of March, 1599, when he left England to assume command (and governorship) in the fifth year of the Nine Years War. English forces had been repeatedly humiliated by Hugh O'Neill, who had drawn out the war with a deft self-presentation in which he always seemed on the verge of submitting. Essex claimed to have had enough of this kind of temporizing war.

Fortunately for him, the government had already decided to vastly expand English forces in Ireland. For most of the sixteenth century, the quantity of men and money provided by England for the control of Ireland had been shockingly inadequate.[14] Prior to 1534 the English garrison (or "retinue") numbered under one thousand, and even afterwards varied between one thousand and twenty-five hundred except during the larger rebellions. On one level the small size of English forces in Ireland had limited their overall capacity to destroy. The Nine Years War, on the other hand, was fundamentally different. In 1599 Essex asked for, and got, a comparatively huge army: sixteen thousand foot and 1,300 horse, many of them "old souldiers" from the wars in the Low Countries, with the promise of two thousand reinforcements every three months to keep the ranks full.[15] The new commitment of resources radically altered English destructive capacity. Strategies now could be pursued that simply had never been possible before. By 1602 "barely six square miles of Ulster lay beyond the range of cavalry raiders from a military outpost," and no garrison was more than a day's march from the other.[16]

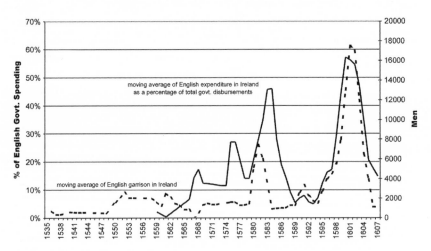

ENGLISH MILITARY AND FINANCIAL CAPACITY IN IRELAND. Graph by author.

In 1599, however, Essex began his campaigning not in Ulster, the root of the rebellion, but instead in Munster, choosing to "shake and sway all the branches" there.[17] In this he violated his orders, forgot his own declaration that "by God, I will beat Tyrone in the field," and ignored much of the advice he had received about how to win the war.[18] The queen expected him to attack O'Neill, and she further ordered Essex not to receive him except "upon simple and single submission"; his life could be granted to him provided he submitted with appropriate humility and gave security for his future loyalty.[19]

Here lies a key point about the English definition of "victory." O'Neill had to be brought to *submit*, and then prevented from rebelling again. This had long been an English goal in their wars with the Irish, punctuated by the occasional colonial campaign of expropriation or the need to expel a foreign invading force (as would happen again in 1601, when the Spanish landed at Kinsale).

Reams of strategic counsel about how to achieve such a submission had been produced over the century. Some of the most interesting advice was written by veteran captains in Ireland for this latest war. Captains Thomas Lee, Nicholas Dawtrey, Humfrey Willis, Carlile, John Baynard, and Thomas Reade advocated what might be called the traditional approach, suggesting a ring of forts around and penetrating into Ulster, the garrisons of which would devastate the countryside between them (both deliberately and naturally in the course of their foraging).[20] Several of the captains bemoaned the cost and ineffectiveness of large campaign armies, which the Irish could easily anticipate, then scorch the English army's predictable lines of advance and ambush them in the many narrow places. Actively patrolling English garrisons (including during the winter) would not only squeeze O'Neill's resources by denying him pasturage for his cattle ("without which they are not able to keep wars"), but could also, if necessary, provide a logistical platform for launching a larger army into the heart of Ulster. After O'Neill's eventual submission, the forts would form a "bridle" on the countryside "to keep the people, who are received to mercy, in their obedience."[21] Although their suggestions for locations varied somewhat, Thomas Lee's plan was the most developed (see map, p. 19). He recommended major provisions bases at Carlingford, Newry, and Kells, which would then support a series of forts extending from Newry to Armagh and on to the key crossing of the Blackwater River, and then westwards via Monaghan and Cloones to connect to the Lough Erne. Once those were established, hemming in O'Neill, the English could mount the long-hoped-for amphibious operations to the

Lough Foyle and to Ballyshannon, which would split Ulster in two and seal O'Donnell out of Connaught.

Lee expected the garrisons to be aggressive. They were not merely intended to defend the borders of the pale; for him (and for the others who shared his thinking) garrisons were an aggressive strategy:

> For it is not the holding of the forces together in gross, nor walking up & down the country, nor living idle in the best towns of the English pale that will affect it. The garrisons should be placed near bordering upon the enemy, the chief commander of the forces will most conveniently be in the middle, having his forces garrisoned on either side, and so to stirring in the night rather than in the day, to enter into the rebels' fastnesses; for it is the night service (well spied and guided) which must end the wars of Ireland, and not fighting with them in the daytime by fits.[22]

This aggressiveness nevertheless retained the strategic goal of submission, not absolute devastation. Dawtrey suggested that once a garrison was placed, a proclamation should be made that "such as are in rebellion against her Majestie and that are willing to cum in, and live as dutifull subjectes . . . shalbe received to mercy and be pardoned." Forts would "da[u]nt them" and make "every man to seeke means how he may make his owne peace," although certain incorrigible rebels were to be killed. After all, Dawtrey argued, "Yt is more glorious to reforme a wicked people than to slay them all."[23]

An important minority continued to recommend restraint and accommodation as the best means of pacification. They feared the war of revenge, the war waged according to the *lex talionis*. No less a thinker than Francis Bacon, in 1602 in the midst of the worst part of the war, admitted the necessity of martial law for a time, but recommended that "carrying an even course between the English and the Irish . . . as if they were one nation . . . is one of the best medicines of state."[24] Even Captain Henry Docwra, usually associated with some of the most frightful aspects of the final conquest of Ulster, made a point of recommending land and protection for the Gaelic lords who helped him, even at the cost of land for himself. He argued that the Irish should not feel driven to such extremities that they would never yield.[25]

As for Elizabeth, she tried to strike a classic balance between promising severity to obstinate rebels and pledging mercy to those who submitted. Her public proclamation announcing the opening of Essex's new campaign in Ireland in March 1599 is worth considering at length. In it she protested her desire for "the peace and tranquilitie of the people of our Dominions" and her preference for "clemencie before any other" attitude. Nevertheless, she continued, "It hath fallen out [that] divers of our Subjects . . . have unnaturally . . . forgotten their allegeance, and (rebelliously taking Armes) have committed many bloody and violent outrages upon our loyall Subjects." She claimed she had tried to avoid bloodshed, asserting that the rebels' consciences would have to agree, but that "after so long patience wee have bene compelled to take resolution, to reduce that kingdome to obedience (which by the Laws of God and Nature is due unto us) by using an extraordinary power and force against them. . . . [and] we shall finde the same successe (which ever it is the pleasure of God to give to Princes rights) against unnaturall rebellions." She pledged to recognize distinctions among rebels; some perhaps had been forced into action. She protested no intent of "utter extirpation and rooting out of that Nation . . . our actions tending onely to reduce a number of unnaturall and barbarous Rebels, and to roote out the Capitall heads of the most notorious Traitours." She concluded by assuring her "duetifull Subjects there, of the great care we have of their preservation, of the abundance of our Clemency and gracious disposition to those that shall deserve mercie."[26]

Elizabeth's public policy thus promised to balance terror and mercy. As always, however, a monarch could only rely on her agents and they in turn on their soldiers. Both groups had their own vision of the proper use of force. And Essex in particular was a man sufficient unto himself. He not only ignored the advice of the captains and the directives of the queen to focus on the "archtraitor" in Ulster; he also intended to ignore a strategy of pacification. He marched into Munster with his own vision of a glorious victory, and he carried with him the sword of martial law.

Still a young thirty-three in the spring of 1599, an experienced if flamboyant veteran, a competitor for the personal favor of Queen Elizabeth, sporting a long red beard and his Order of the Garter, Essex arrived in Ireland to lead the largest army the island had yet seen. Armed with great freedom to impose heavy punishments via martial law, and despite his stated intention to wield a heavy hand, he nevertheless also tried to wage war within the framework of chivalry, military discipline, and a willingness to follow the queen's intent to accept submissions and offer pardons.[27]

At the outset of his campaign, Essex, like other English commanders in the sixteenth century, promulgated a list of orders designed to control the conduct of his soldiers and to clarify the punishments for violating the rules.[28]

THE 2ND EARL OF ESSEX (ROBERT DEVEREUX), BY THOMAS COCKSON. From *A booke containing all such proclamations, as were published during the raigne of the late Queene Elizabeth. Collected together by the industry of Humfrey Dyson, of the city of London publique notary,* STC 7758.3 (London: Bonham Norton and John Bill, [1618]). By permission of the Folger Shakespeare Library.

Essex's articles strictly forbade murders and private quarrels, and mandated that "noe man shall ravishe or force any wooman upon payne of death." If that was wishful thinking, one of the later numbered clauses was a truly desperate stab at restraining a system inclined the other way: "Noe souldier of the armye shall doe violence to any person, or steale, or violently take, or willfully spoyle the goodes of any Irishe good subiect upon payne of death." House burning, corn spoiling, and boat destroying were also forbidden unless ordered. In an attempt to create an environment that would make all these crimes more difficult, Essex forbade anyone from going more than a mile from the camp (without orders) and added that "noe souldier shall breake his order to followe any route or chase, or to seeke any praye or spoyle, except he be commanded."

Although closely resembling other such articles from the era, Essex's version reflects some interesting choices. Other sixteenth-century English regulations designed for service in France or the Netherlands provided more explicit protection for enemy messengers.[29] Essex's articles, and another shorter version produced around the same time in Ireland, made the granting of protections (including for messengers) more difficult. Only those intending full submission were to be allowed entry into the camp.[30] Essex also neglected to say anything about the treatment of prisoners. Normally, according to the other sets of rules and the customs of war at the time, at least some men were to be spared for their value as ransomees. In Ireland even that self-serving reason for preserving the lives of prisoners seemed to have less weight.[31]

With the "rules" set, Essex left Dublin with a field army of three thousand foot and three hundred horse on May 9, joined shortly thereafter by the earl of Ormonde's seven hundred foot and two hundred Irish horsemen.[32] In a campaign dismissed as ineffective by both contemporaries and historians, Essex marched to Stradbally and Ballyknockane (where the men of the English garrison had been killed in 1569), revictualled the fort at Maryborough, forced his way through the pass at Cashel, passed by the town of Clonmel (the paths of English armies really were that predictable), and accepted the surrender (with a pardon) of the rebels holding a castle just downriver, in which he then left a garrison of thirty men. Nearby he engaged in a moderately serious siege, taking the castle of Cahir. He then made a wide loop to the west, swinging by Limerick before returning to Waterford, and then back to Dublin in early July by way of the east coast (see map, p. 19).

In keeping with the usual way of war in Ireland, Essex also devastated the countryside as he passed—presumably concluding that the victims were not "good subjects" as laid out in his rules. Into the deep past of Irish warfare, including wars between Gaelic chiefs, devastation and the theft (or "preying") of cattle had been standard practice. Until the 1590s, however, the armies engaged in such devastation had been quite small. Forces of several hundred men, or even a few thousand, worried about their security and moving along fairly narrow corridors, could spare only so much time to engage in deliberate or incidental destruction. Unfortunately for the inhabitants of Ireland, regardless of ethnicity or background, it was a small island. When Sidney invaded Ulster in 1566, he was surprised at the extent of cultivation near Clogher, and so he stopped his army for one day to engage in dedicated destruction, burning out a twenty-four-mile-wide circle, sparing only the church of the local bishop.[33] A captain of a much smaller force operating in Munster during the Nine Years War claimed he had destroyed a ten-mile-wide strip, having "burned all those parts, and had the killing of many of their churls and poor people."[34] The expansion of English troops in 1599 not only increased their destructive capacity in terms of numbers, but they would shortly begin using winter campaigns for the first time, extending the time at their disposal for destruction.[35]

In addition to devastating the countryside, and engaging in a few small sieges, Essex fought a couple of ugly running battles, especially one near Arklow on the coast. Most of his losses, however, came from disease, desertion, and detachments into garrisons. At one point his central army was down to 1,250 men. When he and his council reassessed their situation in July, they discovered that, after sending major detachments into Munster and Connaught and around the Pale, there remained less than six thousand foot and five hundred horse for the invasion of Ulster, the center of the rebellion. In two months of campaigning, Essex seems to have diffused his power and accomplished very little. Initial impressions of the list of submitted enemies were underwhelming; he recorded in his campaign journal only the submissions of Viscount Mountgarret, Lord Cahir, James Fitzpiers, and John Delahide.

Essex's taking of submissions brings back the issue of victory and its meaning. How did one pacify a countryside of rebellious subjects? The paradigm, Essex's aggressive rhetoric notwithstanding, remained the use of terror *and* mercy. He had granted martial-law powers on a wide scale to local commanders, who presumably followed precedent and set up small circles of punitive or profit-seeking raids around their bases.[36] On the other hand, he pardoned some 1,140 individuals.[37] Some of the pardons covered the whole households of a major figure who had submitted. James Fitzpiers's pardon, for example, included 281 other names. The pardons were also generous: although some excluded pardons for murder, others did not, excluding only debts or intrusions onto royal property (a pardon, after all, could not get mixed up in the ordinary course of civil law—only criminal law). Elizabeth's proclamation had promised this balance of terror and mercy, and, as shown in chapter 1, the cultural vision of how to treat rebellious subjects was well established. In Ireland, however, it had begun to jump out of its traditional groove; persistent martial law seemed to become as or more likely than pardon.[38] In the 1580s, Sir James Croft, Lord Deputy in 1551–52, had complained about the use of martial law, with its granting of nearly unlimited power to English officials. Croft specified that when "we have warres w[i]th subiects, who are to be corrected, chastened and to be reduced to obedience [we must have] regarde . . . to keepe the countrye and people from beinge utterlye spoyled and wasted." Croft went further, pointing to the ignorance of the soldiery, not being instructed by their captains in the "difference there is in making warre uppon subjects rebelling against theire prince and warres to be made against a foreigne [and] publique enemie."[39]

Execution by martial law and the display of the victims' heads was one way to deal with rebels. The other was to accept their submission and offer a pardon. Essex did both, depending on his sense of the power dynamics and the relative incorrigibility of the rebels in question. When he made his choice, he was imagining the people involved as subjects in need of correction and perhaps punishment. Not just a governor, however, he was also a war fighter and soldier, and another line of thinking could see them as soldiers and communities subject to, and protected by, the customs of war. Croft might complain that the captains had not taught their soldiers the difference between subjects and enemies, but it is not clear that even Essex was always sure which standard he was using.

Some of the customs of war that Essex had to deal with were those surrounding the siege and assault of a defended position. The small garrison in Darrilayrie surrendered when summoned, and the men were pardoned. Events at Cahir do not reveal as much about Essex's thinking; the full siege and breach resulted in the rebels sallying out to attack the besiegers. Essex defeated this desperate effort, and a "very few escaped" by swimming. The journal then glides over the details of what happened to the rest. Were they killed in battle? Taken prisoner? Executed? One source suggests all of the above: "What scaped the fyre was slayne with the sworde, & what scaped the sworde was brought to the gallowes & there hanged."[40] Justifications for this level of violence were ready to hand. Rebels who refused to surrender soon enough could be treated as "incorrigible" rebels worthy of hanging, or they could be seen as soldiers who had held out too long and therefore were subject to being denied mercy. Later incidents with combined Spanish and Irish garrisons reveal that the English could make this distinction, though not always consistently. When George Carew reported that the breach in the castle at Rincorran was not yet assailable, but that the Spanish were trying to negotiate a surrender, Mountjoy ordered Carew to accept it, but not for the Irish also in the castle.[41] Carew would later kill all the Irish defenders at Dunboy, who arguably held out after a breach had been made. A smaller garrison taken by Carew shortly after Dunboy made "some defence," but after they "saw the army draw before the castle they sued for their lives, which granted them they yielded it up."[42]

Another theoretical principle in the customs of war was the protection of noncombatants (priests, women, and children), a convention long endorsed by the Catholic Church.[43] Custom had created exceptions (notably the post-siege assault), and it goes without saying that there were many more "exceptions" in practice, but the concept still carried great weight, even in Ireland. The same James Croft, revealing both the weight of the concept and the frequency of its violation, complained about soldiers who "hath slain and destroyed as well the unarmed as armed, even to the plowman that never bare weapon, extending cruelty upon both sexes and upon all ages."[44] One sees both the expectation of protection and its violation when the Irish men of Shillelagh evacuated their homes in 1568 in the face of an attack by Edmund Butler (the Anglo-Irish lord discussed in chapter 1) assuming Butler would follow normal practice and merely burn their houses and leave. Instead his soldiers sought out and killed whomever they could find, primarily women and children.[45] In 1601 Lord Mountjoy allowed the Irish women and children to depart from the besieged Kinsale (despite their potential to add to the Spanish logistical burden if they had remained inside).[46] Arthur Chichester, a notorious spoiler of Ulster in 1602, upon finding five desperate children left starving by his own raids, collected victuals from his solders' knapsacks and left the food with them.[47]

One custom crucial to maintaining these rules and allowing for negotiated submissions was the ability to send messengers between hostile forces under the protection of a passport or a formal truce. The basic necessity of this protection was what led the Spanish at Corunna in 1589 to hang one of their own for violating it. In that same campaign, the English commander, hearing his own messengers threatened, threatened to retaliate.[48] There are countless examples of the English and the Irish allowing the free movement of these kinds of messengers and agreeing to parleys.[49] In 1595 then–Lord Deputy William Russell actually parleyed with the notorious rebel Feaghe MacHugh O'Byrne, who declined to submit but was nevertheless allowed to depart, and the fighting resumed.[50] These agreements were taken very seriously. Lord

Deputy Arthur Grey felt compelled to explain in some detail how the break-down of one parley into violence was not his fault.[51] Essex too observed the custom of parley, but in his case he allowed his notions of chivalry to trump good political sense. After returning to Dublin from his march around Munster, Essex finally moved his now-reduced army into Ulster, where he jockeyed for position against O'Neill's superior numbers. Predictably enough, Essex challenged O'Neill to a duel, which led instead to a half-hour-long private parley on September 7, 1599. The two men reached a truce cemented by the promise of "pledges" (hostages). This unusual private parley immediately created the suspicion that Essex and O'Neill might have colluded. Reports of the incident outraged Elizabeth, and Essex's protestations that O'Neill had been appropriately bareheaded and humble cut no ice with her. Panicked, Essex rushed to England against orders and was eventually denied access to the court. He launched an ill-advised coup, and soon faced the headsman's ax.[52]

In part these customs of war were sustained (to the extent that they were) by the English leadership's imagination of themselves. War remained a raison d'être for the English aristocracy. It was a source of identity and prestige, and, more practically, a means of advancement.[53] But to be all those things, war had to be done "right"—at least at the level of display, and sometimes at more fundamental operational levels. Much of Essex's march around Munster can be seen within the paradigm of previous English "marches" through enemy territories, whether Edward III's, the Black Prince's, Henry V's, or Henry VIII's, all of which were then remembered through a rosy haze. Although those marches all had hard political purpose behind them, and were executed with as much fire and sword as chivalry, the memory of them in sixteenth-century England tended to focus more on glory than on guts. That Essex saw his march in this way is suggested by his knighting of practically everyone involved—a solid eighty-one knights created in one summer.[54] And Essex was hardly the only aristocratic English soldier to shape his choices in accordance with the dictates of the chivalric code and a vision of martial honor. During a 1596 campaign in Leinster, Lord Deputy William Russell, upon being told that Sir Peter Carew had been killed nearby, summoned Captain John Chichester, saying, "It is tolde me that in this place the Quen did lose a verie worthy knight, . . . and I do hope that heare againe hir majestie shall find another as worthie as he"; Russell ordered him to kneel and dubbed him a knight.[55] Edmund Butler, at the tail end of his rebellion (examined in the last chapter), offered to prove his loyalty through a duel.[56] John Perrott, the English "president" of Munster, in 1571 agreed to a tourney of twelve foot and horse each to resolve Fitzmaurice's rebellion, promising a truce until the scheduled day. In the end Fitzmaurice declined to appear.[57] The consternation Perrott's challenge caused among his superiors did not prevent later repetitions. Indeed, although Essex himself had been rebuked by Elizabeth for his "mistress more beautiful" challenge in Rouen (she thought it inappropriate for a "peer of this realm" to challenge a "mere rebel"), when approached by a messenger from the rebel Rorie O'More challenging the English army to a group duel between one hundred men with shield and sword, "to fight it out to the death, without shott or pikes," Essex was so charmed by the offer that he rewarded the messenger, "a naked roague," and promised him a chain of gold if the duel actually took place. O'More apparently never followed up.[58] Considerations of honor also seemed to play a role in Essex's choice of his route of march; he feared the dishonor attendant upon leaving an enemy-held castle unmolested.[59]

Essex's cultural vision of war had prompted him to seek clear-cut martial glory in an environment fraught with more ambiguity than clarity. O'Neill continued at this stage of the war to shift back and forth deftly between rebel and humble supplicant, and he eventually humiliated Essex with yet another truce. This kind of ambiguity contributed to the violence of Essex's campaign, as did the normal customs of war and the growing demonization of the Irish that would eventually render the idea of a mutually honorable duel laughable. But equally crucial to the nature of violence in Ireland were the English military and financial systems there, the role of garrisons, and the fears of soldiers.

Garrisons, Finances, and Fears

In the early spring of 1571, Henry Sidney was ending his second term as the lord deputy and simultaneously surrendering his optimistic hopes for reforming the island, even as he was taking criticism for reform's financial costs.[60] He had hoped that he could establish English law and custom in Connaught, but the Fitzmaurice rebellion had spread there as well and was only now winding down in the face of the war of terror waged by Colonel Humphrey Gilbert. Sir Edward Fitton had been appointed the "president" of the province of Connaught, an office that Sidney hoped, when supplied with its own regional army, would prop up the normal course of English law. So much for the high politics. Lieutenant Thomas Lambyn offered up a completely different view.

Lambyn had been left in command of an English garrison in the major town of Galway in Connaught by his captain, William Collier. Lambyn's was a lonely job, and he was feeling abandoned. Writing in March 1571 he sought to remind Collier "what case I and the rest of your souldiers are in." "We goe so in daunger every daye of our lyves and the cuntrye is so bent agaynste my L. Presydent [Fitton]" that the inhabitants were virtually refusing to feed the men. The soldiers themselves, he said, were at the end of patience. He had taken Collier's advice to "looke veary well unto the souldiers," but he wished "to god that the cuntrye wolde looke better unto us then theye doe." Lambyn did not lack a martial spirit, but he feared being forced out. He was prepared to fight and become the "master of the cuntrye," but only if properly reinforced. Otherwise, he begged, "I praye god sende us shortely away from them."[61]

Lambyn's garrison was one of many hundreds the English deposited all around the country at various times over a century of war. Including only garrisons maintained at royal expense, there were ten such "sundry garrisons and wards" in March, 1569, immediately preceding Sidney's campaign of that year. The two largest were at Maryborough and Philipstown, of forty-four and twenty-four men each, with the others varying between four and thirteen men. And those were the paper strengths.[62] The situation in the spring of 1571 was perhaps even worse: fifteen garrisons in total, three in Connaught, consisting of twenty-one men at Roscommon, ten men at Shrule, and eighteen men in Ballintubber. Lambyn in Galway was probably part of President Fitton's personal force of fifty-two men.[63] A lonely job indeed. Maryborough and Philipstown were the quintessential garrison posts, having been explicitly established and built as marcher forts (originally named Fort Protector and Fort Governor) to protect an expansion of the English Pale in 1548. At least Maryborough came with some comforts that Lambyn probably lacked.[64] A 1560 map of the fort shows a rectangular wall enclosing an area 120 yards by 100 yards, with a blockhouse on one corner and a small castle on the other. No doubt of great importance to the soldiers was the brewery built for them just outside the walls. They also gained a mill in the 1560s. A barracks building sat inside the wall, a wall that promised only a minimum of protection against surprise and almost none against a determined assault. When John Harington saw it in 1599, he described it as "a forte of muche importaunce, but of contemptible strengthe."[65] Harington's judgment notwithstanding, at that moment Captain Francis Rush had just finished holding out for twenty days on nothing but horse meat, having been cut off from provisions by the surrounding rebels.[66]

It was indeed the issue of provisions that made such garrisons a problem, whether in expressly built forts like Maryborough or the wards in existing towns and castles. In wartime the Irish quickly isolated those English garrisons not accessible by sea, and even sea transport of supplies proved unreliable. Attempts to resupply garrisons generated many of the running battles and ambushes of the Irish wars. When not actively under siege, the garrisons usually lived on a "cess" of the surrounding countryside.

The "cess" is a complex and important process that deserves some explanation. One of the most expensive prospects faced by any late medieval or early modern monarch was the maintenance of an army. They typically lacked not only a taxation system capable of supporting a large force but also the bureaucracy to centrally administer and sustain that force. They preferred instead to distribute the burden of maintaining military forces among regional lords, who were then expected to serve the monarch when called. One of the great changes under way in the sixteenth and seventeenth centuries was the increasing replacement of private forces with state armies. For most of the century the Tudor deputies in Ireland relied on a small but decisive royal force supplemented by the private armies of the Anglo-Irish earls, as well as the earls' ability to tap into other Anglo-Irish and Gaelic forces. Private forces therefore both resisted and aided the expansion of English control. Those private forces were sustained by a Gaelic system (long since adopted by the Anglo-Irish) called "coyne and livery." Under coyne and livery, troops were placed in the households of the lords' tenants, who then had to provide them and their animals with food (without reimbursement). Troops thus housed and sustained were always on call to respond to the more or less constant threat of raids by neighboring chiefs.[67]

English policy in Ireland increasingly tried to achieve a similar decentralized maintenance of forces through expanding various limited royal prerogatives into a more general obligation called "cess." Cess was both a noun and a verb, a tax and a process. A general cess, payable in cash or, more often, cattle and food, was used to support the central English garrison.[68] In theory that central army provided protection, rendering private armies sustained by coyne and livery unnecessary and illegal. In practice the use of bulky produce to pay the cess meant that troops were "cessed" (billeted) on households wherever they happened to be. Although, unlike in coyne and livery, soldiers were required to reimburse their hosts for their lodging (and one meal a day) from their pay, the rates were fixed well below market values, took little account of the agricultural seasons, and depended heavily on the honesty of armed men.[69] Farmers on the marches or in newly "pacified" areas supporting garrisons correspondingly suffered. Larger campaign armies grew too large for such a system, and instead depended on provisions (and pay) sent from England.[70] The soldiers in those armies purchased food with their pay from markets set up in camp or from official stores brought from England. When, as often happened, pay or supplies failed to show, the soldiers turned to the countryside, and their captains felt it unjust even to try to stop them.[71] Often the captains lacked the power to stop them even if they chose to; their lives and safety depended on their men.[72]

Lambyn was in part warning Collier about this probability. He was afraid, and the supposedly friendly locals whom he was protecting were refusing to provide the food he believed was owed to his men. If things did not change, the soldiers would take it for themselves and thus increase the likelihood of rebellion. At least Lambyn seemed honest. Some, perhaps most, captains treated their companies like semi-independent fiefdoms whose arms could make them into petty local rulers. Since the supply system depended on squeezing the locals, the captains found it all too easy to squeeze a bit more directly into their own pockets.[73] Bishop William Lyon complained in 1582 that Captains Barkely and Smith demanded cess for 112 men, only maintained 50, did not protect the population from the raids of the Irish, and still did not feel they were getting enough. Their men foraged around the countryside, aided by their horseboys. The ready, almost natural reliance on local farmers as the source of their food created a set of expectations, one that can fairly be called a specific military culture, in which soldiers saw peasants and their produce as their legitimate prey. When peasants resisted, soldiers revealed the extent of this separate military identity, claiming, "Ye are but beggars Rascalls and Traytors and I am a soldier and a gentleman."[74] This attitude persisted, with profound consequences, among professional soldiers in the early years of contact with Native Americans in the new world.

The captains, seneschals, and commanders of small garrisons practically competed at corruption, while the soldiers were prone to "prey" on the countryside rather than cess it. On the other hand, it was equally true that the conditions of soldiers' lives were often hellish.[75] A soldier's official ration was substantial: 1 to 1.5 pounds of biscuit or bread daily, substantially fleshed out by weekly issues of 2 to 2.5 pounds of salt or fresh beef, 2 pounds of cheese, 1 quart of oatmeal, and a large portion of dried cod (for Friday).[76] Not to mention a daily half pint of sack (wine), a daily quart of beer, and on every second day a quarter pint of whiskey.[77] This food allowance substantially exceeded the expectations of a typical peasant in the period, perhaps contributing to the soldiers' sense of privilege.[78] Even in the best of times, however, this ration was frequently adulterated or shorted by captains or by contract victualers taking their cut. A garrison cut off by enemies or an army on the march quickly lost its ration entirely, becoming desperate and often dependent on the countryside—especially on Irish cattle, the main measure of wealth in Gaelic Ireland. Recall Captain Francis Rush and his garrison's resort to horse meat for twenty days in 1599.

Other aspects of campaigning in Ireland contributed to the soldiers' misery. Harington complained that sixty "lusty men" of the Roscommon garrison died from "drinking water, and milk, and vinegar, and aqua vitae, and eating raw beef at midnight, and lying upon wet green corn oftimes, and lying in boots, with heats and colds."[79] Captain Thomas North's company was described as "a most miserable, unfurnished, naked, and hunger-starven band," whose lack of shoes had led to "rotted" feet and legs.[80] The starving garrison in Newry in 1598 mutinied and nearly killed a paymaster sent from Dublin.[81] Worse than victualling problems, however, was disease. Always the great killer of a premodern army, it seemed to the English to be worse in Ireland than elsewhere. It was primarily disease that reduced Sir Henry Docwra's garrison at Lough Foyle in 1600 from four thousand to fifteen hundred within a year.[82] It beggars the imagination and numbs the senses to contemplate the likely environment created as thousands of men, cooped up in close quarters, lay dying from dysenteric diarrhea.

Meanwhile, the soldiers were expected to fight. Lambyn was fortunate in having a purely defensive mission in Galway, and such roles were common enough in every war (and in "peacetime"). Some garrisons, like Maryborough or Philipstown, were located to defend nearby settled or anglicized countryside. Others were situated as a deliberate burden on submitted Gaelic territory in a kind of peacetime "pressure by cess" strategy. As the captains recommended to Essex, however, garrisons could also be given an aggressive wartime role.[83] Deliberately situated in hostile, "rebelling" countryside, the garrison by its very presence would denude the surrounding countryside of food and forage, making enemy military movement more difficult.

Larger garrisons, especially those made possible during the Nine Years War, could do more than merely eat the country out; they could mount longer raids to deliberately devastate the country in an expanding radius. The best example of this strategy were the frequent English efforts to place a large force in the Lough Foyle, dividing Ulster in two and separating the O'Donnells from the O'Neills. An early attempt against Shane O'Neill in 1566 was partly successful, but true success came only with Sir Henry Docwra's expedition in 1600.[84] Docwra landed on the shores of the Lough (essentially an estuary) with four thousand foot and two hundred horse. Over the succeeding months he established a chain of garrisons from Culmore, to his main base at Derry, to Dunnalong, and finally to Lifford.[85] Docwra also planned a whole series of garrisons to connect his main base at Derry to Donegal and then eventually to the key crossing of the Erne River at Ballyshannon.[86] His forces distributed deep into Ulster, and supported by the Gaelic chiefs Cahir O'Dogherty and Niall Garbh O'Donnell against his O'Donnell kinsman, Docwra was able to devastate large sections of the countryside and place even more outlying garrisons. The Lifford garrison, for example, burned Newtown in March 1601, took two hundred cows, and "put to the sword more than a 100 persons, of men, women, and children, this being the second time it hath been preyed within these four months."[87] In all of this raiding, despite Docwra's access to seaborne supply and the careful positioning of his main forts on the river, he could still complain that he should not be held responsible for the killing of civilians when the government's supply was so inadequate.[88]

Docwra was in the relatively privileged position of leading a substantial force in mutually supporting garrisons that could communicate with each other by water. More common were the smaller, isolated garrisons, who had a great deal more to fear. After O'Neill's success at the Battle of Yellow Ford in 1598, he sent troops into Munster and Leinster to stir up trouble. They "spoyled the Country, burnt the Villages, and puld downe the houses and Castles of the English, against whom (especially the female sex) they committed all abominable outrages." In a single month, "the English were murthered, or stripped and banished."[89] Captain Harington speculated that the troops had absorbed Irish beliefs about magic, and further believed that some of the Irish had magical powers. One besieged chief seemed to vanish "by dint of witchery" through solid walls.[90] Another English observer of the Munster and Leinster troubles after 1598 complained of garrisons "not once barkinge or shewinge themselues," or even, being "half naked, and many of them sick, and allmoste starved," allowing rebel forces to march through towns they were intended to defend.[91]

One source of the soldiers' fear was the Gaelic style of war and the very geography of Ireland. Ulster remained largely terra incognita for much of the century. English commanders understood its coastline and the significance of a few key entry points (the Moyry Pass, Ballyshannon, and the Lough Foyle) and had long held Carlingford and Carrickfergus. But the interior remained unclear, if not a blank. In a vivid example of English ignorance of alternate channels of movement, on September 21, 1561, Lord Deputy Sussex claimed that he had Shane O'Neill on the run, fleeing "from wood to wood without offering any skirmish," but then just a few days later O'Neill's forces appeared in Sussex's rear and burned four villages in the English Pale.[92] Even in areas where the English knew their way around, the woods and bogs remained forbidding. The Irish regularly used them as places of ambush, "plashing" the trees together to block the "passes." The lightly equipped Irish kerne readily passed into and out of the bogs. In the figure on page 31 the Irish emerge from the bog to attack and seriously damage Henry Harington's force near Wicklow in 1599. It was these men whom Essex would later decimate for cowardice.[93] A part of Essex's army led by Captain Bosworth and Captain Gardener once pursued the Irish "very rashly into the bog (and one of them being heavy laden with his armour stuck fast in the bog), [where they] were slain before they could be seconded or rescued."[94] Before the campaign had even begun, Captain Thomas Reade had worried over Essex's ignorance of Irish conditions and tried to advise him that the Gaelic "manner of fight will be by skirmishes in passes, bogs, woods, fords, and in all places of advantage. And they hold it no dishonour to run away, for the best sconce and castle for their security is their feet."[95] And who would not fear men who "assayle the Enemy with rude barbarous Cryes, and hope to make them afrayd there-with, as also with their nakednes, and barbarous lookes," men who "never sparing any that yeild to mercy, yea being most bloudy and cruell towards their Captives uppon cold blood, . . . not only mangling the bodyes of their dead Enemyes, but never beleeving them to be fully dead till they have cutt of their heads."[96]

Irishmen in bogs and woods were bad enough, but among his other fears, Lambyn probably worried about the Irishmen in his own ranks. The same men who in other circumstances the English called scorpions, vipers, or cat-erpillars were regularly recruited into the ranks of English companies in large numbers—especially during the Nine Years War.[97] At least those men would have been somewhat visually disguised in the official clothing issues from England.[98] More visibly worrisome were the units of kerne hired into English service or serving with a Gaelic or Anglo-Irish chief nominally on their side. Trust often ran short; even the lord deputy sometimes arranged to keep the allied camps separate from his own.[99] In contrast to the kerne in Irish service, who were often simple men obligated as tenants to serve in a chief's "rising out," the ones in English service—whom the English knew best—were mer-cenaries who blended the worst aspects of Gaelic war with the worst part of being hardened veterans. In 1600 one writer reminded Mountjoy that the kerne were "cruell & bloudy full of revenge in deadly execution, licencious swerers and blasphemers, ravishers of women & murtherers of children."[100] English soldiers told stories of an Irish kerne sent to serve in France who had dueled with a Frenchman and returned swimming across the river with the Frenchman's head in his teeth.[101] These men, who cut their hair long in a forelock (or "glibb"), hiding their eyes, were capable of anything. They had killed Thomas Smith in his own house, boiled him, and fed him to the dogs.[102] They cut off men's heads and set them up with their genitals stuffed in their mouths.[103] In a tight spot, could anyone trust they would not betray the English on the battlefield? It had already happened far too often.[104]

Prejudice and fear go hand in hand, and the deep and broad English prejudices against the Irish are amply documented.[105] Those prejudices seem to have worsened in the sixteenth century and further expanded to encompass the Anglo-Irish, increasingly called the "Old English," who had become a little too Gaelic and were now a lot too Catholic. Most wars require some form of demonizing the enemy, but the supposed "incivility" of the Irish provided a strong foundation for the transformation of "subjects," even the Anglo-Irish, into barbarians. Any longish account from the era, whether written by officials, pamphleteers, soldiers, travelers, or even in policy recommendations, included a section condemning the habits, lifeways, attitudes, superstitions, laws, and now religion, of the "wild Irish." Continuing to emphasize the views of the captains and the soldiers, one finds Captain Barnaby Riche, who had fought in Ireland for decades from the 1570s on, describing the Irish in 1615 as "by nature given to murder, theft, robbery, rape, ravin, and spoil" whose "savage maner of incevylyte . . . is bred in the bone. [T]hey have yt by nature, and so I thynke of ther inhumayne crewelty, that are so apt to rune into rebellyon."[106] Captain Nicholas Dawtrey also believed Irish incivility to be rooted in the bone, suggesting that "an Ape will be an Ape, though he were clad in cloth of gold."[107] Captain Thomas Gainsford repeated many of the standard complaints against the Irish, reminding his seventeenth-century readers that the Irish thought theft no crime; their priests bred illegitimate "strumpet" daughters; they engaged in "strange" hasty marriages; believed being lice-ridden no disgrace, and preferred a savage diet of raw flesh and blood accompanied by whiskey.[108]

Even those captains willing to acknowledge the humanity of the Irish, who could believe there to be no "better souldier under the sune," or that, "tho' I was sente oute to fighte withe some, there did appeare no reason for my not eatinge withe others," increasingly believed that an Irishman's word meant nothing and that the holding of an Irishman's relatives hostage was an equally poor guarantee of submission.[109] This shift was of great importance.

Much of the possibility of restraint in the Irish wars depended on defining victory as the submission and reformation of sept or clan chiefs, not the destruction of the enemy or even of his army. As belief in the reliability of submission faded, so did the hopes for reform and restraint. The same policy recommendation from 1600 that described the kerne as "cruell & bloudy" further discounted the sincerity of any Irish submission and suggested that obedience would only come through keeping "them in continuall subjection."[110] Captain Riche was a little harsher a little earlier. Writing in 1578 he blamed the continual wars in Ireland on "the mercie & lenitie that is used amongest them: and that the onely means to bring the people soonest to conformitie, and the countrie to quietnesse, is without compassion to punishe the offenders, and without either grace or mercie to execute the rebelles, and such as be malefactours."[111] Riche kept up that particular drumbeat in the 1590s, complaining that "pardons and protections" made the Irish bold "to enter into actions of rebellion; for what care they what mischief they commit, when they can still warrant themselves a pardon for a few stolen cows." In the end, he argued that "this reformation must be settled by the sword, not by composition and taking in of the rebels."[112]

And thus returns the paradox of how to treat subjects in rebellion. While English and European traditions allowed for selective justice in the wake of a rebellion in order to showcase the mercy of the monarch and to pacify the countryside, the same traditions allowed for severe, indiscriminate execution of rebels. In Ireland persistent rebellion upset the normal balance between mercy and judicial terror. Applying the methods of judicial terror (martial law) on a wide and sometimes nearly continuous basis undermined English belief in the possibility of the Irish as subjects and surely convinced the Irish that they were not being treated as such in the fullest sense. From an English point of view, however, martial law was nevertheless *law*. Having a commission of martial law excused a very great deal. It cleared consciences and provided certainty.[113] Customs and traditions of restraint in international war, fragile as they were at the best of times, fell apart in the face of the competing tradition of harsh measures for Catholic rebels, who, as it happened, were also widely believed to be barbarians to their very bones. As much as Essex could imagine dueling with Hugh O'Neill (or even with Rorie O'More), a more telling story emerged at the end of his battle at Arklow. Essex's forces had dispersed the Irish troops into the bogs and woods and pursued them some distance. One of the rebel leaders, Phelim McFeaghe, called out from the woods to an Irishman in the English army, and asked him to pass a message on to Essex. McFeaghe "humbly craved leave" to speak with Essex and asked for a pledge of safe return. He almost certainly intended to submit, something that Essex had sought all through his campaign. Furthermore, a contemporaneous English author on military discipline specified that, "should anyone, of any nation, friend or foe, come to camp and desire to speak to any of the officers, they must be escorted to them" and treated "gentlie whatsoever they be."[114] Captain Thomas Lee, too, had recommended strict adherence to the safe conduct of messengers.[115] At first, Essex appeared open to the idea; he replied that if McFeaghe came into Arklow "as a repentant rebel" and tendered "his absolute submission," then "he should have such a safeconduct." But Essex's postscript, combined with a checkered English history of observing safe-conducts kept McFeaghe in the woods: "But if [McFeaghe] sent in any other form, or to any other purpose, he [Essex] would execute the messenger; for he would never suffer his commission to be dishonoured by treating or parleying with rebels."[116]

Essex's governorship was a disaster. Massive expenditure and an unprecedented commitment of military resources had produced little result. Patience for the war was wearing thin on both sides of the Irish Channel when suddenly the danger escalated dramatically with the arrival of the Spanish expedition at Kinsale. Essex's replacement, Lord Mountjoy, immediately turned from suppressing the rebellion in Ulster to defeating the Spanish landing and O'Neill's attempt to reinforce it.

Even after he succeeded, Mountjoy had to return to the problem of pacification. The chorus of voices recommending increasingly harsh measures had risen to a crescendo, and O'Neill's open alliance with the Spanish had erased any trace of his status as "subject" (as it did for all those who joined him in rebellion). The military forces at Mountjoy's disposal remained unprecedented for the wars in Ireland. Although Elizabeth expressed a desire to continue to seek submission to mercy, even (secretly) for Hugh O'Neill himself, on a local level such willingness to accept submissions without proof of active service against other Irish essentially vanished.[117] One Captain Flower, ordered in April 1600 to "burn and spoil" rebels in Munster if they would not submit and give pledges, interpreted his orders broadly. After his march he reported to have "burned all those parts, and had the killing of many of their churls and poor people, leaving not them any one grain of corn within ten miles of our way, wherever we marched, and [we] took a prey of 500 cows, which I caused to be drowned and killed, for that we would not trouble ourselves to drive them in that journey."[118] The calculated use of devastation as a form of political pressure, greatly amplified by the size of Mountjoy's force, was shifting into a strategy of outright starvation.

Fields of corn and herds of cattle had long been targets; now the general population who tended those fields and herds became targets on a larger scale and on a year-round basis. A few captains like Barnaby Riche and William Mostyn had been pushing for this course even before Essex had arrived in Ireland. His failures and the Spanish threat gave such extreme measures more credibility. Mostyn, a twenty-seven-year veteran of the Irish wars, demanded that no "protections or pardons be given" in Ulster except to those who would "promise to kill and draw blood (and do it) upon principal traitors." He further believed the sword a poor instrument compared to famine. "Those not cut off by fire and sword will in a short time be dispatched by famine."[119] For Riche, the time for "gentleness, mercy, and clemency" had passed. He admitted that "Princes should by all means endeavour to advance their subjects in prosperity," but argued tellingly that "these be the Pope's subjects, and therefore your Majesty's protested enemies, and a greater policy for your Majesty to reform them than to enrich them." Although Riche might call it a "reform," he argued that "this reformation must be settled by the sword, not by composition and taking in of the rebels, by pardons, by protections, by putting in of pledges," and "famine must be an especial mean whereby to accomplish it."[120] Using hunger as a weapon was not new: Lord Deputy Arthur Grey de Wilton in 1583 had declared his hope to have driven the Irish to "famish, fight or yeelde." But what Mountjoy, Carew, Docwra, and Chichester achieved in 1602 and 1603 was a horrifying new scale and intensity in its application.[121] They probably believed they had no other choice.

Conclusion

Mountjoy's campaigns and those of his subordinates were a calculated assault on the productive population. Governmental fears and frustrations provided him with the capacity to carry them out. Furthermore, the captains had only the most tenuous control over their soldiers, and they themselves frequently abused the local population (and their own soldiers) in search of profit. Over a century of fighting in Ireland, the soldiers and officers had evolved a military culture in which they simultaneously feared and disparaged the Irish, excusing almost any violence committed against them as necessary for the establishment of order. By the Nine Years War it seems fair to argue that the only restraint remaining on English forces in Ireland was Elizabeth herself. Elizabeth so often repeated that "no extirpation was intended" that one must take her intentions seriously as part of a vision of the Irish as subjects. Even when she ordered a persistent rebel like Shane O'Neill "be utterly extirped," that intention did not extend to all of his followers, much less to a whole region or people. Nor was it applied to every rebel, as suggested by the number of pardons issued both by Sidney in 1569 and by Essex in 1599, including pardons of the principal rebels.[122]

But taking seriously her assertion that there was "no extirpation intended" means also taking seriously her definition of her rights as a monarch and the broader English assumption of cultural superiority. Elizabeth believed in her legitimate right to punish rebels severely. Furthermore, much as Essex was motivated by his own sense of honor, Elizabeth had a costly and demanding sense of her honor as a sovereign prince. In 1597 she informed the earl of Ormonde, then acting as her primary leader against the rebellion, that she would continue to accept appropriately humble submissions from those who came in "with bended knees and hearts humbled." Honor demanded and the law (as she understood it) allowed that the Irish rebels be treated not as "one Prince did treat with another upon even terms of honour or advantage, in using words of peace or war, but of rebellion in them, and mercy in us."[123] Extending mercy in such a case might be politic, but in her eyes it was not morally required. Either way, her conscience was clear.

As the final war seemed to drag on beyond reason, especially under the added threat of Spanish intervention, even Elizabeth's conditional restraint was abandoned, and she approved Mountjoy's tactics of deliberate devastation designed to generate starvation.[124] The fear of Spanish intervention reflected the way the religious schism in Europe had become such a powerful cause of violence. It created an alternative platform of legitimacy that undermined traditional customs of war and allowed for religion to be used as a cause against a monarch, leaving Elizabeth feeling existentially threatened.[125]

Even when Elizabeth might have preferred restraint, her agents were military men empowered by martial law and imbued with a substantial ethnic prejudice against the barbarous Irish. They found it easy (and profitable) to choose means that discarded the limited restraints customary to military practice or to the normal course of law expected for subjects. Even those committed to imposing order had to calculate a way of war against an enemy that waged irregular war and lacked a vulnerable urban infrastructure. The English aristocracy's desire to wage war in an "honorable" way did provide some restraint, but persistent rebellions overwhelmed even the highly developed sense of chivalry of a man like Robert Devereux. The desire of some powerful men, like Robert's father, to found great landed estates at the expense of the Irish (and the Scots in Ireland) greatly aggravated the situation. In the end, the Irish were believed to be less brothers than barbarians, whose land was coveted by others. The poet John Donne, a veteran with Essex of the raid on Cadiz, and a man deeply emotionally affected by his experience of war, nevertheless opined in one of his elegies that

> Sick Ireland is with a strange war possessed
> Like to an Ague; now raging, now at rest;
> Which time will cure: yet it must do her good
> If she were purged, and her dead vain let blood.[126]

There could scarcely be a better epitaph for English good intentions.

The English sense of the Irish as Catholic barbarians wormed its way ever more deeply into their bones over the next generation, as did their belief that the Irish were a conquered people. In 1633 King Charles I's chief advisor and lord lieutenant in Ireland would even say, "Ireland was a conquered territory, and the king could do with them what he pleased."[127] But that same man unwittingly tapped that powerful and volatile English prejudice when he proposed using Catholic Irish troops on the king's behalf in Britain. Sadly for him, it was powerful and volatile enough to thrust him onto a scaffold in London in the spring of 1641 on what would prove to be the eve of civil war.

Part II

Codes, Military Culture, and Clubmen in the English Civil War

In 1642 Charles I, king of England, Scotland, and Ireland, found himself at war with his subjects in all three realms. These wars would shift, change shape, and overlap each other, but it is reasonable to point to the so-called first English civil war, lasting from 1642 to 1646, as the key and core to the outcome of the others. This was the famous one; the war of Parliament versus king, Puritans versus royalists, Roundheads versus cavaliers. It was also a true brothers' war; both sides fought primarily within England and Wales with soldiers drawn mainly from those same places. The state, divided against itself, struggled to raise and sustain two armies, to fight according to the developing European codes and customs of war, and to control and discipline its soldiers in the new collective discipline of battle believed necessary for success, but which also required limiting their off-battlefield violence. Both sides continued to imagine themselves as fellow subjects, and they anticipated eventual reconciliation and a return to civil society. Both sides were also desperate to cultivate the support and loyalty of the countryside. In comparison to concurrent European wars or the wars in Ireland and Scotland, the result was remarkably restrained. The war's violence nevertheless had a lasting impact on the future political history of both England and America.

3

Sir William Waller, 1644

So long as I held any employment in the armies, I constantly endeavoured to express all the civilities I could to those of the adverse party, that so our differences might be kept in a reconcileable condition; and we might still look upon one another, according to Aristotle's rule, as enemies that might live to be friends.

Sir William Waller

The Death of Strafford, 1641

Isolated by the political maneuvers of his enemies, vilified by the London public, and ultimately abandoned by his king, Thomas Wentworth, earl of Strafford, made a noble display on the scaffold. Wearing a black suit, long cape, and cheerful countenance, he attempted to address the tumultuous crowd but was at first shouted down. Crowds at executions in London could never resist a final speech, however, and eventually they relented. He forgave everyone around him, even exchanging the traditional mutual forgiveness with the headsman. He asserted his faithfulness to the good of the kingdom and to the true Protestant religion. He expressed his gratitude to the king who had spared him the half hanging, disembowelment, and quartering usually meted out for treason. He prayed repeatedly, both silently and aloud, swept his hair off his neck, adjusted his head three times on the block, and waved his readiness. The headsman succeeded in one blow and displayed the severed head to the crowd. Depending on the account, the executioner either laughed or said "God save the king." That prayer went unanswered: within eight years the king himself would mount the scaffold.[1]

The crowd in the background on this twelfth day of May, 1641, was more than mere window dressing. The city of London had roused itself for a major spectacle. No mere nobleman, Wentworth had recently been made the earl of Strafford; he was one of King Charles I's leading advisors, and he served as lord lieutenant in Ireland. Around 200,000 people crowded around Tower Hill, perhaps half of London's population, jamming "all the windowes, house topps, trees, steeples and scaffolds" until two of the latter collapsed under the weight. As Charles's advisor, Strafford had long been embroiled in the political struggle with the men in Parliament who had just declared his fate. But it was Strafford's actions in Ireland that most stirred the people of London now gathered to cheer the death of a man they arguably helped kill. Strafford's maneuvers and schemes to raise a Catholic Irish army to deal with Charles's enemies in Scotland, and perhaps even in England itself, had generated a popular frenzy. The crowds had surged around him when he was summoned and charged in January. They had cursed him when he was committed to the Tower. They had petitioned Parliament claiming anywhere from eight thousand to thirty thousand signatures denouncing the continued existence of Strafford's "Irish Popish army" and encouraging Parliament to get on with his trial.[2] The House of Commons successfully impeached him,

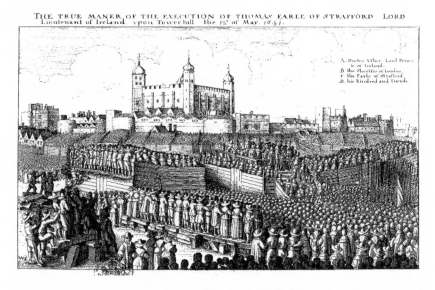

THE EXECUTION OF LORD STRAFFORD. Drawing by Hollar Wenceslaus. *The true maner of the sitting of the Lords & Commons of both Howses of Parliament upon the tryal of Thomas Earle of Strafford, Lord Lieutenant of Ireland*, Wing T2762 (London: s.n., 1641).

but the trial in the House of Lords proved more of a problem. How do you convict a man of high treason when the king still trusts and supports him?[3]

These twisted legalities, the loyalty of some lords to the king, and Strafford's able defense of himself combined to deadlock the proceedings. Strafford's enemies in the House of Commons could not accept deadlock and so resorted to the ancient device of voting to "attaint" Strafford for treason. An attainder vote required merely persuading a majority of "ayes" rather than the stricter standard of proof in a trial. The bill of attainder rapidly passed the Commons. The House of Lords proved harder to convince. While they dithered, and while the king seemed unlikely to sign the bill, impatience in the streets swelled in volume and intensity and soon proved politically decisive. Clergy demanded Strafford's execution in their sermons. Thousands of demonstrators filled the streets around Parliament, literally pressing the peers on their way to deliberate in the house. "Justice and execution," they demanded, promising, "They must have it or would take it."[4] Those lords who emerged from their coaches to promise just that were drowned in cheers. Rumors swirled. Fear spread through the streets and the halls of power that the king would import French or Dutch troops and forcibly spring Strafford from the Tower. The daily crowds outside Parliament grew rougher and more militant. The House of Lords finally agreed to the bill of attainder, and Strafford himself let the king off the hook, offering to go to his death if it would stabilize Charles's hold on his throne. Strafford held little doubt about the role of the crowds in swaying the vote in the House of Lords, and other commentators at the time agreed.

Strafford went to the block well before the start of actual warfare between Charles and Parliament. Nevertheless, his execution highlights key issues about the nature of the brothers' war that followed. The leadership on both sides remained invested in the social structure that had made them leaders in the first place, and they struggled to confine their war within social, cultural, and even "legal" boundaries.[5] Strafford was no barbarian; he was a member of the same social caste and ethnic elite as all other members of Parliament. And therefore his trial had to observe all the legal formalities. At the same time, however, Parliament's defiance of the king necessarily involved redefining the state's relationship to its people. Can a state's interests exist separately from, and even conflict with, the declared interests of the king? That question stretched the laws to their breaking point and beyond, both in Strafford's trial and later, when asked in the context of actual warfare. After all, how could the king's advisor commit treason against the king while doing the king's bidding? English law had long defined treason relatively simply as "crimes against the king, his family and his servants."[6]

But Europe at large, and with it England, was slowly reevaluating the nature of sovereignty and the state. In Alan Orr's interpretation, "treason was not simply a crime against the king's natural person or a breach of allegiance but had increasingly become the unlawful seizure of sovereign or state power."[7] The people themselves, as a corporate body, could now be imagined with the king as a component of the state, even if the more radical notion of the people themselves constituting the state had not yet taken hold. Thus an attack on the people of the state could become treason when done by the king's trusted advisor or, in the end, even by the king. Ideology and events thus conspired to redefine the king's relationship to the state and the state's relationship to the people.[8] Subjecthood and sovereignty again proved key in shaping wartime violence. Under the pressure of war, the shifting definitions of who belonged, who was a traitor, and who should be considered as a subject after some future peace drove calculations of how to win; affected the capacity of a state divided against itself to fight, supply, and control its troops; and even remolded the cultural understanding of war.

Strict legalities aside, Strafford's trial mixed a concern for law with a desire for revenge, thereby foreshadowing the nature of the war to come. His prosecutors may or may not have been innovating in law in the strictest sense, but they surely knew they were pushing its boundaries. The willingness of Parliament to go to such lengths suggests the potential of a civil conflict over the nature of subjecthood and sovereignty to dissolve into retributive, even if marginally legal, justice. Here the legal system did operate, forcing Parliament back onto the device of attainder. Even more, however, the drama in the streets, rather than the legalities of the trial and attainder, revealed the potentially explosive and decisive role of popular attitudes in shaping violence.[9]

Strafford's trial hints at the potential for a brother's war to be restrained and frightful at the same time. The combatants remained deeply invested in some notion of order, and when at war they defined "order" as adherence to a developing European set of codes and customs that outlined acceptable behavior in wartime. The pressure to conform to the codes arose from the traditional military sources: a culture of honor, calculated strategy, fear of retaliation, and so on, but also from the combatants' hopes for eventual reconciliation and a return to civil life. In no way were the codes perfectly followed. The competing sides failed to provide for soldiers, who then turned on the countryside. Military calculations sometimes overrode the dictates of the codes, but equally often concern for popular reaction pressured leaders to control their troops. As the war dragged on, a separate military culture emerged that clashed with the localist cultural values of the English countryside, generating more violence.[10] But, before turning to the specifics of wartime violence, one must begin with how these brothers came to blows.

King versus Parliament

In March 1625 Charles I ascended to the throne not only of England but also of Scotland and Ireland. To a large extent the pressures of managing three different kingdoms, each with its own parliament, its own traditions and apparatuses of government, and worse, its own internal struggle over religion, were responsible for the outbreak of war. Indeed, many scholars now refer to the ensuing conflict as the "British Civil Wars" or the "War of the Three Kingdoms." Beyond this basic agreement about the centrality of multiple kingdoms and religion within them, historians' disagreements over the war's exact causes are legion. The brewing conflict between Charles and the largely Puritan Protestant opposition within England came to a head when the Calvinist "Covenanters" of Scotland in 1638 rejected Charles's religious authority there and rapidly overwhelmed Scottish royalists.[11] Charles raised an army in England and tried to outface or defeat the Scots before the expenses of war forced him to call a parliament and face his critics in England. The campaign sputtered out, a provisional treaty satisfied no one, and Charles returned to England. On Strafford's advice, he reluctantly summoned the English parliament to raise money to pay for a new campaign against Scotland. Charles had avoided calling a parliament for eleven years, and when the members gathered in April 1640, they proved more interested in their grievances with Charles than in Charles's grievances with Scotland. Charles sent the "Short Parliament" home after only three inconclusive weeks.

At the close of the Short Parliament, Strafford made his fateful recommendation to Charles to "Goe on with a vigorous warr . . . loose and obsolved from all rules of gou'ment." And since the English Parliament refused to cooperate with raising money, he suggested that Charles should fight this war with his heavily Catholic army (then in Ireland), which "you may employ here to reduce this kingdome."[12] Meanwhile, Charles struggled in the face of extreme public reluctance and little income to raise troops in England. The Scots, emboldened by this, preempted him by invading England. The Irish army had not arrived, and the Scots defeated Charles's raw army at Newburn on August 27, 1640. The peers of the realm, in council with Charles, pressured him to call Parliament again. This new "Long Parliament" remained in session until purged by force in 1648. The Long Parliament began the attack on Charles's councilors, their eyes quickly fixing on Strafford.

Although it was lethal for Strafford on that fateful day in May 1641, even then the conflict between Charles and Parliament was not yet ready to blossom into war. Ireland proved to be the crucial catalyst. The Gaelic Irish rebelled in October 1641, and, as reported in London, they did so in a way horrible beyond conception. Decades of pent-up resentment erupted in what at least seemed to be an orgy of mutilation, torture, slaughter, and destruction, all tinged with religious overtones. Exaggeration no doubt played a role, but, however overblown, the reports electrified the English public, and their vision of the situation bordered on the apocalyptic.[13] One member of Parliament wrote of the popular fear that the "bloodie murtherers [were] like [to descend] upon us like a swarme of caterpillars." The Puritan artisan Nehemiah Wallington feared that the rebellion was "but one plot against England, for it is England that is that fine, sweet bit which they so long for, and their cruel teeth so much water at."[14] Richard Baxter, writing after the war and explaining its causes, concluded that it came about "above all by the terrible Massacre in Ireland, and the Threatnings of the Rebels to Invade England."[15]

Clearly an army from England would be required to suppress the rebellion, but who would control it? In the midst of the ongoing visceral struggle for authority between Charles and a parliament who had only recently executed Strafford for threatening to use an Irish army, that question proved unanswerable and arguably led directly to war. Parliament feared that any army raised under the king's control would first be used to restore his power in England. As the rhetorical war escalated, both sides maneuvered to secure control of the material and moral resources critical to what more and more people believed was incipient civil war.[16]

The role of Ireland and Scotland in bringing English political conflict to a military head raises an important point. Religious diversity between and among the three kingdoms combined with ethnic preconceptions and memories of old wars to render the Irish and Scottish branches of this "War of Three Kingdoms" significantly more violent than its English component. But from an English point of view, those were not brothers' wars. At best the Scots or Irish were foreign enemies; at worst they were heretics and barbarians.[17] The focus in this chapter is on violence within the English civil war, the royalists versus the parliamentarians, and how they treated each other and the English countryside.[18]

Even considered within this limited English scope, the conflict was complex, its strategies often opaque, and its movements amorphous and nonlinear. Royalist and parliamentarian calculation of what victory meant and how it should be achieved shifted with the seasons. If there was one constant thread, it was each side's need to recruit more soldiers to continue the fight, and therefore the need to defend their best recruiting grounds and to attack the enemy equivalent. Although there were certain geographic consistencies to that process, there were also shifting patterns of enthusiasm, willingness, and even alliance (especially in relations with Ireland and Scotland). For the royalists the person and appeal of the king himself was crucial. His status as the anointed monarch was a powerful draw, and it also lent his cause an undeniable legitimacy in the laws of war as they were then understood.

Christian thought since St. Augustine had wrestled with the conditions under which war would be approved by God, and the dominant opinion, as it was finally summed up by Thomas Aquinas, had settled on three conditions: it must be fought in a just cause (defensive war was always just), it must have a just intention (usually defined as seeking a good end, and peace as an outcome), and it must be declared by a "proper authority." As originally conceived, "authority" was tied to the Catholic Church or to the nominal heir to the Roman imperium—the Holy Roman Emperor. The Reformation, and with it the growing competition of increasingly sovereign princes, had loosened those older definitions. "Proper authority" had become virtually any sovereign, and offensive warfare was increasingly justified if in pursuit of some imagined debt or unsatisfied reparation.[19] Lacking the person of a king, one parliamentarian banner resorted to the simple argument that "That Warre is Just Which is Necessary."[20]

For many, however, even the new, looser definition of a just war still leaned on the semi-mystical person of the king. And so it was only appropriate that Charles, attended by three troops of horse and six hundred infantry (or "foot" in the parlance of the time), formally began the war by raising his standard at Nottingham on August 22, 1642, declaring his opposition to be rebels. Raising the standard was an act of tremendous symbolic and legal significance, a fact dwelled upon at length by a parliamentarian pamphlet published at the time. Among other things, it marked a declaration that the king was at war and could presumptively execute those in arms against him according to martial law.[21] The parliamentarian propagandist, however, dismissed any argument for the right of a prince to make war on his own people, arguing that "War hath its Laws and Ordinances as well as Peace," and that since Charles was violating them, he would find that "God favours just wars and giveth the victory to whom it pleaseth him."[22]

For the most part Charles drew his strength from Wales and the north and west of England, in addition to occasional and contested aid from Ireland and Scotland. Early on he established an almost unassailable base at Oxford but was regularly forced to campaign to hold on to the north and west and could only occasionally truly menace Parliament's most important strongholds in the east and south. Crucially, Parliament held London's loyalty and with it the city's vast manpower and financial resources. In the first years of the war, roughly 1642 and 1643, the royalists frequently had the upper hand on the battlefield, although their failure to take London and their loss of control of the English navy hampered their ability to make battlefield successes decisive and to raise money and import materials.[23]

Both sides quickly confronted the uneasy strategic conundrum of balancing several field armies, who generally went into winter quarters every year, with very large numbers of scattered garrisons, some small, some large.[24] The field armies were expected to decide the war. Their active movements occupied the attention of both sides' leaders, and they believed battle would either force Parliament to submit or bring the king to his knees.[25] Despite this strategic vision of the role of the field armies, the demands for garrisons seemed unavoidable. Although it is a late example, it is noteworthy that royalist forces in June 1645 were divided roughly in half, with 52 percent of their force in two field armies and 48 percent in garrisons in Wales, the west, and the Midlands. One contemporary estimated for that year the royalists had thirty-nine garrisons and Parliament thirty-six in just the midlands and Wales.[26] A modern estimate found as many as 160 garrisons in England in that year.[27]

Aside from a few in major cities or on top of key bridges, garrisons were generally intended to control space—they were like competing ink spots of red and blue on the map, ebbing and flowing over the countryside, sometimes washed away and replaced by the tide of a marching field army.[28] They gathered the income, produce, and recruits from the surrounding area to support themselves and to forward on to their central command. They patrolled the countryside to prevent the enemy from doing the same. They harassed enemy garrisons and made life difficult for passing enemy field armies. A very great deal of the war and its violence occurred within these small skirmishes and patrols.[29] Richard Baxter observed shortly after the war that, aside from a few counties firmly held by one side or the other, "almost all the rest of the Counties had Garrisons and Parties in them on both sides, which caused a War in every County, and I think there where few Parishes where at one time or other Blood had not been shed."[30]

The fate of the field armies in battles or sieges may have been decisive for the outcome of the war, but very often their size and logistical condition were determined by the background noise of the skirmishing and patrolling war.[31] Charles Carlton's sample of over six hundred combat incidents, ranging from the largest battles to very small skirmishes, found that only 17 percent of Parliament's men and 27 percent of royalist men were killed in the war's nine largest battles or sieges. Both sides lost almost half of their dead to incidents in which fewer than 250 men were killed.[32]

There is no room in one short chapter to recount all the events of even the English side of the war. But again, the events of one campaign, and the choices of one leader and his soldiers, can represent the problems of both restraint and violence in the broader war. No leader or campaign is truly typical, but a close analysis of a single march with its accompanying sieges and battles, and especially its interactions with civilians, reveals more about the nature and experience of the war, and the restraints on violence within it, than a fly-over view of the entire war.

Sir William Waller and the Cropredy Bridge Campaign, 1644

Sir William Waller is one of the most important of the forgotten generals of the English civil war. He shepherded a parliamentarian field army through a succession of victories, defeats, and rebounds from defeat for the early part of the war, from 1642 to the end of 1644. His reputation soared and plunged with equal rapidity. On one day a trooper could describe Waller as "resolute and vallorous," but on other days soldiers lost confidence in him and deserted in droves. Waller's military and political career experienced similar ups and downs. At the end of the war, when the New Model Army began to assert its independence from Parliament's control and then finally sought control of the government, Waller, no longer an officer in the army but still a member of Parliament, first resisted and then fled. Waller was thus condemned by both sides in the years after the Restoration, and his role in the war became a footnote.[33]

In many ways, however, Waller's life and beliefs tell us a great deal about the officers in the armies, especially Parliament's. As a young man he joined many of his social class in fighting on the continent. His experience was not extensive, but neither was it idle dabbling. He studied the aristocratic art of fencing and managing a warhorse in Paris. He volunteered with other Englishmen in a Venetian campaign near Trieste in 1617. And finally, he

SIR WILLIAM WALLER. By Peter Rottermond (Rodttermondt), after Cornelius Johnson, 1643. By permission of the National Portrait Gallery, UK.

joined his compatriots in supporting James I's daughter, the queen of Bohemia. He was present when the Holy Roman Emperor defeated the Bohemian forces at the Battle of White Mountain in 1620, a key event in the widening of a regional German conflict into the cataclysm of the Thirty Years War. During that period he not only served alongside his future commander and political rival, the earl of Essex, but also with his future royalist opponent Ralph Hopton.

That such men were present with Waller is no real surprise, but only recently have historians become aware of just how extensively the English aristocracy sought out opportunities to learn the trade of war in the forty years prior to 1642. An older calculation suggested that by the 1620s only one in five of the kingdom's peers had served in combat, but Roger Manning's more comprehensive study has changed that view.[34] Manning argues that a full two-thirds of English noblemen had military experience at the outset of the English Civil War, largely because a reinvigorated sense of the value of military experience had pushed them, like Waller, to fight abroad.[35] Furthermore, a sample of seventy-six officers found that 40.8 percent had fought on the continent, and 9.2 percent more in the recent English campaigns against Scotland in 1638–40.[36] Proportionally fewer gentry and commoners served abroad, although upper-end estimates find 100,000 men from the British isles serving on the continent from 1618 to 1648. Charles Carlton's more moderate estimate suggests at least ten to fifteen thousand Englishmen saw continental service during the Thirty Years War, as well as many more Scotsmen and Irishmen.[37] In addition to this kind of direct experience, Englishmen had been exposed to a voluminous practical military literature in the prewar decades.[38]

One cannot know exactly what Waller or the other veterans experienced, but they surely learned some seemingly contradictory lessons. On one hand, they would have experienced some of the Thirty Years War's now-legendary violence. That war depopulated parts of Germany, spawned a particular genre of wartime-atrocity narrative, inspired great fear in England, and, in some views, permanently changed the way Europeans viewed the cost of war.[39] The Thirty Years War also epitomized the horrors of Europe's religious wars. To be sure, by the end there were Protestant and Catholic sovereigns on both sides struggling over more mundane issues of power and influence, but religion remained at the core of the war, as it would in some ways in England, and its effects could be extreme. Waller himself would have witnessed the impact of religious fervor at the Battle of White Mountain, where a Catholic priest successfully whipped parts of the army into an atrocity-generating frenzy by appealing to their devotion to the cult of the Virgin Mary.[40]

On the other hand, European codes governing the conduct of war, born in the middle ages out of clerical demands to protect noncombatants and from the cosmopolitan nobility's desire to protect themselves from each other, were acquiring ever greater clarity, if not always effect. Chapter 1 outlined some of those developing codes, but they are even more important here, in part because in a brothers' war they had a greater chance of ameliorating wartime violence. Historian Barbara Donagan's work on the English civil war identifies three fundamental levels of authority in setting boundaries on violent behavior in war, all derived from a shared Western European tradition, well known to the English elite, and passingly familiar even to the less literate population. At the highest level, and invoked surprisingly often, were the "laws of god, nature and nations" that provided a moral outline for basic standards of conduct, and which were particularly crucial in providing the legal justifications for going to war. At the lowest level, to be considered later in this chapter, was military law, the specific regulations designed to enforce discipline within the army. The second level, of greatest concern here, and probably the most important in regulating wartime behavior, were the "codes," the "customs and usages," or, as one contemporary described them, the "general sense and practice in all Wars." It goes without saying that such codes were never perfectly observed, but they nevertheless defined a structure of expected behavior that shaped choices and also shaped reactions to their violation. They generated an "etiquette of belligerence."[41] More specifically, the codes regulated several basic areas of conduct: the fate of besieged garrisons and towns, the treatment of prisoners (both officers and men), the treatment of civilians, and communication between enemy forces.

Having and knowing the codes was one thing; enforcing them was another. Adherence generally depended on three beliefs and one material reality: belief in reciprocity, God's providence, and honor, and the ability to pay and feed one's troops. The material factor was relatively straightforward. If, for political reasons, a commander hoped to avoid devastating the countryside and plundering civilians (discussed more extensively in chapter 4), then he had to be able to both pay and feed his army. If paid but unfed, they would plunder for food. If fed but not paid, they would plunder for pay. The mercenary armies of the sixteenth and seventeenth centuries—what John Lynn has called the "aggregate-contract" army—often depended on plunder to sustain them. Their economy, based on pillage, further encouraged a libertine quality to the soldiers' lifestyle, which may have been a major reason why soldiers joined in the first place.[42] Their officers' ability, and even willingness, to control such behavior was limited, and the soldiers' culture sometimes preferred and expected to take plunder even when paid. Reciprocity was similarly straightforward. If one side violated the codes, they could expect the other side to retaliate in kind, although there might be a great deal of discussion, through exchanged threats, over exactly what form retaliation might take. Both sides understood the implications of this discussion, and they feared letting retaliation gain too much momentum.

More abstract beliefs in God and in honor were also crucial and shaped choices. Robert Monro, a Scottish mercenary fighting in Germany in the 1630s, feared that indiscriminate cruelty would rebound on him via reciprocity and God's providence. Like most of his contemporaries, including Waller, Monro believed in an interventionist God.[43] After describing the

cruelties of his enemy, Monro commented that "we ought to learne to for-
beare the like, lest one day we might be used as they used our friends and
Countrymen: for we may be revenged on our enemies crueltie, repaying them
in a Christian manner, without making Beasts of our selves. . . . No truely; it
is just with God, that he misse mercy, that refuseth mercy unto others." One
reason for the success of King Gustavus Adolphus (then leading Swedish
forces in Germany), argued Monro, was that he disciplined his army to prevent them from oppressing the poor, "which made them cry a blessing to his
Majesty and his Army." The enemy, in contrast, "provoked the wrath of God
against themselves and their Army, for their cruelty used in torturing the
poore."[44] Monro surely exaggerated the difference between Gustavus Adolphus and his enemies, but the concern for reciprocity and God's providence
was clear. Waller made a similar postwar comment about his own sins, believing that the destruction of his house in Winchester was God's punishment for his allowing his soldiers to plunder that city in 1642.[45]

Even more important than the fear of God as an enforcement mechanism
was the officers' honor. Certain aspects of the codes, particularly those that
concerned their dealings with each other—promises made to a socially recognizable officer in the enemy army—weighed heavily in their actions.
Monro believed "no punishment more grievous, than the publique ill-will of
all men; especially for just causes. And in my opinion, it is better to be buried
in oblivion, than to be evill spoken of to posteritie."[46] To prove his case,
Monro told a number of stories of officers shamed for their failure to uphold
their reputations, and even of one officer who sacrificed his command and the
civilians in the town to the furies of an enemy storm rather than be hanged
for an insufficient defense—since hanging was a dishonorable death.[47]

This powerful belief in honor was shared by Waller and his contemporary
officers on both sides in England. The concept of a gentleman's honor encompassed a whole range of often conflicting notions. At one level it was tied to
a revived, almost quixotic, notion of chivalry that encouraged officers to challenge their opponents to single combat (much like the earl of Essex had in
Ireland), or that led younger officers to engage the enemy precipitously in
defiance of orders and even military logic. Waller benefited at the battle of
Cheriton in the summer of 1644 when a glory-seeking royalist infantry officer,
against orders, left his strong defensive position and charged down a slope.[48]
This chivalric vision of honor also led some commanders to challenge an
enemy army to meet in an open field—although it is not clear whether any
such challenge led directly to battle. Waller, at least, found the idea amusing.
In 1643 he was pursuing a royalist army whose commander sent him a message

suggesting that they "might fight no more in holes, but in the champagne [open field]." Waller jokingly agreed and even offered powder if the royalists were running short. The battle of Roundway Down followed shortly after.[49]

On another level, honor meant maintaining the visible marks of one's social position. This concern for status goes to the heart of honor, which, in the end, was defined by the appearance of appropriate behavior in the eyes of others—it was (and for many still is) about reputation. But what behaviors enhanced or degraded one's reputation? It was in exactly this period in English and European history that those definitions were shifting. Where honor for Essex in 1599 had been defined primarily by his "wilfulness, stead-fastness and assertiveness," by the 1640s maintaining reputation and status had also begun to include showing "wisdom, temperance and godliness." The former qualities did not disappear as markers of status, but they seemed increasingly tempered by expectations of more restrained and moderate behavior, and in particular behavior in service of the sovereign and not merely in one's own interest. By these new standards, one of the key markers of honorable behavior was the inviolability of one's word. In terms of wartime violence, an officer's word of honor gave force to many of the codes of war, since they often operated upon their promises to each other (less so their promises to civilians). Sieges could end with a negotiated surrender; prisoners were exchanged and treated according to promises made and covenants reached; safe-conducts to enemy messengers were protected by the commander's signature (and promise).[50]

A gentleman of his era, Waller was as touchy as any about his honor. He compared his reputation to his eye, both "such tender things, as being touched, Nature bids me defend the one, Honour the other."[51] But he was also a classically educated man of affairs and a Puritan deeply committed to his faith. His writings display an encyclopedic familiarity with a variety of ancient authors and, unsurprisingly, an intimate knowledge of the Bible. Waller referred to the two sets of authorities almost equally often, but it is clear that he shaped his life based on his Protestant, predestinarian faith. Puritan belief in an interventionist God, however, did not encourage passivity. Waller believed that a man had to act in the world, and in so acting he became God's instrument.[52] He was not as radical in his beliefs as some, and he was insistent on certain concessions to daily reality, protesting, for example, against the appointment of officers based on religious orthodoxy alone.[53] That said, during the course of the war he fought hard for the cause and was intolerant of neutralism. In 1643 he forced the surrendered population of Hereford, in return for protection for their lives and goods, to take an oath to defend the king and Parliament both, and more specifically to assist Essex's parliamentary army.[54] In 1644 he issued an edict denouncing "detestable" neutralists, saying, "We shall take them for nothing more than enemies to the State and men accordingly to be proceeded against."[55]

This broad combination of experiences and sometimes contradictory beliefs shaped Sir William's military choices as a parliamentary general. He fought a number of successful campaigns in 1642 and early 1643. His rapid movements and night marches soon earned him the sobriquet of "William the Conqueror."[56] Unfortunately, he also incurred the jealousy of Lord Essex, then the senior parliamentary general, and the one with whom Waller would have to work most closely in the ensuing years. He also suffered some serious defeats in the summer of 1643, but his reputation was strong enough to secure him an appointment as the major general of the associated counties of Hampshire, Sussex, Surrey, and Kent in November.

After several minor actions that fall, including a failed siege of Basing House in December 1643, Waller, like most generals of the era, went into winter

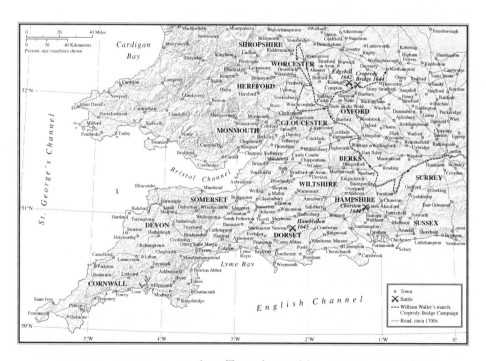

SOUTHWEST ENGLAND IN THE 1640S. The road network here is approximate to the early seventeenth century. Adapted from Great Britain, *A Map of XVII Century England* (Southampton: Ordnance Survey Office, 1930). Map by Justin Morrill, the-m-factory.com.

quarters, and both sides plotted their moves for the start of the next campaign season. As spring began Parliament had five main armies in the field, not counting smaller local forces. Waller's was one, having been built anew in London, from where he began moving south toward Winchester. Meanwhile the king held an extensive central position in the Midlands centered on Oxford, surrounded by garrisons at Reading, Abingdon, Banbury, Wallingford, and Faringdon, and supplemented by a further circuit of smaller posts.[57]

At the start of the spring's campaigning, Waller once again faced the army of Sir Ralph Hopton, a fellow veteran of the German wars and his opponent for much of the previous year. Probably as a result of their previous acquaintance, Waller and Hopton were more comfortable than usual communicating with each other as enemies. Maintaining the contractual etiquette of belligerence established by the codes of war required this kind of communication between opponents, and in a civil war that kind of contact was easier than in the wars with barbarians. Chapter 2 discussed an incident during Essex's 1599 march through Munster in which an Irish rebel named Phelim McFeaghe had been forced to shout the offer of parley, in Irish, to an Irishmen in Essex's army, and he had to do so in the midst of the chaos of battle.[58] Communication had not always been that difficult in Ireland, but between Waller and Hopton one finds an easy and cordial exchange of letters. Waller professed his "unchangeable" affections and averred that "hostilitie itself cannot violate my friendship to your person; but I must be true to the cause wherein I serve." Waller even claimed to hold a "perfect hate" for this "war without an enemie." Waller concluded by setting forth the utopian ideal of honorable war: "We are both on the stage and must act those parts that are assigned to us in this Tragedy, but let us do it in way of honour, and without personal animositie." Waller and Hopton were perhaps more genial and voluble than most, but this kind of correspondence was not uncommon, as commanders used the platform of social equality and the binding qualities of honor to open these channels of communication.[59] Drummers and trumpeters constantly appear in the narratives of campaigns, and especially sieges, carrying offers of battle, pardon, or surrender. Without the inviolability of those messages (and messengers), and without the mutual understanding of the codes that lay beneath them, hope for restraint dimmed.

Waller defeated Hopton at Cheriton on March 29, 1644, and Hopton's diminished army fell back to join the king's forces at Oxford. Parliament now sensed an opportunity to threaten the king and ordered Waller and Essex to trap him. Suspecting such a plan was afoot, Charles recalled the garrison at Reading, added them to his army, and even raised two regiments of scholars and their servants from the university. Sir Edward Walker, Charles's secretary, admitted that the defeat at Cheriton forced Charles to switch from an offensive to a defensive war. It is probably more accurate to say that it forced Charles's army into full evasion mode, as he tried to buy time for his other field armies either to win their own battles and relieve the pressure on him or to join him.[60]

In stripping the garrison from Reading, Charles confronted one of the major strategic problems faced by combatants in the English Civil War: how to balance forces between garrisons and field armies. Ultimate victory was thought to depend upon a decisive battlefield confrontation. In fact, there had been widespread surprise when the first major battle of the war at Edgehill had not decided the war. Walker, in an account annotated and corrected by Charles, weighed the problem of the Reading garrison versus the overall royalist calculation of victory: "All our consultations tending to nothing but how to make the enemy fight, which we not being able without disadvantage to do," they were now forced either to join the Reading garrison to the field army or abandon Reading to a siege they could neither prevent nor lift.[61] Charles hoped that with just enough reinforcements he could defeat Waller's and Essex's converging armies in battle one at a time.[62] But Walker also noted the need to prevent Waller and the other parliamentarian forces from expanding their recruiting base. Although Charles could not yet face them in open battle, he could march, and he could use the tidal effect of a field army to efface parliamentary garrisons in his path and plant new royalist ones.

This calculated need for garrisons greatly increased the capacity of an army to inflict violence. A field army was a tremendous imposition on the countryside, but it moved in relatively narrow corridors and its stay was usually temporary. A cautionary note is in order, however; armies larger than a few thousand men rarely moved in the neat arrows seen on maps. They nearly always spread out on parallel roads, with detachments often sent even further afield. This was especially true at the end of each day as the army spread out into evening quarters, potentially occupying several separate villages. Nevertheless, armies came and went, while garrisons were more or less permanent impositions, widely scattered, and designed to extract resources from the countryside.

As for Waller, his goal (and Parliament's) at this stage of the war was simply to capture the king. Waller thought he should "follow the King wherever an army can march," believing "the war can never end if the King be in any part of the land and not at the Parliament."[63] This calculation meshed perfectly with Waller's vision of war "without an Enemie"—two armies circling each other, seeking only that battle which would rejoin king with Parliament, either as prisoner or ruler. Furthermore, Waller would seek to "express all the civilities I could to those of the adverse party, that so our differences might be kept in a reconcileable condition."[64] But Waller faced the same necessity of controlling the countryside, and he too (or other forces on his behalf) fought the garrison war. Facing the same conundrum farther west that summer, Essex complained that he needed money, and unless Parliament sent some he would have to divide his army to collect it locally. With two royalist field armies nearby, he could neither hinder them nor divide his army any further than he already had for the placing of local garrisons. And since the royalists were free to move, they were successfully raising "great forces," although they were doing so, he said, by "terrifying [the country] with menaces."[65] Dry strategic calculation could also be overruled by the desires and prejudices of the soldiers, as had happened in Ireland. Later, as the king marched farther westward, Waller worried that some of his men were "much dejected" by the direction of the march, and he feared one of his regiments would completely break up, "many of the soldiers having already run away, and more likely to follow."[66]

As Essex and Waller attempted to converge on the king at Oxford, Waller struggled with managing his typically cobbled-together army of veterans, conscripts, former royalists, militia, and probably some outright mercenaries.[67] A brothers' war without clear sectional boundaries and with only limited ideological commitment by the rank and file produced a complex pattern of soldier recruitment, heavily dependent on coercion and therefore plagued by desertion and changing sides. By 1644 Parliament had some more or less "regular" regiments of long-serving troops, who usually remained in service with the same colonel. Many of those men had been conscripted into service in a manner much like that discussed in chapter 1, but now in much larger numbers. Quotas were issued and volunteers were preferred, but local officials tasked with recruitment often had to coerce. In this war without boundaries, both sides might tap the same contested regions for men, although in general the majority of recruits tended to come from the "safer," more firmly controlled areas. Both sides also recruited heavily among prisoners taken from the enemy. Detailed records do not exist, but there is little doubt that some men changed sides repeatedly. All but sixteen of 1,100 survivors at Cirencester in 1643 switched to the royalist side, and in May 1644, after the taking of Lincoln, over seven hundred royalists in turn "chearfully desire[d] to serve the Parliament."[68] Furthermore, compared to English armies in sixteenth-century Ireland, there was much greater voluntary service by the gentry and nobility, who in turn recruited men by exploiting the old loyalty of a tenant to his landlord. The elite could promise to raise a regiment and then seek, with more than a modicum of social pressure, "volunteers" from the men living on their estates. This system reinforced the normal local hierarchy in a time of great upheaval, helped preserve the social order, and in some cases was preferred by the soldiers.[69] Cities were also particularly crucial sources of men: Parliament-controlled London enormously so, especially its large and relatively prepared "trained bands" of select militia, expanded during the war with "auxiliary" regiments.[70]

Waller complained that his army was never really his own. Parliament frequently sent or recalled the "city and country regiments," and since they had "no dependence upon" him they "would not follow [him] further then pleased themselves."[71] Indeed, Waller's London regiments had departed for home shortly after Cheriton, and he received a new batch on April 17 for his pursuit of the king. They could fight well, and had done so at the battle of Newbury the previous September.[72] But these new regiments from South-wark, the Tower Hamlets, and Westminster gave Waller plenty of trouble, and many deserted during the pursuit of Charles, wishing "for nothing more [than] to see the smoke of their owne Chimneys," and almost all the rest disheartened and brought "to their old song of home, home," after the battle at Cropredy.[73] Waller was not alone in feeling oppressed by this system, and it was partly for this reason that Parliament "new modeled" its army in 1645 into a more professional standing force.

Synchronized Collective Discipline

Even prior to the formation of the New Model Army, Waller was doing his best to bring his odd mixture of troops under control. A key to that process was a clear system of military law, and Waller issued his own supplement to the existing body of such law on March 13, 1644.[74] By that year the basic corpus of military law had been firmly established, and Waller's amendments and clarifications changed little, but his issuance of them provides an oppor-tunity to consider how military law had developed since the sixteenth cen-tury. Looking at the development of military law provides an insight into two key issues. The first is how such codes reflected the increasingly common calculation by early modern European military leaders that the most desir-able characteristic in a soldier was obedience to orders within a comprehen-sive system of collective discipline. The second is how military law also reveals the extent to which an army's leadership concerned itself (or did not) with the potential violence of the army against civilians, and how they codified aspects of the customary codes of war into specific punishable regulations. We will deal with each of these issues in turn.

Chapter 2 explored the earlier development of English "articles" of military law. In that era the articles were essentially temporary, established by a single commander for the duration of a campaign. In contrast, the assorted printed codes from the English Civil War, both royalist and parliamentarian, were remarkably homogenous; furthermore, many of them were directly derived from similar articles drawn up by the Swedish and Dutch armies over the preceding decades.[75] The vast majority of the regulations were related to strictly military matters: the consequences of striking or contradicting an officer, challenging a fellow soldier or officer to a duel, falling asleep on sentry duty, and so on. The regulations also included clauses about immoral behavior, regulating blasphemy, fornication, drunkenness, and other such sins.

More germane to our purpose were the various regulations that might control troops and limit the frightfulness of war. Some articles sought to preserve the resources of the countryside from uncontrolled plunder for the use of the army. Soldiers were thus forbidden to "waste, spoyle, or extort any mony, or goods from any of Our Subjects" or "to cut down any fruit-Trees, . . . or to spoyle any standing Corne in the eare or in grasse," or "burne any House or Barne, or burne or spoyle any stacke of Corne, Hay or Straw, or any Ship, Boat or Carriage, or any thing which may serve for the provision of the Army."[76] This list of prohibitions was of course conditional; if ordered, the soldiers should indeed destroy, but otherwise such things were potentially useful and not to be idly wiped out.

But there were also a few somewhat more idealistic, if not entirely altruistic, rules. Clauses 133 and 134 of Charles's orders prohibited destroying hospitals, churches, schools, and mills, unless ordered—even during the storming of a city. Nor should any man ever "tyrannise over any Churchmen, aged men, or women, or Maids, or Children, unless they first take Armes against them." Parliamentary regulations were similar, and even more detailed, and both sides tended to step up the number of proclamations against plundering over the course of war. This suggests that not only was the problem growing, but so was the strength of the formal desire to control and manage it.[77] The regulations also included clauses governing the treatment of prisoners and the surrender of fortresses, subjects to be explored in more detail later.

Of course the real question with any set of regulations is whether or not they are being enforced. The records on that subject are much less clear than the black-and-white certainty of the printed rules. Most historians agree that

Parliament's generals tried harder than the royalists to control their troops' violence, especially against civilians, but even their efforts were far from uniform or perfect.[78] Waller, at least, tried. Only two sets of court-martial records survive from the entire war, and one of them happens to be from this period of Waller's pursuit of the king in the summer of 1644 and somewhat after. Waller's court met regularly, except during the most intensive periods of pursuit. Forty-one officers and men faced formal charges, while warrants were issued for others. Most charges (nineteen individuals) were for mutiny, desertion ("running from the colors"), or some other form of military disobedience. The court also tried eight men for plundering civilians, and one of them was additionally charged with abusing a prisoner. The punishments for plundering varied from being tied neck and heels or hung by the wrists (both painful positions enforced for a specified period of time) to running the gauntlet, and even to death (three of the eight accused of plundering were sentenced to death).[79]

This organized effort to impose a disciplinary code on the soldiers reflected a spreading European vision of a new kind of discipline necessary for success on the battlefield. Medieval commanders had also sought the obedience of their men, but the ideological core of a medieval army—the set of behaviors they imagined as most central to the winning of a battle—was the cooperation of highly skilled individuals. Their solidarity, commitment, and skill would carry the day. In saying this I am specifically describing the *imagined* battlefield.[80] Much of medieval warfare revolved around sieges and destructive raids; battle was relatively rare. When it occurred, it was planned, scouted, and prepared for by men who made war their life's business. I am not arguing that medieval commanders were "amateurs" about to begin a process of professionalization; indeed, "chivalrous" and "professional" are adjectives that can describe the same person (medieval or otherwise). But medieval military leaders nevertheless imagined the decisive battlefield element to be the skill of the individual knight, honed in the tournament and living for war.[81] These were men who, in the words of the medieval chronicler Jean Froissart, could neither "achieve perfect honour nor worldly renown without prowess."[82]

The imagined reliance on individual prowess began to shift in the late middle ages as part of an ongoing reemphasis on infantry. In the fourteenth century more than one kind of infantry began to emerge and successfully oppose the knight, but of these kinds it was the pike that had the real long-term impact. The pike formation would require a new kind of collective discipline, but that need emerged only gradually. The first consistently successful pikemen were the Swiss, but their success was rooted in the organic communalism of their social organization rather than in a new form of discipline.[83] Early efforts to copy their success retained strong aspects of individualism, and even libertinism, as mercenary pikemen aspired to the model of knights, whose social class they could not otherwise emulate. By the beginning of the sixteenth century, however, there was an increasing recognition of the need for a new kind of collective discipline.[84]

While the pike began the move toward collective discipline, gunpowder sealed it. Early gunpowder weapons required a synchronization of effort that allowed the slow-loading, unreliable weapons to send a wall of lead in the desired direction. It took some time for Europeans to understand this need, and as it became clearer, they increasingly turned to ancient Roman models of synchronized maneuver and insisted on ever-more carefully drilled and trained armies—not just skilled individuals.[85] Part of this shift resulted from the recognition that pikemen and shot were socially expansible categories of soldier (unlike the skilled medieval knight), and elite leaders presumed that such men could be effective only if tightly controlled and directed by the "naturally" martial elite—who increasingly served as "officers" rather than warriors. In one sense, gunpowder did not "require" synchronization of effort; rather, a cultural prejudice about those who would wield it suggested that synchronization was the only way to make it effective. In addition, shot had to be combined with pikes for their mutual protection, further increasing the need for synchronicity and slowly imposing greater discipline on the pikemen as well (and slowing the aggressive charge originally pioneered by the Swiss). Army components now had to march in sync, fire in sync, and maneuver in sync with other component parts. It was thus the emergence of a technology within a particular vision of social hierarchy, and then filtered through an inherited Roman model, that defined how an army should act. On this new imagined battlefield, firmly laid out by the end of the sixteenth century, individualism had no place, aside from occasional quixotic challenges to single combat by elite commanders. In one sense, this "military revolution's" most important change was in the mind.[86]

The importance of this shift cannot be overstated. In essence, the sixteenth century saw the demand for skilled, but hopefully obedient, individuals working together, replaced by a demand for units of soldiers shaped by *synchronized collective discipline*. The difference between the two is subtle but significant, especially in its long-term implications for the nature of military culture. We can see the shift clearly by comparing some of the representative martial literature from the late middle ages to sixteenth-century military manuals. The late fourteenth-century *Tree of Battles* by Honoré Bouvet (also known as Bonet), an influential and widely distributed encapsulation of much of medieval thought on war, provides a starting point. It was admittedly not a practical military manual; it was more a guide for knights explaining their martial obligations and the nature of "right conduct." Its sections dedicated to on-the-battlefield behavior certainly emphasized obedience, but in a highly individualistic sense. Bouvet, in fact, seemed concerned about excessive individual "boldness," which needed to be reined in. He described a good general as one who properly arrayed the troops and knew how to conduct battle, but there was no sense of him being a trainer or discipliner of troops. As for the "virtues" that would win battles, Bouvet emphasized strength, justice, temperance, and wisdom—all virtues of the individual knight that in their right application would contribute to military success.[87] Similarly, Geoffroi de Charny's mid-fourteenth-century *Book of Chivalry* dealt primarily with issues of right conduct, but, in terms of the necessary virtues for successful war, he too emphasized almost entirely individual skill (in battle), with only passing praise for knowing how to deploy troops and a condemnation of undisciplined boldness.[88] Richard Kaeuper's study of the chivalric literature found an "utterly tireless, almost obsessional emphasis placed on personal prowess as the key chivalric trait."[89]

Perhaps the clearest evidence emerges in the way Christine de Pizan's 1410 *Book of Deeds of Arms and of Chivalry* characterized the late Roman writer Vegetius's call for a commander who, in Vegetius's original words, should "train his soldiers diligently," who would find victory only through "drill-at-arms [*armorum exercitio*] . . . and military expertise [*usuque militiae*]" acquired by "constant exercises [*adsiduo exercitio*]" and by constant practice of moving into and out of precisely regulated and linear formations. Vegetius concluded, as part of his most famous line about preparing for war, "He who wants victory, let him train soldiers diligently [*milites inbuat diligenter*]."[90] Pizan transmuted this advice into "He who likes victory must be adept with arms . . . [he] should fight with competence." And she concluded that Rome's success had come from "skill at arms and by the instruction of their troops." When she glossed Vegetius on the subject of formations, she acknowledged the necessity for proper order and distance, but she omitted the sense of the necessity of group practice or the strict linear quality of the Roman formations. In a section of her own creation describing early fifteenth-century practice, she designated a key role for units composed of men "most eager to fight, and [who] are expert in their skill with arms."[91] Pizan's vision of Roman success, and the chivalric tradition in general, focused on individual prowess and a generalized sense of (or hope for) obedience.

Early modern authors, however, impelled by the need for synchronized movement, began to understand Vegetius closer to his original vision of Roman discipline. To choose but one example, Matthew Sutcliffe (whose work, incidentally, was supported by the second earl of Essex), writing in 1593 in a much more practical and long-winded manner than Bouvet or Pizan, believed success on the battlefield depended on a commander who "redress[ed] . . . disorders . . . and by sharpe punishment maintein[ed] true militarie discipline," since "with exercise of armes and observance of true discipline of war great enterprieses most happily are atchieved." For Sutcliffe the synchronization of parts was key: without "array & order . . . armes have no use." "As well can an armie march or fight being out of array," he continued, "as a body doe the functions of the body, having the partes out of frame," and success required "instruction and exercise."[92]

With the late sixteenth-century adoption of synchronized volley fire (in addition to the already necessary synchronization of pike with shot), the perceived necessity of battlefield discipline and training only accelerated.[93] Using a gun in the seventeenth century required no great skill, but using it in an army did require uniform, synchronized, imposed, and mechanistic discipline. Others have explored the implications for technology and military success, but the implications for the Western world's vision of discipline and what it meant in terms of wartime violence were equally profound. The sixteenth century did not change the long-held ideological desirability of a virtuous soldier who obeyed orders not to plunder or kill peasants, but it now imagined such a comprehensively disciplined soldier as absolutely necessary for military success.

This new calculation of the necessity of discipline was one reason for the increasing clarity and standardization of codes of military law (another reason was the increase in the number of social inferiors composing the soldiery). To have truly disciplined soldiers meant that all aspects of their behavior had to be obedient to command—including their violence off the battlefield. This period also saw a vast proliferation in printed manuals, of which Sutcliffe's was one, designed to guide officers in the conduct of their units, all emphasizing the necessity of drill, uniformity of movement, and obedience. Most of these manuals did not yet dwell on off-battlefield violence the way military law did, but many were beginning to incorporate advice about how to manage soldiers' behavior in camp and garrison, and especially to limit their unauthorized movement (much of the worry here was about desertion rather than plunder, but as a restraint on frightfulness the effect was similar).[94] Parliament's Puritan side was not content to rely solely on law or manuals, and encouraged the publication of "soldiers' catechisms" and sermons that added the force of religion to the problem of shaping and controlling soldiers' behavior.[95]

This comprehensive approach to discipline and drill required a higher ratio of officers to soldiers. English companies in Ireland in the sixteenth century typically had one captain, one or two petty captains, and perhaps a sergeant and a drummer running companies ranging from fifty to two hundred men (usually around a hundred).[96] The army as such lacked any larger organization above the company level (other than the army itself). By the very end of the sixteenth century, spurred by experience in the Low Country wars, where new forms of drill were being pioneered, English armies began to look more familiar to a modern eye.[97] Companies of "foot" or troops of "horse" were banded together in regiments. A regiment of foot was designed to have ten companies totaling 1,200 men. With a rank structure reflecting and reinforcing the social hierarchy, the colonel of the regiment had his own company with two hundred men, the lieutenant colonel had 160, the major 140, and the six captains had one hundred men each. Each of the captains' companies in turn had one lieutenant, one ensign, one "gentleman," two sergeants, and three corporals (plus a drummer and/or a trumpeter). The regiment added a small staff composed of a quartermaster, provost, surgeon and mate, carriage master, drum major, and chaplain.[98] Although still a very low ratio of officers to men compared to modern armies, it was a substantial increase in the level of control over soldiers compared to the sixteenth century.

Obviously, none of these changes produced perfect restraint, especially as the desire to control clashed with the state's failure of capacity to keep their men paid and supplied. Following Waller's army as he pursued the king into the west provides a glimpse into how these conflicts between capacity and control unfolded.

Waller pursues the King

With a deft series of feints and countermarches Charles slipped out of the net at Oxford, leaving a substantial garrison there to defend it. To all appearances he was heading for his reliable recruiting grounds in Wales. Essex, the senior parliamentary commander, ordered Waller on June 6 to continue pursuing Charles, while Essex headed into the southwest to raise the siege at Lyme. Waller protested, arguing that the west was his province, but he submitted to Essex's authority, gnashed his teeth, complained to Parliament, hanged a plunderer from his army, and forged on in pursuit. As a first step, however, Waller bowed to the pressures of the garrison war and his need for money and recruits. So he turned his sights on the royalist garrison at Sudeley Castle.

A trusted royalist commander, 250 to 300 men, and two cannon defended Sudeley.[99] In the kind of act typical of much of the war's destruction, they hastily burned a nearby barn to clear their fields of fire and prevent the attackers from using it for cover.[100] Waller initially approached accompanied only by cavalry, and the garrison spurned his demand for surrender. Waller then made his own preparations, brought up his foot, and selected the "forlorn hopes" who would lead the assault (a moment of extreme personal significance for the men so chosen). Seeing the infantry, the garrison commander asked for a parley and offered his surrender but demanded the full "honors of war": the right to march away, fully armed, with a guarantee of safety as far as Worcester. Waller had barely begun the siege, and he saw no need to offer such generous terms. He offered them only "fair quarter," in this case meaning he would spare their lives, imprison them, and guarantee that the officers would have horses.[101] Unwilling to accept the dishonor of capture, the garrison's negotiators spurred their horses back to the castle, and the commander fired a volley of shot from the walls in defiance. Waller calmly began his bombardment. Fortunately for him, one of his shot shattered the body of the castle's chief gunner. This entirely unnerved the defenders, mostly pressed men, "who had never seen the warres before." They belabored the garrison commander with their fears, even physically threatening him. To Waller's surprise, and to the great relief of the men selected for the "forlorn hope," the garrison surrendered the next morning.

In one sense this was a non-event, so why dwell on it? Admittedly, Sudeley's surrender was unexpectedly quick. Typical sieges might last days, or even weeks, and the terms could vary depending on both sides' assessment of the movements of armies around them, and of how desperate the garrison might be. The decision to surrender was fraught with potentially fatal consequences. If a commander surrendered too soon, he faced recriminations from his own side, ranging from loss of honor to execution.[102] But "too soon" proved highly subjective. Parliament's first set of military regulations demanded that commanders only surrender "in extremity."[103] This was an imprecise term at best; the commander's definition of extremity might clash with his soldiers'—as had happened at Sudeley. In a 1643 revision of the regulations, Parliament expanded and clarified the definitions: the garrison had to be out of food, with no relief army expected and the physical fall of the fortress imminent. The rules now even prescribed decimation for soldiers who forced their commander to surrender.[104] Waiting "too long," however, could have even worse consequences. As discussed briefly in chapter 2, exasperated soldiers forced to assault a recalcitrant garrison were granted license to destroy, plunder, and kill. The English civil war saw grotesque examples of what happened in those circumstances.[105] The Thirty Years War, the Eighty Years War in the Netherlands, and the concurrent campaigns in Ireland offered even worse instances.[106]

Despite the vagaries of definitions, the basic outlines were clear to all—and this system of negotiated surrender was a great preserver of life and property. What was remarkable about the English civil war was how often a siege ended in a negotiated surrender. Charles Carlton analyzed a sample of twenty-five major sieges, in which the defenders successfully held out in twelve, negotiated a surrender in eleven, and suffered a storm in only two.[107] On a number of occasions the terms of surrender were violated, and the besiegers killed or plundered the defenders as they tried to march away. But frequently these incidents were instigated by the soldiers, often over the violent efforts of their officers to stop them. Since the leadership on both sides recognized that an effort at control had been made, the system remained strained but intact.[108]

Part of Waller's haul at Sudeley was some 250 to 300 soldiers and twenty-two officers captured. The fate of prisoners who survived the process of surrender is of enormous interest in evaluating wartime violence, because it was then that armed men made choices about what to do with helpless enemies. The situation speaks volumes about the frightfulness of a particular war. Both sides in the English civil war regularly took hundreds or even thousands of prisoners through the surrender of fortresses, or when regiments gave up the fight on the battlefield. At the outset of the war, both sides claimed the other to be rebels, with all the attendant violent implications for the treatment of prisoners (discussed in chapter 1). In December 1642, however, the royalists threatened civil trials for treason for three prisoners, and Parliament immediately threatened to do the same with prisoners they already held. The royalists relented. This early-established reciprocity led both sides to accept the other as a "lawful enemy," to be treated according to the then-current European customs of war for the treatment of prisoners and, indeed, the codes more generally.[109] There were exceptions where both sides felt that a prisoner had been executed outside the bounds of normal practice, but they tended to become causes célèbres, and threats of reciprocal treatment generally served to discourage escalation.[110] Under preceding systems, including well into the sixteenth century, prisoners were seen as a source of income for those who took them and for the army commander. English articles of war from the preceding century, and even Charles's wartime regulations, laid out specific rules for who could ransom whom and under what restrictions.[111] Parliament's article on the subject reflected the more current view of prisoners, merely, that no one should be killed who had thrown down his arms.[112]

After successfully surrendering, a prisoner's fate was governed more by emerging custom than by written regulation, and that custom was less and less about profit. First, common soldiers were often recruited on the spot into the victor's ranks—a process eased in a brothers' war by a shared language, but also fairly common in the continent's wars.[113] Those refusing, or those not offered the chance, were marched away under guard and confined, generally in conditions ranging from inadequate to horrific.[114] But practical problems played a role here. Prisoner hauls were often so large that the logistics of guarding and housing the prisoners proved insurmountable, especially for forces barely capable of feeding themselves. Prisoners also could be exchanged, relieving the burden of care and returning soldiers to the army they were at least nominally interested in fighting for. Exchange was particularly common for officers but was also regularly practiced for common soldiers. Waller offered an exchange in the summer of 1643 when the royalists held eighteen soldiers and a corporal, versus his three captains, two ensigns, one quartermaster, and sixty-four private soldiers.[115] The exchange of prisoner lists was another of those circumstances requiring regular and relatively free communication between enemies. Officer prisoners were frequently given their parole and were expected to obey its terms, which always prohibited them from fighting again (until exchanged) and usually confined their movements. Gentlemen who violated their parole, or were so accused (often very publicly as a form of propaganda), went to great lengths to have their case publicly adjudicated.[116] Thus in the treatment of both officers and men, reciprocity and honor operated to keep the system intact. In a brothers' war, where everyone's definition of honor was roughly the same, that was a relatively simple task.

Perhaps the most striking and measurable fact about the taking of prisoners was the ratio of prisoners taken to men killed—a ratio which suggests a significant level of restraint on the battlefield. Carlton's database of 645 combat incidents (from very small to large) contains suggestive statistics. In his sample, Parliament's armies suffered a ratio of those killed to taken prisoner of nearly 1:1, while the royalists' ratio was 1:1.66. At first glance these numbers might seem a bit bloody, but they combine losses suffered in victories and in defeat. Generally the victors had few if any prisoners taken, which exaggerates the number of dead in the sample. The string of royalist defeats late in the war, which generated large prisoner hauls, weight their ratio toward prisoners. To mention a few nonsystematic examples that convey a better sense of the ratio of killed to captured for the defeated side, at Roundway Down parliamentary forces suffered six hundred dead to eight hundred prisoner (1:1.33); at Adwalton Moor it was five hundred to fourteen hundred parliamentarians (1:2.8); and at Boylston Church (a smaller engagement) the royalists lost seventeen dead and three hundred prisoners (1:17.7).[117]

Part 4 will discuss how these numbers compare to a later era, but even these ratios attest to a built-in restraint in the era's imagination of warfare. Prisoners were not "the enemy"; they were a resource to be utilized either for recruiting or exchange (as well as being a highly public signifier of victory). Robert Monro, the Scottish officer who fought for Sweden in the Thirty Years War, encouraged this attitude, noting that Gustavus Adolphus "with clemency doth follow the example of the ancient Romans, who, of all victories, thought that victory best, which least was stained with bloud, having given quarters and service to three thousand Emperiall Souldiers [i.e., he recruited them], without drawing one drop of blood."[118] This differed from wars with "barbarians." So long as a barbarian prisoner was seen as a valuable recruit, this same attitude might persist, but if he could not be ransomed or exchanged, or had no value as a recruit, his life held much less value.

With Sudeley taken, Waller picked up stakes, hanged a deserter, and marched after the king. He raced ahead, hoping to cut Charles off from reaching Wales, but Charles unexpectedly faced about and rushed back toward Oxford.[119] On the way Charles demonstrated one of the conundrums faced by communities during the Civil War. Passing through Evesham on June 16, and irritated at the town's previous "alacrity in the reception" of his pursuer, Charles fined the town £200 and "compelled them to deliver a thousand pair of shoes for the use of the soldiers."[120] As he neared Oxford he reunited with its garrison; thus strengthened, the king again hoped for battle. Essex was far gone into the west, and now the king's army and Waller's were roughly comparable. Charles "resolved no longer to live upon his own quarters . . . but to visit the enemy's country . . . where he would stay and expect Waller."[121]

Meanwhile, Waller, eventually discovering Charles's deception, hesitated for a few days, but finally moved again in pursuit.[122] As Waller moved closer to Charles's army in the countryside north of Oxford, practically within shouting distance of Edgehill, site of the first major battle of the war, both leaders began calculating where they could fight a battle with the greatest advantage.[123] In the end, as was often the case, they fought on a field that neither would have chosen. Marching in parallel on opposite sides of the River Cherwell, Waller suddenly launched a charge across the Cropredy Bridge, hoping to cut off the royalist rear, which had fallen behind the main body. The see-saw battle that followed does not lend itself to neat maps with blocks of troops and arrows, but in essence Waller missed the rear and hit the middle.

Waller's horse hurled themselves at the column of royalist troops. Like most cavalry charges of the war, there was a certain element of rash desperation involved. Inexperienced cavalry tended to fire their wildly inaccurate pistols from too far away. Whether charging or standing on the defense, the best option was usually to hold fire until literally at arm's length.[124] The fighting in such circumstances quickly became confused and chaotic; the mass of men and horse surged back and forth until some began to flee, and the chase was on. At Cropredy the charge of Waller's horse initially overcame the royalists' hasty defense, but they were eventually pushed back across the river, abandoning their artillery train to the royalists. Waller's regular infantry regiment from Kent and the London Tower Hamlet's trained band lowered their pikes, held the bridge, and prevented the retreat from becoming a rout.

As was often the case, at the end of the day the two armies were still facing each other, and they tested each other's positions the next morning. Eventually the royalists sent two messengers across to Waller. One proposed an exchange of prisoners.[125] The other came directly from the king, and represents another classic moment in a war between brothers. The king, "desirous to avoid the Destruction of His Subjects, . . . [sent] a Message of Grace and Pardon to all the Officers and Souldiers of that Army (if they would lay down their Arms and submit)." Waller refused to announce the king's message.

Despite the face-off the day after the battle, Waller's troops, especially the London militiamen, quickly proved they had no more stomach for fighting. In this sense, in its effect on the minds of the soldiers, battle could be decisive. Aided by the interposing river, Waller proved competent enough to hold the army together during the fight, even in retreat. The next day, however, sensing their defeat (and probably the loss of the cannon in particular), his soldiers began to disappear. Even without hearing the king's promise of pardon, ninety-two men slipped across the river to join the royalists, apologizing for their rebellion. As soon as Waller began to move again, he was "extremely plagued by the mutinies of the City [London] Brigade, who are grown to that height of disorder that I have no help [hope] to retain them, being come to their old song of home, home."[126] Waller had heard that song before, and when more London men arrived a few days later, four hundred men of one of the old London regiments immediately departed. The new London men were little better, being "so mutinous and uncommandable that there is no hope of their stay. . . . Yesterday they were like to have killed their Major-General."[127] Fortunately for Parliament, their northern army had almost simultaneously won a much more significant victory at Marston Moor, which helped check the spread of disaffection.

Conclusion

In many ways Waller's campaign in the summer of 1644 exemplified all that was "regular" and conventional about the violence of the English Civil War. It saw sieges ending in negotiated surrender and battles between regular forces ending with an exchange of prisoners and an offer of pardon. It was generally restrained by the nature of combat, by mutually understood codes of conduct, and by the officers' sense of honorable action. The cultural and utilitarian system of codes and the regulations for controlling soldiers were under constant strain, and military law was a fragile reed indeed. But both sides admitted the problems of control and proved capable of acknowledging to each other that sometimes soldiers could not be restrained. A full-fledged war of retaliation never ensued. This is not to deny that this war, like any, involved substantial destruction. We have seen one small aspect in the burning of the building at Sudeley Castle. When a fortress included a whole city, the destruction was much more extensive.[128] The garrison war was destructive as well, but the intent there was always for productive (if painful to the inhabitants) control of the countryside, not its destruction. Of overwhelming importance was that within England the calculation of victory never involved a strategy of destruction, much less the strategy of famine developed at the end of the sixteenth century in Ireland.

But this brothers' war has another lesson to teach. Waller's march did *not* represent the utopian ideal of an army moving and fighting without extraneous, or "collateral," violence. There were many breakdowns in discipline, and a major aspect of Waller's march not yet touched upon here was the plundering of the countryside by the marching armies and by the garrisons surrounding them. Examining plundering and its consequences in the English civil war highlights two issues. The first was the failure of the state, or more accurately, the inability of the state, to pay and supply its troops. The cost of artillery, fortifications, and guns notwithstanding, far and away the greatest drain on an early modern state's budget was military pay and provisions.[129] The greatest drain also proved the easiest to skimp. Such skimping in turn helped foster a divergent "soldier's culture" that granted license to the troops, in their minds and sometimes with the concurrence of their commanders, to take what they needed from the countryside. This was a common enough problem in any war in the period, but in a civil war this phenomenon also generated two further escalating effects. First, both sides drew resources from the same population, which greatly increased the impact of plundering, at least in certain regions. Furthermore, a brothers' war generated widespread (although not universal) intolerance of neutralism. Charles's £200 fine on Evesham and Waller's proclamation against neutrality were only two examples of the way a society at war with itself demanded allegiance.[130] A peasant in an international war who did not take up arms might be plundered, and might even suffer from a deliberate strategy of devastation designed to hinder the enemy, but he or she was not attacked merely for trying to stay out of the way. The military culture of legitimized plunder, enhanced by the civil-war context and the intolerance of neutralism, eventually produced a backlash in an English countryside wedded to a culture of localism. "Neutralism" in the English civil war reflected the ability of the populace to affect the military capacity of the state. Most dramatically, neutralism emerged on the battlefield itself as the "clubmen" movement. The following chapter considers the intersection of all these issues—control of plundering, soldiers' culture, and the power of neutralism to affect calculation—but it must begin with the role of popular attitudes in the actual outbreak of the war.

4

The Clubmen, 1645

There is nothing accompanying this service hath more afflicted me than to see these insolencies that are sometimes committed by the soldiers and not have power wholly to restrain them. I know that the soldiers' plunder is put into a bottomless bag. . . . Our reputation is extremely lost thereby with the common people, who for the most part judge our cause by the demeanour of our army.

Col. Brereton, 1645

Out of Wiltshire we are advertised, that a party of the Kings forces went lately from the Devizes to plunder Collingborn, but the Country folk to prevent, killed one of them, and three Horses, and drove back the rest; the Country likewise lost a man.

Kingdomes Weekly Intelligencer, 1645

London and the "People," *1642*

On January 3, 1642, Charles I made a fateful decision.[1] Fearing the rising tide of resistance led by the so-called Junto in Parliament, he prepared articles of high treason against five members of the House of Commons (including William Waller's future cavalry commander) and one peer in the House of Lords. Parliament ignored the king's indictment, and Charles proved unwilling to wait. On January 4, anticipating a potentially hostile public reaction, he ordered the mayor of London to search houses in the City for excessive arms in the possession of "persons of mean quality," and to discover why some cannon had been moved inside the City. Summoning his guard and the pensioners, described by Parliament as a "great multitude of men armed in a warlike manner," the king breached centuries of tradition of parliamentary privilege by marching into the House of Commons and placing himself in the Speaker's chair. The reaction astonished him. Warned beforehand, the six in question had fled. The rest of the House greeted him in silence, then after a brief exchange sped him on his way, crying "Privilege! Privilege!"

Parliament met the next day but almost immediately adjourned to meet deeper in the City, at the Guildhall, presuming on the friendly protection of the populace. Even as Parliament was adjourning, the king himself arrived at the Guildhall to address the City's Common Council and explain his warrant for the arrest of the members of Parliament. The turbulent, divisive meeting went nowhere. As the king walked out, however, the sympathies of the "ruder people" in the outer hall became clear: they besieged him with cries of "privileges of Parliament." Fearing this surge of popular resentment, one royalist expressed the hope that "we find not that we have flattered ourselves with an imaginary strength and party in the City and elsewhere, which will fall away if need should be."[2] The king did have some local support—the trained bands of Middlesex stood guard outside his court at Whitehall. But on the night of January 6, the trained bands of London, fearing an attack by royal supporters, turned out around the City. They shut the gates to the City and prepared to defend its walls. Instead, Charles packed up his family and left London, not to return until he was dragged back under arrest.

Almost a year earlier, Strafford's execution had showcased the intensity, passion, and public involvement in the brewing conflict. Much of that passion proved to be invested in the public's vision of the meaning of subjecthood, and particularly the constitutional role and privileges of Parliament. It is hard to imagine "privileges of Parliament" as a popular rallying cry that would help start a war, but so it did. Constitutional conflict over a host of issues boiled down to a question of authority versus privilege, as did the internal wars of so many other contemporary European states.[3] Charles's ambitions for autocratic control were relatively modest compared to other European monarchs, but in this instance questions of elite privilege were also bound up with emotive issues of religion and English "liberties." Accordingly, Parliament's cry of privilege received enthusiastic popular support, especially in the key location of London at the critical moment in January 1642.

The opinions of the wider populace mattered before and well after January 1642. Their power to push events forward and sway outcomes was evident from the 200,000 who cheered Strafford's execution and the 100,000 who attended his funeral to the much larger number of people throughout

England who were deeply and viscerally afraid of Catholicism, especially if forcibly thrust into the country by bloody Irish hands. Perhaps even more influential was the people's "latent" power to sway decisions through elite fears of what they might do. The collapse of censorship in 1641 had helped spur a massive expansion of public access to political debate. Pamphlets poured off the presses, and both sides learned to play to the crowd. Parliament's famous "grand remonstrance," which listed Charles's supposed abuses and had precipitated the crisis of privilege in January, was in essence a gigantic effort to mold public opinion.[4] This is not to enter into the debate over whether social or class discontent itself led to revolution, but rather to emphasize the undeniable latent power of public opinion to shape events. It always threatened to become active, and political and military leaders had to take it into account.

That potential power was made abundantly clear at the very outset of the fighting. As soon as both sides had managed to raise sufficient troops and equipment to wage war, they clashed at the Battle of Edgehill on October 23, 1642. The battle was initially inconclusive. Charles's army recovered more quickly, and he pushed on toward London with great hopes of forcing his way back into the City and ending the rebellion. Reports of enthusiastic royalist plundering, and especially the violent sack of Brentford, however, stiffened the resolve of the initially downcast London populace. The London trained bands turned out in unheard-of numbers, eventually increasing parliamentary forces to over 27,000 men in the immediate vicinity of London. The bulk of that army mustered at Turnham Green and there confronted Charles on November 13. With no hope of defeating this startlingly large army, the king turned away, losing his first and best chance to retake London.[5]

Turnham Green demonstrated that this civil war, like other brothers' wars in this book, depended to some extent on the energy and motivation of the wider public, not just at the outbreak of violence, but also in the long hard fight that followed. Frequently, that energy failed.[6] The decline in public enthusiasm limited state capacity, producing fewer troops and less money. Sometimes the public not only lost its enthusiasm but also reacted violently to the presence of troops at all, especially troops out of control. Here the latent power of the populace became active, and that power in turn reinforced thoughtful commanders' desires to control their troops. But even the best-intentioned of commanders, men like Sir William Waller, must have occasionally felt that their efforts at control were dwarfed by the problems of capacity and the culture of the soldiers.

Plunder, Garrisons, and Military Culture

Chapter 3 described "Waller" pursuing the king in the summer of 1644, but it was his soldiers who did the marching.[7] And despite Waller's numerous proclamations against plunder and his regular courts martial (and ensuing executions), it was those soldiers who faced the daily necessity of finding both food and "quarter" with the local people. For them, a military culture fostered by a common experience in the ranks, combined with a specific religious outlook, actually encouraged plundering in some circumstances.[8]

Consider marching. Every day on the march began with the painful process of the soldiers assembling on the road in their units, each man in his proper place in the column. On hot days the dust of the scramble was insufferable; on wet ones the mud was ankle deep and unshakeable. It always seemed that some other unit's delay left them standing, waiting to start, but eventually the mass of men began to move. The horse had almost certainly already set off. The horsemen were indifferent scouts for the most part and had been known to plunder their own infantry, but at least they went first, taking their dust with them and slightly lessening the likelihood that the infantry would be ambushed. So the day's slog began. Over the course of the Cropredy Bridge campaign the men marched almost continuously, on thirty-six of the sixty-nine days between May 23 and July 31, 1644, covering nearly five hundred miles.[9] The battle itself took place on June 29, but battles did not end the torturous cycle of walking, finding quarters, sleeping (most nights), rising, and walking again. At least battle would wreathe them in the acrid smell of gunpowder, replacing the pervasive stink of men who did not bathe often in peacetime—much less now. A thirteen-pound musket almost five feet long, or, weighing the same and even more awkward, a sixteen-foot pike, rubbing the same spot on the shoulder with every step. A musket rest for the musketeers (an object abandoned early in the war). Armor down to the thighs for the pikemen. Sword and helmet for all. A minimum of food and alcohol (and perhaps water) in a knapsack for the day's march. Perhaps worst of all, for the musketeers, the constant clicking and clacking of the myriad little bottles of gunpowder swinging from their over-the-shoulder bandolier. Step, clink. Step, clink. The daily distance might not seem so long except for the endless morning process of getting started, and, at day's end, the equally endless process of being directed to their quarters for the night, or almost as often, having to find their own. The hurried chase after the king had meant twenty-one nights in the open, without tents. One night the rain poured so hard the army could not move.

A PIKEMAN AND A MUSKETEER. This image was part of a single sheet publication that sought to summarize all the motions of the pikeman and musketeer from a widespread genre of instruction books with many diagrams. Gervase Markham, *A Schoole for Young Souldiers containing in briefe the whole discipline of warre*, STC 17387 (London: Richard Higginbotham, 1616).

But this was the middle of England, and it was common for a column to pass through one or more villages every day. Sometimes they had to march through villages already full of quartered soldiers, pushing on another mile or two to find better prospects. The men expected shelter and food. If they paid the residents, it was "quarter," and if they didn't, it was "free quarter," but soldiers never seemed to note the difference.[10] What they did know was that villagers were at least supposed to bring them food. All too often, however, the people's larders were bare, as the royalists marched a day or more ahead, eating up the countryside. For the three-day pause outside Wordsley the villagers brought mutton, bread, cheese, and beer. But quarters varied: a barn sheltering two or three hundred; the salt cellars at Salwich that were so dry the soldiers in turn drank "the Towne drie"; the church in Tewkesbury; the fields near Stowe, and so on. On this march the army was relatively small and the quartering process relatively brisk; when Waller and Essex had joined forces during the approaches to Oxford earlier in the summer, the process had been tedious, extended, and dangerous. An army dispersed into quarters was vulnerable to the sudden descent of the enemy cavalry, who might come in to "beat up the quarters" or even compete for the space.

Vulnerability, fear, hunger, thirst, and frustration all sought an outlet: someone to blame and someone to provide. Fortunately for the soldiers, it was usually easy to find the "malignants" (open royalists or just Catholics) in the villages as the men passed through or ate. There was no point in sparing them any pains, as Parliament would probably confiscate ("sequester") their property anyway, and the officers were less likely to interfere when the victims were malignants. It was even easier to have good Protestant villagers point out the property of the papist priests. Puritan preachers and occasionally even their own commanders encouraged or inflamed the men's anti-Catholic leanings, and besides, altar rails made excellent kindling, and priestly vestments restored and brightened up worn-out clothing, or made fine handkerchiefs.[11] The officers usually noticed attempts to plunder the wealthy and powerful of either side, and regularly pronounced the law, threatened justice, and even hanged an occasional plunderer (usually the violent ones), but it was a simple matter to sneak into the houses of small folk at night for "whole great loaves and cheeses" as well as "meat and money." And, in the dark of night, did the peasants' religious preference really matter as much as the soldiers' stomachs? Sometimes, for no given reason, commanders selected a malignant's house and licensed the soldiers to pillage it "to the bare walls." For men eager for some compensation (they received only seventy-seven days' pay for that year's worth of service) and fed on rumors of the enemy's barbarity, that kind of order was easy to follow. And those rumors were ugly: the enemy mutilating the dead, their bodies "stripped, stabbed, and slashed . . . in a most barbarous manner." Enemy prisoners followed the column tied with match cord, two by two. There they were vulnerable, and vulnerability and rumor were a potent mix. Revenge was always possible; if not now, then when they were next at mercy.

A marching army's need for quarter was only one problem civilians faced. More pervasive and permanent was the problem of the garrison. As suggested in chapter 3, a garrison's function was to control the countryside. More specifically, garrisons were to extract provisions, money, and recruits from the surrounding territory, and to prevent the enemy doing the same. That process of extraction was always hard on the locals, but it could also be violent, largely because the garrisons usually operated beyond the control of a central command.[12] A garrison commander had enormous freedom to use the armed men under his control to enrich himself. Historian Philip Tennant has provided an intimate look at the inner workings of the garrison at Compton House in the summer of 1644, quite close to Cropredy bridge, and not far from Sudeley Castle, captured by Waller early in his pursuit of Charles.[13]

Initially garrisoned by royalists as part of their extended ring of strongholds around Oxford, their 150 men left little imprint on local records until the king stirred from Oxford, and Waller's pursuit generated a burst of local activity. The Compton garrison stung the edges of Essex's army, and the roundheads decided to deal with them. Colonel William Purefoy, a rigorous Puritan from that region, rounded up troops from nearby parliamentarian garrisons, laid siege to Compton House, and fairly quickly induced the garrison to accept quarter for their lives and civil usage.

As the new parliamentary garrison settled in under the command of William's nephew, Major George Purefoy, they began the process of allocating logistical responsibilities to the surrounding villages in an area that grew to encompass dozens of parishes. The garrison conscripted labor and materials to repair and improve the fortifications; carts were at a premium, as was bedding, and even chamber pots were requisitioned. The village constables were tapped for all of these tasks, and they felt the pressure intensely, torn between the force of arms and their strong ties to the local community. Meanwhile, Purefoy proved an apt student of the plunder-for-profit game. At one point he even imprisoned a local constable, William Calloway, until he produced eight hundredweight of butter and seven of cheese. Purefoy imposed taxes on much of the surrounding countryside. He was supposed to use those taxes to pay for the food and materials scoured from the countryside, but many of the villagers found obtaining that payment more than a little difficult. They understood that trouble, however, as a symptom of a broken system, and felt it within their rights to complain. One complainant carefully noted the non-military items stolen from him, including a "riding coat, waistcoats and petticoats, boys' hats and stockings, silver clasps for a bible, and a pair of sifters tipped with silver." He also pointed out that George Purefoy was present as all this was taken, making the offense much worse—it was not just soldiers "out of control," but under an officer's supervision. Complaints did not seem to go far nor reform Purefoy. He warned the town of Shutford to produce its arrears of taxes or he would "plunder yor towne and hang yor Constable." He apparently even descended into kidnapping travelers for ransom and appropriating the salary of a local schoolmaster. Few locals were exempt, but known supporters of the king were singled out for repeated plundering raids.

Thus the men in Purefoy's garrison at Compton settled in for two years of living off the produce of others, with only the occasional serious threat from the royalists. It was the kind of life one could get used to, a direct result of the inadequacy of the early modern state, and one of the wellsprings of a divergent military culture.

The root of free quarter and the garrison war lay in the inability of the early modern state to raise sufficient pay for its troops or to manage the delivery of supplies to an army on the move. Major urban garrisons were less of a problem, because they were at the center of an existing market network that had always brought produce to the city. Supply problems were common across Europe and had led other seventeenth-century European powers to rely on military contractors who accepted a ruler's commission to raise and run an army. After receiving an initial cash payment or credit to begin recruitment, the contractor financed and profited from his army by using it to extract cash from the countryside they traversed, sometimes through more or less regular "taxation," sometimes through somewhat less regular "contributions," and sometimes through outright plunder for profit.[14] Statesmen were explicit and calculating about the use of plunder to balance the revenue shortfall. France's Cardinal Richelieu knew that the cost of entering the Thirty Years War would exceed the state's revenue, and a subsequent royal council admitted that they could not pay the soldiers, but predicted that "they can be maintained without giving them a lot of money if instead we give them more freedom . . . to live off the country, although not so much that they become completely undisciplined and disobedient."[15] A perilous balancing act indeed. A wiser English officer, quoted in the first epigram opening this chapter, admitted in 1645 that this kind of plunder for pay provided no real benefit to the state, since the "Souldiers plunder is putt into a bottomless bagg: the State looses it: [and] the Souldier accompts it not for pay."[16] To be fair, however, it was often only this promise or potential of plunder that kept the army in the field.

The sixteenth and seventeenth centuries saw the costs of an army far outpace the peacetime revenues of English monarchs—revenues traditionally based primarily on their personal land holdings. Historian Azar Gat argues that under normal circumstances, a state can only sustain 1 percent of its population in a permanent armed force. This percentage can expand, but only for a time: in war, or through drastic social reorganization such as that in seventeenth-century Sweden or eighteenth-century Prussia. Herbert Langer estimates that forces operating inside the Holy Roman Empire during the Thirty Years War equaled 2.5–3 percent of the population.[17] England's prewar armed force was minuscule, well below Gat's 1 percent threshold. The civil war, then, was all the more shocking to Englishmen, as it led to two separate "states" tapping the same population to support as much as 2.35 percent of the country.[18]

As for finances, both sides, but especially Parliament, rapidly innovated new ways of raising income in wartime. By many measures, especially compared to pre-war state revenues, they were wildly successful, and both sides managed to avoid outright contracting along continental lines. The nature of a brothers' war, however, worsened the problem by dividing the available tax base of the country against itself.[19] Costs spiraled and experimental revenue streams failed to meet the need.[20] The soldiers' pay was supposed to cover their food when they quartered on the countryside, but more often than not the soldiers had no pay to offer, even if they had wanted to.[21] Parliament's financial system for paying the army through 1644 was decentralized and ad hoc.[22] Waller's army in 1644 received only seventy-seven days worth of pay. Essex similarly complained in the same summer that, despite the arrival of £20,000, the failure to provide "continuous pay" meant that, despite the best efforts of the officers, "the soldiers cannot be kept from plundering."[23]

Without centralized pay, the English frequently fell back on the same system that came into being in Germany, a system somewhat disingenuously called "contribution." The alternative German expression, *Brandschatzung*, was somewhat more honest; not directly translatable, it might best be rendered as "torch payment," meaning a payment to avoid being burned out. Essentially, under the contribution system towns paid off marching armies or nearby garrisons with cash in order to avoid the less certain and possibly violent alternative of plunder.[24] As unseemly as the contribution system sounds, one scholar has suggested that in the Thirty Years War it was relatively successful in limiting violence, and may have been deliberately restricted to be proportionate to the means of a given village.[25] In England the subject has undergone extensive scrutiny, and a surprising amount of order is discernible, even to the point of royalists and parliamentary forces reducing their exactions when they knew that both sides were demanding funds from the same region. Stephen Porter's study of the "fire raid" finds many threats of burning if contributions were not paid, especially against towns in contested areas, but relatively few actual burnings.[26] This supposed restraint would have been opaque to the villagers, however, while its threatened and sometimes actual violence was quite clear. A regiment of soldiers on the doorstep demanding money was a far cry from an ordered taxation system administered by familiar local officials. From the villagers' point of view, the military tax was often levied randomly, whenever an army hove into view; worse, it could be demanded by competing armies in rapid succession.[27]

The failure of capacity—that is, the failure to provide for the army through a regular and centralized system of pay and supply—meant that the more improvised solutions of quarter, contribution, and the garrison war were unavoidable. All three, combined with the frequent lapses in pay, generated uncontrolled plunder. Scholars of the wars on the continent refer to this as a "pillage economy"—soldiers sustaining themselves and their marching families on the property and produce of the peasants around them. Large numbers of women followed the mercenary contract armies on the continent, generating not only more hungry mouths, but in fact, according to one historian, stimulating and managing the pillage economy. The officers, though they might deprecate the necessity of pillage, confessed their powerlessness to stop it. Colonel Robert Monro, campaigning in Germany, admitted the necessity of taking booty "onely in respect of the circumstances," but nevertheless it created a "guilt . . . worse than the punishment." Monro may have felt guilty, but it is less clear that soldiers did, even in England, where the problem of contract armies never became as dire as in Germany or Italy. Henry Slingsby, a royalist officer, was more matter of fact. When, in late 1642, the soldiers began "to enquire after their pay," and finding there was no "treasure or treasurer" nearby, he confessed that "the souldiers must be the Collectors & in the mean time live upon free Billett, which caus'd great waste to made, especially where the horse came, and put the countryman at a great charge, so great as not to be imagin'd." Regrettable, to be sure, but, he concluded, "Soldiers must be satisfied."[28] For the soldiers, demanding quarters, taking food, and even plunder, became a part of their way of life, their culture.

The soldiers' experience of marching, suffering, fighting, and needing together, welded them into groups of men with a new set of collective values.[29] With a few exceptions, these men did not begin the war as professional soldiers. They were civilians called into the field by enthusiasm or coercion, and there they found a new kind of lived experience that demanded new behaviors. Every collection of human beings learns behaviors appropriate to the survival of the group. Soldiers learned and shared those lessons quickly, intensely, and without the moderating elements of stable neighbors and a full family network.

This creation, existence, and transmission of values within a discrete group precisely fits the definition of "culture" discussed in the introduction. To a large degree, culture changes in response to changes in material conditions, and soldiers at war were thrown into dramatically different material conditions. In the normal course of life in villages and towns, the shared grievances or problems that altered values were experienced privately at first and then slowly shared in public arenas like taverns and churches, where possible solutions were explored. For soldiers the whole process was speeded up; the rate of cultural "improvisation" and dissemination soared, although within a framework of precedent accumulated during previous wars. Furthermore, frequently soldiers generated those new shared values in defiance of the army hierarchy. Soldiers sat around campfires and explained to each other that one had to do what one had to do, and the devil take the hindmost. In other words, within the larger "military culture," there stirred a partially separate "soldiers' culture." This soldiers' culture represented a collection of values generated by their shared learning, communicated by them to new recruits, and in the early modern era, not yet dominated by social authority.

This was hardly Europe's first war; there were already many elements of a military culture (and even the soldiers' culture) in place, much of it familiar to Englishmen via direct experience, sermons, military manuals, and common gossip. In many ways the codes of conduct and the customs of war were themselves a part of this culture. Immersion in the military culture, and judging one's conduct against those values, rather than against broader societal values, granted a kind of license to soldiers to act in ways that were otherwise unacceptable.

In one way, however, societal culture already granted license to the soldier at war—being "at war" at a minimum granted license to kill in battle. Even that simple admission was one reason that Western society had struggled to define and create legal conditions for going to war in the first place. Christian theology essentially demanded that the license to kill be limited by rules. St. Augustine had provided the original basic justification in the needs of the state and its charge to defend the people under its care. The rest of the *jus ad bellum* (the rules of going to war) were mere refinements, but were still designed to limit the granting of a license to kill.[30] Waller himself argued within this worldview, saying after the war that he had simply acted in self-defense.[31] Re-immersed in a peacetime society, he found he had to touch base with that most fundamental justification for his wartime actions. Similarly, ministers on both sides, building on a strong pre-war tradition of just-war thinking and energized by a sense of embattled Protestantism, published a wealth of material and uttered a host of sermons designed to assuage whatever doubts soldiers might have about the rightness of their cause.[32] Matthias Milward preached to the London militia in 1641, reminding his listeners what they had been "taught by others" about the three causes of "undertaking a war": "Necessary defence, due revenge, and reparation of damage."[33] Another preacher suggested that Christ only forbade "private revenge and resistance," not a just war, even if against the king.[34] Another relied on the developing role of the state in the *jus ad bellum* to point out that "a right Commission makes the Warre it self lawfull to the Souldier, although it were undertaken by the Prince upon unjust grounds: for the Subjects duty is, to mind his owne call rather then the Cause."[35]

Societal culture provided only the basic definitions of just war that provided the license to kill while at war. Once soldiers were in the crucible of war, it was military culture that regulated when one killed, how one killed, and what one was supposed to do in the many other situations that did not confront a peacetime population. How did military culture grant license for the soldiers' many other acts of violence beyond killing on the battlefield? Some military rituals and customs provided a kind of license: it was easier, for example, to execute prisoners after a hasty trial to provide a legal cover.[36] The right to plunder civilians in the aftermath of an assault was seen by the soldiers as an entitlement, something they were owed by virtue of their sacrifices and sufferings, much like the plunder literally and legitimately taken off the battlefield. That sense of license was explicit, as was a corresponding sense of restraint. When a group of nuns in Mechelen in the Netherlands in 1580 pleaded with invading soldiers for a quick death, the men spared them entirely, saying the "fury was over and they no longer had license."[37]

Returning to the English civil war, one senses this entitlement in the experience of the earl of Chesterfield's wife. In late 1642 Parliament's forces attacked the earl's home and a small royalist garrison there, eventually forcing them to flee. The victorious officers then tried to convince the earl's wife to pay their men half a crown each for drink to keep them from sacking her house. She claimed to be out of money. When the officers offered to lend her the money for that purpose, she "refractoroly and willfully said, that shee would not give them one penny." One can imagine the soldiers standing nearby hearing this exchange, shrugging their shoulders, and then plundering the house as their just due after an assault. And so they did (although the officers protected her personal chamber and the goods therein).[38]

Everyone admitted that plunder was part of a soldier's reward, but the soldiers had also come to value the *process* of plunder as entertainment. Lady Chesterfield had refused to pay, but when Essex offered his own men twelve shillings to prevent them from plundering Reading in April 1643, where Essex had just agreed to surrender terms preserving the town, the enticement failed. The soldiers hoped for more, and for the "fun" of collecting it themselves.[39] Reading was sacked. This kind of license was a recruiting and retention tool for military service. General George Monck, writing after a lifetime of military service, believed that two things "cause men to be desirous of this [military] Profession; the first is Emulation of Honour; the next is, the hopes they have by Licence to do Evil."[40]

The separate military culture, and the license it granted to soldiers to open their "bottomless bags," did not sit well with England's broader societal culture.

Military advice written in 1593 showed both calculation and military culture in its assumptions about the role of villagers, suggesting that "if the country where our army passeth doe not furnish us [willingly] with victualles, the same is utterly to be ruinated, and burned. Which if the countrey people do perceive, either for feare or for hope, they will succour us."[41] That advice was written by an Englishman about fighting abroad, but by 1642 Englishmen had spent twenty years being terrified by the horrors of the Thirty Years War, and they were deeply concerned about the spread of such conduct to their island. The 1641 rebellion in Ireland had crystallized and intensified those fears. In part this was a generalized fear of war and of Catholicism, but there was also a specific fear of the behavior of armies and of soldiers.[42] Armed men under central authority were seen as contradictory to the English culture of localism and liberties.

Thus we return to the power of the political public, those men who sat on the sidelines but whose opinion mattered in a brothers' war. Political and military leaders on both sides felt compelled by military necessity to pursue the constitutional high ground in the initial outbreak of war, and then during wartime they competed for the moral high ground. As Parliament's Colonel Brereton worried, "The common people . . . for the most part judge our cause by the demeanour of our army."[43] The consequences of failing to assuage public expectations could be severe. Turnham Green and the thousands of London militiamen may have been decisive, but there were other less organized, but still militant reactions to the nature of the war and the license of soldiers. The most militant reaction was that of the "Clubmen."

The Clubmen, 1645

Few places more powerfully evoke a sense of uninterrupted habitation and tradition than the stone circle at Stonehenge. In the spring of 1645, on that windswept plain rising above Salisbury in the county of Wiltshire, and nearly simultaneously on other ancient hills in Wiltshire and neighboring Dorset, thousands of unhappy farmers and villagers gathered. There they pledged a solemn vow to each other to resist the depredations of both armies.[44] The men and women of the south and west of England suspected that they had "tasted the Miseries of this unnatural intestine War" more deeply than the rest of England. Indeed, a map of England showing garrisons, the tracks of armies, or the locations of major battles would have shown a heavy concentration of activity in their counties (the other major zone was Yorkshire and the Scottish marches).[45] Some commanders had been more careful than others at controlling their men. Some, like the royalist Lord George Goring, were notorious for not doing so. But even the best-intentioned officers admitted control was impossible if they could not pay their troops. Accused of letting his troops run wild in Cheshire, Colonel Brereton offered this typical explanation: "I have not had power to hould the Reines of Discipline as otherwise have bin Convenient, when extreame want of all necessaries have inflamed the Souldiers discontents to an unmaisterable height."[46]

Worse, peace negotiations at Uxbridge during the winter of 1644–45 had raised false hopes of impending reconciliation between the king and Parliament.[47] The negotiations collapsed, dashing hopes and igniting long-simmering resentments, producing a movement and even an organization. Although it was primarily an upwelling of the common folk's anger, lesser gentry, priests, justices of the peace, and lawyers provided a tempering of social authority and helped articulate widely shared frustrations.[48] As grievance led to action, the people followed the time-honored pattern of their ancestors. Uprisings against authority in the countryside had nearly always been led by at least a minor cross-section of the local power structure—sometimes the lesser folk pushing their superiors into action, sometimes the latter trying to manipulate the country to their own ends. They usually outlined their grievances in writing, attempting thereby to achieve a local consensus about their anger and their goals, and to make sure that the authority to whom they appealed understood the context of their actions.[49]

The Dorset and Wiltshire men and women issued their statements in late May and June of 1645.[50] There were some interesting regional differences between their statements and those issued by other counties, but many of the basic ideas were similar.[51] They appealed to "our ancient Laws, and Liberties," fearing that they were "altogether swallowed up in the Arbitrary power of the Sword." They affirmed their Protestantism and their loyalty to an England ruled jointly by king and Parliament. But finding their "enjoyment of our Religion, Liberties, or properties" threatened by the soldiery of both sides, they resolved to resist "all Plunderers, and all other unlawfull violence whatsoever."

The statements and their articles of association laid out a practical framework for defending themselves. Traditional local officials, the constables and tithingmen, would select watchmen who would regularly walk their precincts. They would not interfere with peaceable soldiers on the roads, but if they encountered violence they would either intervene or raise the men needed to put a stop to it.[52] Their rhetoric carefully branded such violent men as those who merely "pretend themselves to be Souldiers," whom they would take, disarm, examine, and then convey to their home army for punishment.

A number of the statements made a crucial distinction between plunder and quarter. They would respect a "lawfull Warrant" as well as quarter "demanded according to Order Martiall" (provided the soldier behaved "himselfe fairely").

Despite this seeming acknowledgment of military necessity, in practice the Clubmen were both more creative and more militant. They negotiated agreements with garrisons on both sides to pay them a regular stipend, provided they would not plunder or requisition any further. Although this sounds like typical "contribution," one must imagine in this case that it was more of a bargain than the usual outright armed extortion. In one instance they even persuaded two enemy garrisons to meet together for a drinking party. When faced with a sizeable threat, they arranged systems of signals with bells and messengers to muster forces reported to range from four thousand to twenty thousand. Once assembled, they marked their hats with white ribbons and marched under banners and flags, also colored a peaceful white. At least one banner, however, added a challenge: "If you offer to plunder or take our cattel, Be assured we will bid you battel." They came to their musters armed with their eponymous clubs, but also with muskets and pikes. They killed the occasional violent straggler from the armies and promised the competing local commanders that they were prepared to defend themselves. The royalist garrison at Devizes proved a particular target for the Wiltshire Clubmen. In early June they refused to pay further contributions and then attacked groups of soldiers who emerged, killing several and capturing twenty horses.[53]

At first the Dorset and Wiltshire Clubmen focused their attention on the royalist garrisons in their vicinity. They were true neutralists, however: when Lord Thomas Fairfax arrived with the New Model Army in late June to relieve Goring's siege of Taunton, the Clubmen quickly turned on his unruly soldiers.[54] Fairfax's approach took him through the heart of a country "in Arms under the nation of Club-men."[55] In Salisbury (Wiltshire) one of his columns was taken aback by the city's militant confidence. Meanwhile the Clubmen in Dorset "forced the quarters" of another parliamentarian force, with several killed on both sides.[56] They also interrupted the army's communications and threatened messengers and even foraging parties.[57]

The Clubmen soon filled Fairfax's thoughts and altered his plans.[58] He met with several Clubmen leaders, solemnly promising them he would control his troops and punish disorders, but he still considered them a threat. During his July 25 council of war, the argument varied between advancing against Goring in the west or first teaching "better manners" to the Clubmen at their rear. Some in the council preferred to deal with the Clubmen first; ironically, Parliament's committee for the war expressed a similar concern about the Clubmen's potential that same day.[59] In the end the council recommended continuing westward, and Fairfax initially agreed. The army marched ten miles west the next day, but Fairfax's worries overcame him. He changed his mind and returned to capture Bath and deal with the Clubmen.[60]

On August 1st an opportunity presented itself. Learning that the Dorset, Wiltshire, and Somerset Clubmen had planned a joint rendezvous, Fairfax dispatched one thousand horse to break up their meeting. They captured some fifty of the leaders, which quickly led to a massive outpouring of Clubmen into the field. Fairfax sent Cromwell to deal with them; he found ten thousand men and their colors on a hill near Shaftesbury. He persuaded them to go home, but not far away he found two to four thousand more militant Clubmen dug into an ancient fort on Hambledon Hill. After they fired on Cromwell's first parley attempt, and two more attempts did not get much further, Cromwell lost patience and ordered a general assault. The Clubmen threw back his men, killing one or two and wounding several more. A flanking force finally crawled up the back of the hill and the surrounded Clubmen broke and fled; some few were killed (Cromwell claimed only twelve).[61]

Cromwell's violent dispersal of the Clubmen on Hambledon Hill was perhaps the most dramatic moment of the movement, but the movement did not end there. It spread to other counties over the remaining months of 1645, and Clubmen continued to appear even in those counties where Fairfax and Cromwell had seemingly broken their power.[62] But more and more often, they proved amenable to cooperating with Parliament, taking their revenge on Goring, and increasingly acknowledging the improved discipline and pay of Fairfax's New Model Army.[63] Declining royalist fortunes also hastened this drift toward Parliament. A variety of Clubmen detachments marched under Fairfax's orders in August, and that fall one group under a parliamentarian colonel captured a hundred royalist horse. One parliamentarian optimist predicted that soon their army would include eight thousand Clubmen. A least two "regiments" of Clubmen working with Fairfax were destroyed by Goring in October.[64] They also continued to act on their own, attacking the royalist garrison still at Devizes.[65]

In the end the military significance of the Clubmen movement failed to measure up to royalist or parliamentarian fears. Even so it was not negligible. At a time when only three armies in England exceeded twelve thousand men, the threat of anywhere from three thousand to a stunning twenty thousand Clubmen was not to be taken lightly. Charles Carlton downplays their significance, noting that probably only six hundred Clubmen were killed during 1645. However, compared to his own calculation of yearly casualties, simply in terms of relative scale those six hundred men were equivalent to 16 percent of all the parliamentarian dead in 1645, and greatly exceeded the two hundred parliamentarians killed in the decisive battle of Naseby earlier that summer.[66]

The Clubmen are also significant for what they suggest about the relationship between the state and its people. The Clubmen emerged as a result of a clash of cultures between English localism and a soldiers' culture that was created by a failure of control, a failure caused, in turn, by the limited capacity of the state to pay its troops. But in another sense the Clubmen movement represented a militant form of localist resistance that was the very reason for limited state capacity in the first place. If that seems circular, it is. It was a cycle that reinforced itself. Early modern European states engaged in a more or less constant struggle to extend their power in the face of tenacious resistance up and down the social scale. This restraint on capacity generated violence because it meant that armies often went unpaid, but it also limited the ultimate capacity of the state to wage frightful war. It did so by restraining how far a ruler might go for fear of his own people's reaction, both in using force against his own people and in demanding money and recruits for wars against other states.[67]

Kings struggled against localism in their efforts to expand state capacity, but they also confronted a broader European trend in which "peace" was increasingly seen as something valuable and desirable. According to J. R. Hale, the expansion of occupations and the expectation of luxury in this era both changed the attitude of the normally martial aristocratic elite and reinforced the already "in-built pacificity" of the urban merchant elite. He concludes that "peace . . . was now more generally seen as a positive, attainable, prolongable, above all profitable and interesting phase of national life."[68]

Furthermore, generals in the field conceded that angry villagers could prove troublesome. Colonel Monro repeatedly mentioned incidents of violence between soldiers and the aroused peasants in Germany, who, on one occasion, "cruelly used our Souldiers (that went aside to plunder) in cutting off their noses and eares, hands and feete, pulling out their eyes, with sundry other cruelties." He further repeatedly recommended working hard to avoid "violence to be used to Boores [farmers]."[69]

In England these trends were more powerful than elsewhere. The preceding century had seen the rise of "a comparatively large, well informed, experienced 'public,' with a keen appetite for news, and capable of political action and influence."[70] Every effort at conscription in England in the preceding century had met with massive resistance that both limited the number of recruits and made the process more expensive.[71] Furthermore, as a brothers' war, the English civil war had from its beginning generated a powerful neutralist "faction" rooted in more than self-interested resistance to conscription. That movement gained strength after the fighting broke out, especially when no immediate conclusion seemed to be on the horizon. Even though popular opinion in London had urged Parliament to break with the king, a competing neutralist impulse nevertheless generated a powerful peace movement there in December 1642.[72]

Partisan sources tended to equate neutralism with supporting the other side, and this was indeed often the initial reaction to the Clubmen.[73] But historians' deeper analysis has found many men deliberately avoiding the choice between king and Parliament. Instead, neutralism emerged from a profound sense of localism. It is not enough to say, as Lawrence Stone famously asserted, that the civil war was "striking" for the "almost total passivity of the rural masses." In fact, the men of twenty-two counties produced neutrality pacts that tried to demilitarize their locales. And there were a variety of other public declarations, statements, and pacts, all designed to protect local communities from the armies of both sides.[74] Like English civil war generals, historians must take into account the possibility that neutralism could become active. That potential to be active, as neutralists or as partisans, was always present in the calculation of the leadership.[75] The Clubmen were just the most visible manifestation of an active localism that could reshape strategic calculation. Perhaps even more common was the way one's own side's localism forced alterations in central strategy through demands for the local use of local money and local troops. In one of many examples, the residents of Derbyshire convinced the parliamentary committee for the war to disobey Lord Fairfax's orders and to keep their regiments of horse within reach, since they felt threatened by nearby royalist garrisons. Communities demanded that their side protect them, even at the cost of the overall war effort.[76]

Neutralism and even peace campaigning were possible reactions to the strains of war and political division, but another was hostility to the side deemed the worse offender. It was this kind of reaction that led Londoners to oppose Charles at Turnham Green after hearing of royalist plundering at Brentford. The King recognized the problem in regular speeches in which he tried to assure the countryside of his attempts to control the soldiers.[77] Unfortunately, many of his officers seemed less concerned. One parliamentarian observer admitted that although some of the King's "Gentry and Superiour Officers" were attentive to controlling their soldiers, there was "no Security to the Country, while the multitude [of officers] did what they list." He believed "this was kept from the knowledge of the King," but he also was certain that it was their disorders that "filled the Armies and Garrisons of the Parliament with sober, pious Men. Thousands had no mind to meddle with the Wars, but greatly desired to live peaceably at home, when the Rage of Soldiers and Drunkards would not suffer them." Parliament's armies were far from perfect, but on the whole their reputation was generally better, although any given local garrison commander, like Purefoy at Compton, could change that equation. The royalists may have acquired their reputation for plundering in part because the area of England under their control shrank in late 1644 and 1645. Charles Carlton goes so far as to argue that plundering lost them the war, although other historians have disagreed with such a radical conclusion. It is clear, however, that a concern for public opinion touted in a culture of localism shaped military calculations, especially with regard to the use of indiscriminate violence.[78] Furthermore, the English civil war did not lack examples of spontaneous local resistance to the forces of either side. This resistance seems to have been less brutal than those described by Monro in Germany, but the possibility of villagers' collective violence surely always lurked in the minds of soldiers who were regularly quartered in small groups in small villages.[79]

Active neutralism, peace movements, the effect of violence on recruitment, resistance to taxation and conscription, spontaneous peasant revenge, and organized Clubmen all speak to the power of the populace at large to affect strategic calculation and state capacity. But spontaneous peasant revenge in particular held the potential to greatly increase the horrors of war; indeed, there

PEASANTS' REVENGE. This image is part of a series by Jacques Callot called *The Miseries of War*, published widely in the seventeenth century and intended to convey the miseries of soldiers and civilians. Here he shows oppressed peasants turning on soldiers. All of his images depict the Thirty Years' War. Jacques Callot, *Les miseres et les mal-heurs de la guerre* (Paris: [Israel Henriet], 1633), pl. 17. From the Robert Charles Lawrence Fergusson Collection. Reproduced by permission of the Society of the Cincinnati, Washington, D.C.

is little doubt that something of the sort occurred in Germany during the Thirty Years War. In this sense military culture quite literally clashed with the culture of localism. Soldiers saw themselves as beings apart, no doubt encouraged by the "imperious carriage" of their captains. Their profession legitimated their violence. Villagers were *supposed* to bring them food and lodge them. For example, when royalist troops trapped inside York in 1644 found themselves without money, they were distributed into billets in the town. If a townsman lacked the resources to provide for a soldier, it made little difference, "for the soldier knew him that was appoint'd to pay him, & if he refus'd the soldier lays hands on him or any thing he had."[80] Soldiers with this vision of their own status and power reacted with outrage and a special kind of furious violence to the attacks of peasants.[81] Furthermore, the organized Clubmen and their more spontaneous brethren had no investment in the traditional codes of war and no intimate familiarity with how they worked. It is even possible that this unfamiliarity explains their firing on Cromwell's first attempt to parley—not to mention the solitary armed Clubman who, when confronted by Cromwell and his troops, actually tried to get off a shot rather than lay down his musket as instructed.[82] As fragile as the codes of war were, soldiers' vengeance could be terrible when the codes were violated without even a passing regard for the rituals and forms soldiers held dear. Fortunately, in England the war was comparatively short, and the Clubmen's reliance on local social leaders made it easier for them to open channels of communication to army leaders. Parliament proved relatively tolerant of the Clubmen, and the height of the movement remained confined to a few months when the first civil war was almost over.

Conclusion

Most historians of the English civil war have long since discarded the old idea of a purely gentleman's war, devoid of atrocity and relatively confined in its destructive impact. But they have also generally agreed that it was not nearly as horrific as it could have been. The simultaneous experience in Scotland, Ireland, and of course Germany, provides clear counterpoints to how much uglier the war could have become. William Waller's campaign showed how conformity to a mutually understood set of rules, enforced by deeply embedded beliefs in God and in honor, and combined with a widespread desire for eventual reconciliation, kept some key restraints in place. But the codes of war also provided a license for violence, particularly in the aftermath of an assault on a fortified town. In Barbara Donagan's apt phrase, "Permissible cruelties existed side by side with humane protections."[83] The escalating violence of the second and third civil wars can also partly, if not largely, be explained by anger and frustration at the sense that the defeated royalists had violated the conditions of their surrender at the end of the first war.[84] Parole violators (and the parliamentarians who changed sides in the later wars) were seen to merit much less mercy. The conditions of war and the bureaucratic weakness of the emergent state (especially one divided against itself) also helped generate a divergent soldiers' culture in which the men felt a license to plunder as a substitute for absent food and pay. The longer soldiers lived in that culture, the more separate they felt, and the more license they took. Military calculation also played its role. The new need for collective discipline to achieve battlefield success hardened and elaborated the system of military law, although at this stage disciplinary systems had not yet caught up with the problems of supply and the resultant soldiers' culture. Military calculation also suggested the need to control the space in which human and financial resources could be mobilized. That need seemed to mandate a system of garrisons, and those garrisons inevitably lived off the countryside. This was enough of a problem in itself, but some garrisons, like Purefoy's, morphed into profit-making enterprises, with all the resultant outrages.

Scholars have long recognized the impact of an active and informed citizenry on the nature of warfare. There is less agreement on when and where such a citizenry existed, what "popular" feeling means, whom the term "popular" includes, and how such a movement affects power or authority. The Clubmen are a reminder that military men often worried about the effect of their armies on a countryside that might rise against them. Similarly, there is a growing historical sense of the resistance of European common folk to the impositions of their royal masters, which in turn limited the emergent states' military and financial capacity. Most of the international wars of the seventeenth century were fought by mercenaries, but the brothers' war in England demanded an appeal to the long unused, or barely used, citizenry as soldiers.[85] In that appeal to the obligations of subjecthood existed the possibility both for greater restraint and for greater frightfulness.

The Clubmen, and Englishmen in general, understood their subjecthood within specific circumscribed boundaries. They had "ancient liberties." They expected social authority to be wielded by men whom they knew and, often, whom they had chosen. This cultural perspective defined their response to the violence and the demands of the armies around them.

Elite leaders on the national stage understood that potential power and had to build their military and political calculations around it. They had to appeal to the public, generate propaganda in vast quantities, tap emotions, feed prejudices, and polish the public image of their army's virtue. In England that process produced restraint, in part because officers on both sides adhered to codes of war, enforced by elite notions of honor, to a greater extent than in the contemporary international wars fought by mercenaries. The calculated appeal to the public, in England, if not the British Isles more generally, gave the codes more teeth and more political purpose. This calculation, however, also limited the capacity of either side to raise men and money. Local cooperation could only be pushed so far. Finally, this calculation also all but excluded the use of a deliberate strategy of devastation, much less the strategy of famine used in Ireland at the beginning of the seventeenth century.[86]

Despite the division between royalists and parliamentarians, both sides continued to acknowledge the English public as a "political public": a public whose opinion had to be taken into account. After the war, that public would remember soldiers' actions, and that memory, along with Cromwell's military rule, intensified their dislike of armies. Having experienced the effect of the soldiers' culture firsthand, they wanted no further part of it. At the very least they did not want to have to see it. Much of the military establishment after the 1640s would be kept in Ireland or abroad, out of the sight of the political public.[87]

A political public only exists, however, when respect for its future participation in the polity exists. The Irish tossed over the walls by Sidney's men in 1569 had been placed outside that public, at least in the minds of the soldiers and, over time, increasingly in the minds of their masters. A similar imaginative problem surely existed in the minds of soldiers in Virginia in 1610 who found themselves encumbered with a Paspahegh queen and her children as prisoners. Their commander had ordered that no prisoners be taken, but the queen and her children had been spared by a subordinate, and now he could not bring himself to kill them. His soldiers disagreed. While rowing back to Jamestown, the men finally forced their commander to kill the children "by Throweinge them overboard and shoteinge owtt their Braynes in the water." The soldiers continued to grumble about the queen's survival, and eventually a superior officer ordered her burned alive. The commander had "seene so mutche Bloodshedd thatt day" that now in "Cowldbloode" he refused to burn her. He ordered another officer to execute her quickly with a sword while he himself turned away.[88] It was only 1610; the English new world enterprise had barely begun, but already there seemed little room for Indians as fellow subjects or even as honorable enemies.

Part III

Native American Warfare

Countless wars with Native Americans profoundly shaped American culture and society. Understanding the often ugly violence of the Anglo-Indian wars requires careful attention to the combatants on both sides. We have seen warfare through English eyes in Ireland and in the English civil war. We now must examine warfare from a Native American perspective. Lacking a written tradition of "laws of war," they nevertheless had powerful and systematic restraints built into their ways of warfare. Some of those restraints derived from a social organization that limited the scope of mobilization and thus the capacity to destroy on a large or continuous scale. Other restraints derived from a flexible vision of war that accommodated a gradation of intensities suited for particular purposes. War did not always have to be all-out, although Indians were perfectly capable of making it so. Unfortunately, European and Indian ways of warfare, and their systems of restraint, proved to be not only different but dissonant. Mutual misunderstanding of beliefs and practices designed to restrain war's violence instead unleashed it. In that dissonant and violent environment, fragile hopes for incorporating the Indians as coequal civil subjects dissipated even as Indians themselves resisted those hopes. Over time Englishmen and Americans came to see Indians not merely as uncivil, but as true barbarians, fit targets for the most extreme forms of war.

5

Wingina, Ralph Lane, and the Roanoke Colony of 1586

This mocking (between their great ones) is a great kindling of Warres amongst them: yet I have known some of their chiefest say, what should I hazard the lives of my precious Subjects, them and theirs to kindle a Fire, which no man knowes how farre, and how long it will burne, for the barking of a Dog.

Roger Williams, 1643

The assault would be in the night when [the Pequots] are commonly more secure and at home, by which advantage the English, being armed, may enter the houses and do what execution they please. [And] before the assault be given, an ambush be laid behind them . . . to prevent their flight"

Narragansett advice to the English, 1637

Wingina and Lane

Wingina had reasons to worry and nearly as many reasons to hope.[1] He and his Roanoke people lived comfortably on the rich resources of land and sea on the coast and islands of Wingandacoa, now North Carolina. No absolute monarch, he nevertheless held some influence over a wide region inside Pamlico Sound. A number of villages deferred to him, acknowledging his ability to distribute food and other goods among them, especially the bits of raw copper that had traveled across many miles and through many hands before arriving in the coastal tidewater. He enjoyed friendly relations with his nearest neighbors, especially the powerful chief Menatonon of the Choanoakes. Their loose alliance helped protect him and his people against the Iroquoian-speaking Mangoaks further inland. Although he had recently warred against several peoples to the south, he was now at peace with them. He also had contacts among the Chesepians to the north and the Croatans who lived on the southern islands of the Outer Banks. Indeed, Manteo of the Croatans had been visiting Wingina's people the previous year when something entirely unexpected had happened—the source now of his new hopes and worries. An English ship had appeared, seeking trade and information. Wingina knew of these white men, and he knew they had goods that could greatly increase his prestige and influence if he controlled their dispersal among his people. At the time of the Englishmen's visit, Wingina had been recovering from a battle wound, and he had not met with them personally. But his brother Granganimeo had spent several days in their company. Language difficulties had kept Granganimeo from determining exactly what the English wanted, but their trading had been useful and may have marked the beginning of a mutually beneficial relationship. At the end of the English visit, one of Wingina's young men, Wanchese, and the visiting Croatan, Manteo, had left in the English ship.[2] Wingina hoped they would return with knowledge he could use, knowledge that would allow him to manage the potential benefits of this relationship. But he also feared there might be dangers here.

Manteo and Wanchese sailed to England with Arthur Barlowe in 1584, arriving at a time when England and Spain were hurtling toward open war. Spain was Europe's undisputed great power, flush with New World silver and determined to see Catholicism triumph in post-Reformation Europe. Protestant Queen Elizabeth of England had become increasingly militant, encouraging piratical attacks on Spain's treasure fleets and aiding others in their fights with Spain. The stage was set for some of the most dramatic, even mythic, events in English history. Almost as an afterthought (certainly as a sideshow), some of her courtiers, including her favorite, Sir Walter Ralegh (often now spelled Raleigh), expanded their colonizing vision from Ireland to encompass the newfound lands of America.

England had been expending its colonizing energies in Ireland, and so came late to the New World game. The Spanish had been enlarging and consolidating their hold on the Caribbean islands and the former Aztec and Incan empires for three generations, and they laid a nominal claim to virtually the entire hemisphere. The English adventurers were interlopers, challenging the mightiest monarch in Europe with impossibly grandiose hopes. Those hopes pointed in at least two not necessarily compatible directions.[3] On the one hand, they hoped to establish bases to contest Spanish supremacy

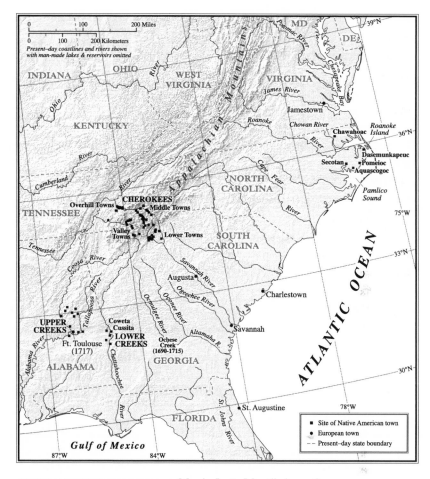

SOUTHEASTERN INDIAN TOWNS. Map by Justin Morrill, the-m-factory.com.

in the New World and thereby profit. On the other, the leaders of the expeditions hoped to lay personal claims to vast new estates, much as some of them had done in Ireland. Sir Humphrey Gilbert, the colonizer of Ireland discussed in chapter 1, was one of the most famous of these Ireland–New World crossover colonists. Notorious in Ireland for his brutal suppression of the Desmond rebellion, and spotlighted by historians for lining the path to his campaign tent with the heads of his Irish enemies, he had lately attempted to plant a colony in Newfoundland. The effort disastrously failed, costing Gilbert his life. His half brother and fellow adventurer in Ireland, Walter Ralegh, took up the mantle, but he turned his eye to more propitious climes further south.[4] Gilbert's failure and the extraordinary Spanish successes suggested the need for planning, not only to avoid Gilbert's mistakes but also to avoid Spanish retribution. A key player in both goals would be the natives. They too would have to be planned for.

Barlowe's voyage to Wingina's territory had been part of that planning. His reconnaissance, followed by a year's worth of language exchange with Manteo and Wanchese, laid the groundwork for a new expedition—this one designed to establish a more permanent presence in Wingandacoa. Ralegh ignored the local name, and named the region Virginia in honor of Elizabeth. Ralegh had been preparing continuously since Barlowe's return, and, among other recruits, he had brought Ralph Lane into the project. By 1585 Lane was a veteran soldier and relentless proposer of military schemes. He petitioned for and received an appointment in Ireland, but was soon recalled to Ralegh's service. Lane's first brief service in Ireland generated few records, but when he returned there in 1592 after his adventures in Virginia, he became the muster-master general of English forces in the opening years of the Nine Years War. During that time he was regularly accused of corruption and malfeasance by a government-appointed inspector. His own letters from that period are defensive and proud, but they indicate at least a passing interest in protecting the countryside from "the spoil as well of the soldier as of the rebel."[5]

When Lane sailed for Virginia his head was full of orders, plans, and not a little hope of personal gain. His instructions, his problems, and his actions reveal much about the way English cultural attitudes and modes of military calculation would shape the violence of English-Indian relations for the next three hundred years.

The overall command of the 1585 expedition went to Sir Richard Grenville; Lane had command of the soldiers and the specific task of controlling their behavior. Spain's own priests had accused its agents of inflicting terrifying atrocities on the natives, and the English hoped to profit from the supposed native resentment. English colonial promoter Richard Hakluyt urged the planners to build a project "most void of blood," and a military planning document, much influenced by Hakluyt's work, laid out exceptional punishments for colonists who treated Indians violently. The punishment for raping an Indian was to be death; for forcing Indian labor, three months' confinement; even entering an Indian house without permission meant six months' imprisonment or labor. But Hakluyt also acknowledged the hard reality of possible, but to his mind unjustifiable, Indian opposition. England was going there for "the glory of God" and the "increase of the force of the Christians." If the Indians resisted "[we] who seeke but juste and lawfull trafficke," then "we will proceed with extremitie, conquer, fortifie and plant . . . and in the end bring them all in subjection and to civilitie."[6]

Much of this rhetoric rings true to English goals and experiences in Ireland, whose people also were to be made subjects and brought to English civility. Unfortunately for the Indians on the receiving end of Lane's plans, many of the men recruited to the project were also veterans of Ireland. Such men expected to live off the country if official supplies failed; to compel obedience; to preempt presumed betrayal with betrayal; and to assume resistance was in fact rebellion.

Grenville's fleet arrived off the Outer Banks of North Carolina on June 25, 1585, and explored the islands and the mainland around Pamlico Sound. Grenville landed Lane and his men on Roanoke Island inside the banks and then departed in September. Aside from a few letters in August and September, much of the next few months are a documentary blank until a report from Lane outlining his interactions with various local Indian peoples beginning in March 1586, and covering the period until his departure in June.

The whole story of Lane's colony is not necessary here, but within it there is yet another "march" to follow. Lane's march displays some of his and his men's assumptions and fears, fears similar to those held by other early English colonists, and shows how they erupted into violence. Ralegh and Grenville tasked Lane with exploring the mainland, and the lure of mineral wealth on a Spanish scale drove him and his men to comply eagerly. Lane contacted Wingina shortly after his arrival and found him only too happy to provide Lane a base at Roanoke. There they would be conveniently close for Wingina's purposes, but also separated by a broad sound from his main town at Dasemunkepeuc on the mainland.

Limited English resources immediately played a role in defining their relationship. Storms during the transatlantic voyage had destroyed much of the preserved food and seed for the colony and Lane found himself dependent on Wingina's people. But the Roanokes were struggling through a period of drought and climatic uncertainty; they complained of little rainfall in the spring of 1586, and during the 1587–89 period, the years of the famous Lost Colony, the region underwent the worst two-year drought in eight hundred years.[7] Nevertheless, the Roanokes, and Lane's control over his men, kept the English alive over that first winter: out of 108 men only four died— an almost unheard of survival rate for first winters in the New World.[8] Spring, however, brought the usual pinched rations until the summer's bounty began. But spring was also the campaigning season, and with its arrival, Lane began to explore again. Unfortunately, short rations and active campaigning had always proved a troublesome mix.

Wingina encouraged Lane's plans to explore inland. Accounts of the events that followed are confusing, reported only by Lane, and commented on only briefly by Thomas Hariot, who opined that Lane became excessively paranoid and aggressive toward the end of the colony's existence.[9] The Indians' true intentions must remain obscure, but what Lane claimed to believe is clear. Lane intended to explore, and reports of strange minerals and the possibility of a distant sea had focused his hopes on mineral wealth and a passage to the western ocean. But he did not abandon the more immediate practical problem of seeking a good harbor for a permanent settlement. At Wingina's urging, and convinced by him that the Choanokes and Mangoaks were preparing to attack the English settlement, Lane sailed up the Chowan River, seeking the Choanoke chief, Menatonon. Perhaps legitimately fearing attack, Lane surprised Menatonon in his own village, and in a very conquistador-like manner, immediately seized the chief, a man crippled but "very grave and wise," bound him, and then pressed him for information over two days.[10]

Menatonon told Lane a great deal. He showed him pearls from the vast bay to the north (the Chesapeake) and suggested that white men had been in the bay before dealing with the Indian king there. He told him of a great sea at the head of the Morotico (Roanoke) River. Menatonon probably added to the chorus of voices who had already told Lane about strange minerals further inland. And finally Menatonon told Lane, who by now was thoroughly impressed with Menatonon's knowledge and cooperation, that it was Wingina who intended to destroy the English. Each of these pieces of information played into the calculations and choices Lane made in the next few weeks.

Regarding his dreams of controlling the bay, Lane could only lay out a hypothetical model of what he would do if a new supply of men arrived. He would proceed further up the Chowan and there build a "sconse," a kind of raised, ditched, and palisaded fort, to guard his boats. He would then march overland, building additional sconces every two days' journey, preferably in the midst of some convenient cornfield, adequate to feed a small garrison of fifteen to twenty men. And so he would continue until arriving at the bay, where he would build another, larger fort as the center of the colony. His soldierly calculations are a near perfect reflection of his experience in Ireland: chains of forts with surprisingly small garrisons, dependent for food on dominating a nearby farming community, connected ultimately to the sea and thus to England.

Lane would never get to execute that plan, and for the moment he focused on continuing inland in quest of the strange mineral. He released Menatonon in exchange for a ransom and took his son as a "pledge" (to use the term then current in Ireland for the same process) to send back to Roanoke for safekeeping, along with some of his men and the larger boat. Lane himself headed west up the Roanoke River with forty men in two rowboats. As they rowed through the territory of the Mangoaks, from whom he hoped to capture prisoners to guide him to the mines of the interior, Lane grew frustrated. The Indians withdrew from the banks of the river, taking their food with them. In Lane's thinking, reinforced by Menatonon's report, this confirmed Wingina's conspiracy. The veteran soldier expected to find food as part of his march, in much the same way as he hoped to build his sconces in local cornfields. Three days of rowing west produced no food and no contact. With only two days' food left in the boats, and facing a long row back east to Roanoke, he put the question to his soldiers: continue at the risk of starving or turn around? Eager to pursue the least rumor of upstream riches, the men voted overwhelmingly to continue, and to eat their dogs at the last. And so it happened. Two more days of rowing yielded only the sight of distant Indian fires and, finally, a surprising volley of arrows into the boats. Lane pursued, but in what would become the defining pattern of white-Indian warfare, the Indians "wooded themselves" and disappeared. Lane feared that even if he caught up with them, he would find none of their food. Lane and his men turned about and rowed hurriedly downriver, relying on stews of their dogs and sassafras leaves.

Lane's safe return to Roanoke seems to have startled Wingina, who may have expected the Choanokes to kill him. This, at least, is what Lane believed. The Roanokes reverted to a cooperative posture, planting fields and building fish weirs for the English colonists. Lane remained suspicious, and he feared the hungry gap between the sowing and the harvest. Heightening the tension, the man most convinced of the value of helping the English, Granganimeo, had died during the winter, and in April the other English-inclined Native leader, Ensenore, also died. In a potentially threatening sign, Wingina changed his name to Pemisapan, a word implying watchfulness.[11] Lane became increasingly convinced that Wingina/Pemisapan intended his destruction, encouraged in this belief by Menatonon's son, Skiko, who went back and forth between Lane and Pemisapan. Finally, Lane launched what he intended to be a preemptive night attack on Pemisapan's village. His boats rowing from Roanoke to the mainland chanced upon a canoe, but the noise they made killing and decapitating the two Indians therein alerted those on shore, and

Lane called off the attack. In a move not unlike the first earl of Essex's dealings with Brian McPhelim O'Neill, Lane then requested a meeting with Pemisapan, under truce, to discuss the situation. They met in Pemisapan's village, and as they stood together in a small group, Lane cried the signal "Christ our Victory" and his men fired. In the smoke and chaos, Pemisapan briefly played dead, but then jumped up and ran off into the trees. Lane's Irish "boy" (reminiscent of the horseboys from chapter 1) shot him as he fled. Another of Lane's Irish servants followed him into the woods and returned with Pemisapan's head.

There is a great deal in Lane's story that should sound familiar, both from the wars in sixteenth-century Ireland and from the popular image of white-Indian warfare in North America. At least a quarter of Lane's men were experienced soldiers, men who Lane referred to as the "wylde menn of myne owene nactione."[12] The others were assorted "experts" gathered for specific investigations of New World resources (apothecaries, miners, and the like) who lacked a cohesive identity with which to oppose the military men—even if they had wanted to. Soldiers expected to take food from civilians—this was, as suggested in Parts 1 and 2, an assumption of the era's military culture. The English were unprepared for the food scarcity of late spring, due partly to their losses from storms, and they were frustrated by the Indians' ability to disappear into the wilderness. As in Ireland, but further exacerbated by the width of the Atlantic, English agents in the New World interpreted their instructions broadly, and instructions to behave beneficently toward the Indians ran against the grain of Lane's assumptions. Accustomed to the side switching of Irish rebels, pardoned one year and rebels again the next, Lane expected the same kind of double dealing from the Indians. Did he see them as barbarians? Perhaps, but he had hardly had time to build up a whole network of racial prejudices. He, like many other early English travelers, expressed what seemed to be sincere admiration for some of the natives—their carriage, their physique, their capabilities, and their hospitality all impressed.[13] Stories of Indian barbarity would have filtered into the English imagination via Spanish reports, but they do not seem responsible for Lane's choices.

Instead, Lane acted true to both his sets of instructions: to be beneficent, but also to be prepared to impose an English vision of civility. English rulers had an inclusive vision of civilizing Indians from barbarians into brothers, but they framed those hopes in terms of political sovereignty and their own assumed cultural superiority. Sovereignty required control and order; cultural superiority justified it. Most historians have emphasized English legal rationalizations designed to justify dispossessing the Indians, but from the very beginning of the colonial era there existed a parallel belief that it was only a matter of time before the Indians could be made civil. Furthermore, in the early years of exploration and settlement, the practical needs of overextended and fragile bases demanded a certain level of friendly cooperation.[14] Like Lane, Thomas Hariot optimistically reported of the Roanoke Indians that "in respect of us they are a people poore," but they have their own ingenuity and "excellencie of wit," and it is likely they will "desire our friendships and love, and have the greater respect for pleasing and obeying us. Whereby may be hoped, if meanes of good government bee used, that they may in short time be brought to civilitie, and the imbracing of true religion."[15] The Jamestown colony received similar instructions about incorporation. William Strachey reported that Governor Gates in Virginia, after witnessing the Kecoughtan Indians kill one of his men, was "not a little trouble[d]," as he had "since his first landing in the country (how justly soever provoked) would not by any means be wrought to a violent proceeding against [the Indians] for all the practices of villainy with which they daily endangered our men, thinking it possible by a more tractable course to win them to a better condition." Revealingly, however, the contemporary English commentator Samuel Purchas wrote in the margin of this account: "Can a leopard change his spots? Can a savage, remaining a savage, be civil? Were not we ourselves made and not born civil in our progenitors' days? And were not Caesars's Britons as brutish as Virginians? The Roman swords were best teachers of civility to this and other countries near us."[16] This self-justifying reference to their own savage, pre-Christian past gave "reform" a hard edge: reform *required* conquest, and indeed Gates destroyed and occupied Kecoughtan. But in theory, at least, it was a form of conquest that should have been forward looking, toward eventual reconciliation and incorporation.

But the English were not the only party in this relationship. Native Americans, like the Irish before them, all too naturally, and often all too effectively, resisted English sovereignty and conversion, and they fought even more desperately to control their traditional means of subsistence, most especially their land.[17] Constant strife over land in the ensuing generations contributed to the slow development of deeper and deeper prejudices that played into the colonists' use of violence, and may even explain its later contours. But at this early stage, Lane's unhesitating seizure of prisoners, ransoming of chiefs, and preemptive attacks, complete with decapitations, arose naturally from his European military experience, combined with a powerful sense of vulnerability, perched as he was alone on the edge of the ocean.

Lane's story also hints at some of the differences between English and Indian visions and practices of war. Nothing in Lane's brief experience in Ireland encouraged him to show benign restraint now, while commanding soldiers on the edge of the world. He did, however, work within a "grammar" of war that gave meaning to violent action. Those meanings frequently had been contested in Ireland, and violence there escalated. Shared meanings in the brothers' war of the 1640s rendered its violence mutually comprehensible, and thus more generally restrained. Surprisingly, in Lane's story, as violent as it was from our perspective, the grammar can still be seen at work—at least from his perspective. He held pledges against future conduct; he took ransoms and released prisoners, and most of all, he sent signals about strength while in a fearfully shaky position indeed. But could the Indians understand Lane's language of war, and would Lane understand theirs?

To make sense of the extreme violence that became the norm between whites and Indians in North America, we need a more complete understanding of how Indians imagined war, an understanding that moves beyond English prejudice or even the contest for land (without ignoring either). What were Indian systems of restraint? How did they imagine victory, and what was their social capacity for wielding violence? Much of the rest of this chapter and the next shifts the lens away from the English and examines these questions from a Native perspective. Understanding Native systems and cultures of war will show how deeply Indian systems of restraint did not mesh or synchronize with the European systems examined in the first two parts.

Native American Society and Warfare

When Lane's men sailed up the Chowan River, like other early Europeans arriving on the eastern seaboard, they confronted a silent and apparently empty wall of trees, occasionally punctuated by a clearing and a settlement. The European explorers had only vague and usually inaccurate notions of life in and beyond the trees—vaguer and more inaccurate even than their knowledge of the bogs and byways of Ulster. Although Lane and his successors would spend the next three hundred years trying to understand, profit from, and dominate the peoples they encountered, they could never grasp those peoples' tremendous complexity and variety. They saw the forest, but only an exceptional few ever saw the trees. That lack of comprehension in turn mars our modern effort to understand Native American society from the contact and colonial eras. Using primarily European sources, some oral traditions, and a smattering of archaeological evidence, we can construct only a pale silhouette of the real nature of Native American societies and Native American war.

The problem becomes more acute when one asks if and how those societies changed as a result of the arrival of Europeans. If knowledge of the contact era is but a shadow of reality, then the understanding of precontact Indian societies is the shadow of a shadow. Virtually every conclusion about Indian social and cultural structures of war must be tempered by this cautionary note. Are even the earliest seventeenth-century descriptions of Indian war a reflection of societies that had already markedly changed? We know, for example, that even prior to the *Mayflower*, European-introduced diseases had wiped out a significant percentage of certain coastal New England groups.[18] Similarly, along the St. Lawrence River and in the southeast, regions that sixteenth-century European travelers Jacques Cartier and Hernando de Soto found to be heavily populated, were described by seventeenth-century travelers as barren or nearly deserted.[19] The magnitude of the depopulation certainly had some consequence on warfare—whether to limit it or expand it is not clear.[20]

There is one more caveat. There was of course no such thing as "Native American society." Societies varied from Maine to Florida, or even from one valley to the next. But the same caution applies to the phrase "European society." Without downplaying the dangers of overgeneralization, one can draw some conclusions about Native American societies in the eastern woodlands because they shared many characteristics, especially in war. The following broad-brush descriptions move beyond Wingina and the Roanokes, later into the colonial period, and reflect some of the changes that came with European contact.[21]

During the colonial era Native Americans typically lived in villages or towns ranging from a few families to several hundred or, in a few cases, several thousand people. They centered their lives and allegiance on an extended kinship group or clan, which often, although not always, followed the maternal line. Ethnically and linguistically related clans often occupied the same region and formed larger associations, finding in such cooperation both strength in numbers and a wider array of marriage partners. These long-standing associations meant that each town included members of several clans, which cemented inter-town ties. Individual towns, however, retained a high degree of political autonomy. These clusters of towns are traditionally called "tribes" (e.g., Seneca, Cherokee, or Narragansett), but they remained fractious, independent entities who might act together out of choice but were quite capable of refusing to do so.[22] Rarely was there any truly autocratic tribal leader, however much the Europeans hoped and pressed for such a figure. In part because of the legacy of the European efforts to construct a unified tribal system, this book avoids "tribe" and instead uses the English version of the Indians' most common word for themselves: the "people."[23]

Compared to European states, Native American political structures of the colonial era were relatively egalitarian. In the precontact era, and even during the earliest periods of contact, it is clear that many Native societies operated as chiefdoms. Political power was centralized in the hands of chiefs who, like Wingina, managed the flow of prestige goods and may even have collected and stored food for redistribution. This basis for hierarchy proved vulnerable to the massive shocks of disease, depopulation, and forced migration suffered up and down the east coast by the mid-seventeenth century. Although traces of the chief-centered, stratified system remained, colonial-era societies operated through consensus and divided authority.[24] This was even truer in times of war. The decision-making process was presided over by a council of elders, often split into peace chiefs and war chiefs (discussed in more detail in chapter 6), and in some societies further checked or confirmed by a more or less formal council of women. Individual sachems or chiefs still held some forms of leadership, as well as claims to land or tribute, which they could pass down to heirs (often through the maternal line). Mobilization for war, however, depended not on chiefly power but instead on the combination of group consensus, the reputation of the leader proposing an offensive, and individual warriors' desire for prestige.

Subsistence for most of the region depended on a combination of farming, gathering (especially among coastal or riparian groups), and hunting. Depending on the level of farming activity, different groups might be more or less sedentary, perhaps leaving the town in smaller groups for the winter hunting season. Some peoples sent out winter hunting parties without ever completely abandoning the towns. Over long periods, they might exhaust the soil around the town and then relocate within a local region. The need for periodic relocation and for dependable hunting territories meant that every group had a much more expansive need for land than Europeans assumed at first glance.

To this basic political and economic portrait one must add war, an activity fundamental to Native American culture in a variety of ways. Young men could gain status and authority through demonstrated courage and aptitude in war. Groups could gain or protect territory at the expense of others, incorporate prisoners into their population as kin or as labor, and impose tribute on other peoples. Deeply enmeshed within these material motives were the require-ments of revenge; above all, blood demanded blood. The killing of a member of one group mandated revenge on the other side. The rewards and requirements of war were so thoroughly entwined in Indian societies that irrespective of the arrival of the Europeans war and violent conflict was nearly endemic through-out much of the eastern seaboard and beyond.

That last assertion about "endemic war" demands deeper scrutiny. It is difficult to be sure exactly how Wingina, Menatonon, or other Native Amer-icans thought about war at the outset of European contact. The evidence is confined to archaeology, ethnographic analogy with other similar societies, and early European accounts, which, as discussed, may already have come too late to capture a true precontact vision of war. Worse, especially for under-standing the Roanokes, or even the Powhatans who greeted and warred with the English colonists at Jamestown, the first English colonists arrived at the geographic nexus of two clearly different patterns of precontact Native American warfare.

The preceding several centuries in the American Southeast and Midwest had seen the flowering and then decline of the so-called Mississippian chief-doms. Compared to the usual stereotype of North American Indian society during the colonial era, these chiefdoms were larger scale, urban, agricultural, and hierarchical.[25] They competed with each other for control of labor, trade routes, and territory. Since a chief's power and position in part depended on acquiring prestige and controlling local sacred space, attacks on an enemy chiefdom sought to plunder prestige goods and burn or otherwise defile ritual centers in order to undermine the enemy chief's capacity to effectively mobi-lize his own population. Spiritual power could be gained not only from the taking of prestige goods but also through taking human body parts as "tro-phies," which in turn transferred their personal spiritual power to their taker. In addition, it seems clear that Mississippian warriors took some prisoners as slaves. This combination of motives and social organization produced a form of warfare featuring open battles and large-scale sieges in addition to ongoing smaller-scale raids and ambushes. The first Spanish explorers encountered this style of warfare in the early sixteenth century, but Mississippian societies were then already on the decline, breaking off into smaller and smaller polities. The Spanish arrival likely hastened that process, and by the time solid docu-mentary evidence begins again in the late seventeenth century, the Native peoples of the Southeast had shifted into the form of warfare practiced by the Indians of the seaboard and in the Northeast.

A MISSISSIPPIAN WARRIOR. This pre-contact image shows a successful warrior returning with the severed head of his enemy. Mississippian warrior, Castalian Springs Mound Group, Tenn, 1250–1350 AD. Courtesy, National Museum of the American Indian, Smithsonian Institution, Image D150853, photo by Walter Larrimore.

The precontact evidence for the "Northeastern" pattern of war is less clear, but recent work in anthropology and archaeology suggests that it too could be highly lethal, but that lethality depended heavily on surprise. Northeastern societies lacked the centralized hierarchy of the Mississippians, so their style of warfare strove to avoid risk. Major losses on a campaign could have a significant demographic impact at home, something that their more egalitarian political structure would not tolerate. Furthermore, the balance of offensive and defensive technology limited the capacity for killing and destruction—provided major surprise was avoided. Many towns in the Northeast were fortified (although, oddly, few to none in precontact New England), and those fortifications were usually resistant to enemy raids. But surprise could turn deadly, as witnessed by Jacques Cartier in one of the first contacts in the sixteenth-century Northeast. In 1533 the Toudaman Indians surprised a party of some two hundred sleeping Iroquoians inside a temporary palisade. They set fire to the palisade and then killed all but five of the inhabitants as they rushed out.[26] A few years before the first English arrival in Roanoke, the Werowance Piemacum, one of Wingina's neighbors, apparently invited many men and women of Secotan into his village to a feast under some form of truce and then surprised and killed the men, taking the women and children prisoner.[27] The more usual balance of offense and defense also existed on the individual level. Indians in the Northeast frequently wore armor, and even

PICTOGRAPH OF IROQUOIS WARRIOR WITH PRISONER. From a French copy of an Iroquois drawing, ca. 1666. From Colonies, CııA2, folio 263, Archives Nationales, Archives d'Outre Mer, Aix en Provence, France.

helmets, that were generally proof against most arrows. Finally, since the usual goal of going to war was to take prisoners, warfare was characterized by the unexpected attack. Warriors swooped in on an isolated group, suppressing resistance with arrow fire, closing in to disable the men with clubs, seizing and binding the women and children, and then quickly returning home, seeking to avoid pursuit. In this world, cordage for binding prisoners was a crucial part of a warrior's toolkit.[28]

With the collapse of the Mississippian chiefdoms and the arrival of European technologies, Native American warfare across the eastern part of the continent rapidly homogenized. Although raids and ambushes had always played a major role, now they increasingly dominated warfare. It is important to emphasize, however, that Europeans neither taught Native Americans about lethality in war, nor introduced the concept of war that sought the total destruction of an enemy.[29] They already knew. In a later era, admittedly well postcontact, the Narragansetts' advice to the English in 1637 was clear and stark: "The assault would be in the night when they [the Pequots] are commonly more secure and at home, by which advantage the English, being armed, may enter the houses and do what execution they please."[30]

Blood Feud and the Levels of War

Chapter 6 will detail the techniques and restraints of postcontact Native American warfare, but for the moment there is a more fundamental problem to discuss: what was "war"? Used with regard to Native American societies, that word is an all too convenient catch-all for a whole spectrum of conflict, representing different motives and functions. The broadest, and slipperiest to interpret, function for war was its use to administer political "lessons" in proper relationships between groups. Most scholars acknowledge that Indians fought war for such political purposes, but it is less clear to what extremes Native Americans would press war in a political cause. For example, did political adjustments through warfare include outright conquest? Two other purposes of war are more clearly understood. A great deal of Native American conflict was filtered through the demand for blood revenge coupled with the expectation of achieving personal status through war. The mandate that relatives take blood revenge for the killing of one of their own was arguably the single greatest factor in patterning violent relationships both within and between peoples. Furthermore, young men in virtually all Native American societies looked to success in war not only to assert adulthood but also to increase their status within the group. Warriors returned home bearing enemy scalps or prisoners as their individual possessions, and even though such prizes were redistributed upon arrival in the village for the benefit of all, for that moment, possessing trophies redounded solely to their own credit as men.[31]

These three motives for war were intimately intertwined. A war that arose from material or political causes might frequently take the specific form of the blood feud. To put it another way, the recruitment of individuals for a series of raids that might have political or material consequences still relied on blood feud and status rhetoric to motivate young warriors.[32] The obvious problem with these latter two motives for war is their potential endlessness. There are always new young men in need of proving themselves, and each act of revenge typically begets a fresh desire for revenge from the other side. In fact, however, although it was extraordinarily awkward to contain, the seemingly endless loop of the blood feud possessed both structure and restraint.

The basic principle of blood debt was simple.[33] The killing of any person, accidental or otherwise, placed an obligation on the dead person's kin to exact revenge on the killer's people. This obligation had two crucial, open-ended components. The timing and intent of the original killing were irrelevant—an accidental killing, a deliberate murder (as a European would define that term), or death in battle all equally mandated vengeance.[34] The other open-ended component was the lack of specificity in the target for revenge. Any member of the killer's people would do—in the case of an intragroup killing, the killer's "people" meant the killer himself and his relatives. If the killer were from outside one's own people (the tribe), any member of his people would suffice (anthropologists call this "social substitutability").

In terms of warfare, the key issue arose when a death was caused by, or blamed on, another people. In some cases a blood gift could be offered to calm relations. Such a diplomatic overture might avoid the outbreak of war, but success depended on the willingness of the clan council and the dead person's relatives to accept it.[35] Furthermore, if revenge was taken and a new killing occurred, one side could suggest that the second killing had balanced the first, but convincing the other side to accept that balancing proved difficult. The Cherokees of the town of Keowee, for example, threatened the Catawbas that a failure to accept a recent killing as squaring the balance would lead to escalation:

> If the Catawbas continue to take Revenge, we will not only go against them Ourselves, but draw the whole People of our Nation against them: but if they are satisfied we are also, for as they began first, and laid our People in heaps, we have kill'd two of them, and laid them on our own; and now we are satisfied, if they will be so; but if they are not, we will soon go against them, as we think nothing of them, and as it was intirely their own Fault.[36]

This difficulty of "balancing blood" arose from the nature of a war party. The party ostensibly mobilized for revenge but was composed of individuals hoping for status, especially young men, who had no particular desire to limit their attacks to one person. Once in contact with the enemy, individual needs for prestige led them to take as many scalps or prisoners as they reasonably could, because those prizes conferred status.

There followed an obvious response. Any deaths inflicted by the first war party created a need for a new retaliatory strike. The potential endlessness of this cycle was the weak link within Native American restraints on war: the cultural mandate for revenge proved extremely difficult to overcome. Historian John Reid summarized the problem: "A [Cherokee] warrior who had recently lost a brother in a Creek raid might tell a Creek peace delegation that he would bury the bloody hatchet after he had taken one Creek scalp. If he succeeded, it would be for the brother of his Creek victim to decide if the war would continue. Peace negotiations therefore were largely promises to forgive and forget."[37] In some cases the original source of enmity between two groups might be lost in the depths of time, but endemic raiding back and forth continued nonetheless, as each new killing reinvigorated the blood debt.

Even within this revenge-based system, however, restraints existed. The process of mobilization limited the size of the avenging party and thus its capacity for destruction. Decisions for war were reached by consensus, and Native American leaders lacked the power to coerce participation.[38] As indicated in the just-discussed Cherokee-Catawba conflict, mobilizing warriors from outside the family or from members of other towns required extra effort. Unless the dead person was a prominent figure, his or her death was unlikely to stimulate a multi-town mobilization. An influential leader with a significant following could possibly expand the pool of recruits, but only as long as his reputation held out.

While a lack of coercive political structures limited the scale of war, the Indian ideology of revenge was itself another fundamental limitation. The revenge motive did not carry with it the motivation to pursue the wholesale destruction of the enemy people—a few scalps and prisoners would suffice. This "tit-for-tat" understanding of feud was unfocused in its targeting since any victim from the enemy clan would do, but it was limited in its scale. This limited notion of revenge differed dramatically from the European ideology of revenge in war. The Europeans also believed in retaliation, but they were much more thoroughly lethal. The European ideology of revenge presumed that an original violation of norms, however minor, authorized a no-holds-barred retaliation.[39] For Europeans, extreme retaliation was necessary to enforce the codes of war. When the Powhatans attacked the Virginia colony in 1622, the Virginian response was fired by a vision of catastrophic retaliation and then sustained by a greed for land. The sources clearly suggest that some leading Virginia colonists were glad of the excuse to clear the land of Indians for their own use, but ideologically they justified their hugely destructive violence by citing the Indians' "treacherousness." Edward Waterhouse, an agent for the colony, summed up the Virginians' explanations: "Our hands, which before were tied with gentlenesse and faire usage, are now set at liberty by the treacherous violence of the Savages, not untying the Knot, but cutting it: So that we . . . may now by right of Warre, and law of Nations, invade the Country, and destroy them who sought to destroy us."[40] In this light, the blood-revenge system seems much less destructive.

Turning from the blood feuds to wars with political and economic causes, it seems clear that Indians distinguished between "grand" and "little" war.[41] On occasion, sufficient motivation existed for large parties of warriors, perhaps six hundred to a thousand, to attack their enemies "in the name of the tribe" rather than as part of a limited blood feud, apparently in the hope of inflicting damage well beyond the norms of the little war—the tit-for-tat war—practiced in the blood feud.[42] Expeditions such as these were at least intended to force a reformation in relationships between groups, if not to conquer territory or impose tribute. They attested to the power and vitality of the attackers and sought to coerce their enemy into modifying their behavior in some way. This interpretation has been applied to many of the larger attacks on the European settlements, but it also applies to significant inter-Indian wars.[43]

Furthermore, Native American societies did not necessarily recognize a strict duality between little and grand war. The little war of the blood feud could escalate into larger attacks otherwise lacking the political or economic intent of "grand" war.[44] Conversely, a people could launch a long series of small-scale "little" war–style raids, but with a larger political calculation in mind, attempting in effect to conquer by harassment. The Narragansetts, for example, advised the English in 1637 to attack the Pequots over the "space of three weeks or a month, [and] that there should be a falling off and a retreat, as if you were departed, and a falling on again within three or four days." It was even possible to resolve disputes through single combat between champions; truly a "little war," but one with larger political import.[45]

Nor did Native Americans draw a stark line between war and peace (perhaps in part due to the endemic nature of blood-revenge needs). Peace did not exist as a formally declared state so much as it was achieved through the mollification of hostile feelings.[46] In societies lacking coercive central authority, and deeply invested in the blood feud, it proved impossible to control the hostile feelings of every individual. There also existed an intermediate level of conflict between peace and outright war. In that state, significant levels of violence could be in play, perhaps intended to carry a political message, or to let off young men's steam without a full commitment to war. Essentially, there was another point on the spectrum between the poles of peace and war, which, for lack of a better phrase, I call the "not quite war" level of conflict; in early Jamestown John Smith called it "sometime peace and warre twice in a day."[47] The "not quite war" was politically motivated as an effort to affect relations between groups, but it employed only the small-scale methods of "little war." If Native leaders determined that peace was a better alternative, they could disavow a raid as having been "only" little war, that is, it had been simply about a personal grudge.[48] Furthermore, the "not quite war" functioned well within the imperial context. Helped along by the autonomy of individual towns within the larger groups, it allowed Indian headmen to "capitalize on [their] decentralized, kin-based politics to cultivate connections with rival colonies and so avoid dependence on a single European power."[49]

Some examples may clarify this concept. The Cherokees engaged in a round of "not quite war" with the English in 1758. Nominally allied with the British during the French and Indian War, Cherokee warriors returning home from Pennsylvania were attacked and killed by Virginia militiamen in a dispute over stolen horses. As a matter of course, the relatives and townsmen of the dead sought blood revenge, aided by anti-British, pro-French factions within the Cherokees. Small groups of warriors raided the white frontier in Virginia and North Carolina throughout the summer and fall of 1759. Strikingly, at the same time all remained peaceful around the two British forts in the midst of the Cherokee towns. Trade and relations there continued, if with greater worry and tension. There was no sense of the two peoples being "at war," even while some Cherokees were undeniably wreaking havoc on English settlements elsewhere. Confronted diplomatically, the Cherokee leadership downplayed the violence, describing it as young men out of control. The conflict thus was both simple blood debt and political at the same time, all while the Cherokees tried to avoid a broadening of the conflict.[50]

In another example of such behavior, a Creek party attacked the English trader John Sharp's home near the Cherokee town of Nayowee in 1724. The Creeks approached his house, fired several volleys, wounding Sharp in the leg, and then plundered his house. All of this was done within sight of the Cherokee town, whose residents declined to interfere. Sharp nevertheless survived. Historian Tom Hatley interprets Sharp's survival as proof that the Creeks were merely making a point, "a symbolic statement about the strength of Creek men to the onlooking Cherokee villagers."[51] They could just as well have been making a violent point to the English in general and to English traders specifically, but at any rate Sharp was neither killed nor captured, nor were the Creeks then "at war" with the English.

Once "at war," whether at the "grand," "little," or even the "not quite war" level, a number of factors built into Native American society tended to limit the extent of the conflict's lethality and overall destructiveness—for both the attacker and the defender. These structural restraints were not absolute, and they may or may not have originated as deliberate efforts to limit war, but there is little doubt that they served that function. Following up on what happened after the Creek attack on Sharp provides a better sense of how they worked. As in the earlier chapters, it is easier to perceive the details of restraint and violence by turning to a single campaign as an example. The attack on Sharp spurred a protest from the government of South Carolina to the Creeks. Tobias Fitch traveled to Creek country in 1725 to deliver the protest and demand an apology, and one result was a Creek expedition to attack the Yamasees. If that sounds like an indirect chain of events, it was. The genesis, execution, and results of that expedition serve as an example of *how* Indians went to war, *what* exactly they did, and what restraints functioned during the process.

6

Old Brims and Chipacasi, 1725

The [Iroquois] Warriors sought fame ... in desultory excursions against the Cherokee and Catawbas. . . . [They] left home in parties from two hundred to ten . . . [and traveled until they] came upon the Head Waters of Holston, along the Banks of which the Cherokee Hunters were frequently scattered;—these they often surprised, killing and taking them prisoners. At other times, they proceeded to the Villages, but only in small parties to prevent discovery. . . . When [they], had gained Scalps or Prisoners, they fled to where their comrades awaited their return, to support them in case they might be surprised by superior force. . . . Many of these Parties were overtaken [during their march home], others triumphantly returned with Scalps & Prisoners.

Journal of Major John Norton

They repeat this Request [for cheaper goods] again & again & say unless Goods particularly powder be sold Cheaper they must disperse themselves & that the Sachems can no longer keep up their authority over the Young men.

Five Nations of the Iroquois to the Governor of New York, June 1711

Old Brims

The August Georgia sun beating down on the plaza at Coweta did not improve Old Brims's mood.[1] As he sat and listened to Tobias Fitch, South Carolina's messenger to the Creek towns, he could feel his anger and disgust grow. Brims had long been a *mico*, or leader, among the "Lower" Creeks (see map, p. 123). He may even have been in Coweta when the South Carolinians made one of their first contacts with the Creeks in the late 1670s. Coweta had become a gateway on the trading path between the South Carolinians and the Creeks, and that had strengthened Coweta against its Spanish-supplied enemies to the south and enhanced the town's status among other Creeks. In the 1690s Brims had joined a migration of some of his people to Ochese Creek on the Ocmulgee River, in order to live closer to the English. Up through 1713, Brims and many Creek men aided the English during their various wars with the Spanish and other Indians. In 1711 the English even gave Brims a commission designating him an emperor. The title meant little to his fellow Creeks, but it did mean that the English tended to deal with him first.

Nevertheless, the Creek-English relationship had its troubles. The Creeks on the Ocmulgee River found the trading system increasingly disruptive as corrupt English traders drove them into debt. One trader even killed a relative of Brims in 1715. So when that same spring the Yamasees of South Carolina launched an all-out war on the English, many Creeks joined in what nearly became a pan-Indian alliance.[2] Brims tried to limit Creek involvement, but the pressures of war pushed him and his fellows on Ochese Creek back to Coweta and the other towns on the Chattahoochee. Then in 1717 Frenchmen appeared and built a trading post and fort named Toulouse in the heart of the Upper Creek towns (near modern-day Montgomery, Alabama). A French source for European goods lying to the west favored other towns over Coweta, but Brims proved adept at managing the flow of change. He quickly reestablished friendly ties and trade with the English using his son Ouletta as an emissary to Charlestown.

Aging, and now called Old Brims, he enjoyed a reputation and influence among the Creeks enhanced by his long relationship with the English. In 1722 the English governor renewed Brims's commission, affirming his favored status and designating his son Ouletta as his heir. Maintaining that position, however, was growing ever more complicated. As Brims sat listening to Fitch in 1725, some forty-five other chiefs from various Lower Creek towns were also present. Some favored the French, while others wanted to reduce their reliance on Europeans all together. Brims knew very well that he could neither speak for nor control all of them. Brims's son Ouletta had died while on a raid into Florida, and now the question of who would receive Brims's English commission after his death was up in the air. Worse, what Fitch and the English did not seem to understand was that he and his people had other problems besides supplying the English with deer-skins. Just ten years ago, in the midst of the Yamasee War, the Cherokees to the north had murdered a diplomatic party of twelve prominent Creeks.[3]

Now here was Fitch demanding that the Creek war with the Cherokees must end.

Old Brims had finally had enough. The blood of his kinsmen killed in the Cherokee country still called out to him and his people for revenge. To Fitch's claim that some of the other Creeks might make peace with the Cherokee, Brims replied that they "may do As they please But we have Nothing of Makeing a peace with The Cherokeys. For them men that was killed by the Cherokeys of Mine When the White people were there is not over w[it]h Me as yet, nor never shall be While there is a Cowwataid [Lower Creek] Liveing."[4] This anger, and Creek raids on the Cherokees, which in turn required Cherokee retaliation, kept their war alive until 1752.[5]

Brims recognized that he needed to offer the English something, and he wanted them to agree to his choice for his successor. He knew this would be tricky, because his choice was a man named "Seepeycoffee" (Chipacasi), who had been notoriously pro-French and pro-Spanish in the preceding years. Chipacasi would have to do something momentous to prove his loyalty, and so he and Brims agreed to Fitch's request to send a war party against the Yamasees, who continued to struggle against the South Carolinians from their new homes in northern Florida.[6]

Creek Society and Chipacasi's Raid

To make sense of Brims's choices and Chipacasi's raid, one must look back into the history of the Creeks and of Brims. The Creek Indians were Musk-ogean-speaking descendants of the peoples encountered by Hernando de Soto during his wanderings through Georgia and Alabama in the 1530s and 1540s. They lived in several different clusters of towns along the Chatta-hoochee (the English would call them the "Lower Creeks"), Tallapoosa, and Coosa Rivers (the "Upper Creeks"). By the early eighteenth century they had become in effect a multiethnic confederation of a number of different peoples. That confederation was built around an original ethnic, linguistic, and still dominant core, but now included "daughter towns" of other groups, including the Alabamas, Koasatis, and Hitchitis, and eventually some Apalachees, Shawnees, Yuchis, Yamasees, Natchez, and Chickasaws.

In the seventeenth century, pressed from the east and south by other Indians armed by the Virginians or Spanish respectively, the Creeks quickly turned to the English arriving in Charlestown, South Carolina as potential allies. The Creeks' success at confederation, incorporation, and alliance made them one of the most powerful peoples in the southeast by the turn of the eighteenth century and for many decades to follow.

This history of movement, agglomeration, and confederation was reflected in the presence of forty-five other headmen at the council with Brims and Fitch in 1725. It made for complicated internal politics. In fact, Fitch had just come from another conference with the Upper Creeks attended by sixty headmen. Such broad representation was necessary because each town in the confederacy was an autonomous agent. The confederacy acted together primarily in defense, as a grouping of natural allies who turned to each other for support and for recruits. It was all too common, however, not only for a town to act on its own, but for ambitious individuals to act even further afield. To further complicate matters, there were additional layers of political organization and cooperation between the town and confederacy level.[7]

All of these dynamics challenged English understanding. In one of the councils Fitch brought up the attack on Sharp's trading post discussed at the end of chapter 5. The leader of the Creeks who had shot Sharp was a man named Gogel Eyes. When Fitch confronted him about the attack, Gogel Eyes apparently tried to excuse the incident as the actions of young men out of control. This claim of "not quite war," or at least of young men's desire for status overwhelming more politic hopes for peace, proved unacceptable to the more hierarchically minded English (notwithstanding the parallel phenomenon of white frontier settlers generating Indian conflict in defiance of their own political leaders). Fitch complained, "That Excuse will not do." Gogel Eyes, he said, was "a man in years and ought To know better," and since he had been leading the men who attacked Sharp, he "Should have prevented th[e]ir Rogush proceedings."[8] What looked to Fitch like an evasion of responsibility was actually something quite different. Gogel Eyes' claim that the attack on Sharp was an instance of "young men out of control" reflected the tangles of Creek political structure and the Creeks' general lack of coercive social authority (or "control" in the terms of this book). It was also an effort to limit the act's implications by defining it as not an act of war.

Having said that Creek leaders lacked coercive social authority, what are we to make of Brims's supposed title of king, or even "emperor"? This is also a complex matter. The designation was entirely invented by the English. When they first arrived Brims was already a significant player among the Creeks; over time, however, as the English increasingly funneled their diplomatic gifts through him, Brims gained in stature and authority not only in his town but within the confederation. Native leaders gained prestige not by hoarding wealth but by redistributing it, and he used his privileged access to

English goods to enhance his power and authority over other Creeks and even over other Creek towns. Furthermore, it is likely that Brims was a war chief, and the South Carolinians' favoring of him likely diminished the influence of the Coweta peace chiefs.

Most Indian peoples on the eastern seaboard distinguished between a "civil" chief who dealt with domestic issues and a "war" chief who dealt with outsiders and played a leading role in battle. Furthermore, Native councils tended to be made up of older men, respected for their wisdom and already possessed of war honors.[9] The Creeks went further, and had "white," or peace, chiefs take an oath to be devoted always to the white path of peace and never to shed human blood (unlike the "red" chiefs, who had responsibility for war).[10] Over the course of the colonial era, the power of the war chiefs tended to rise. European traders and diplomats almost invariably conducted negotiations with war chiefs, who received the diplomatic gifts and were assumed to be the real authority of the group.[11] European bestowal of titles like "emperor" or "king" did not make it so, but over time this process weakened the influence of the civil and religious authority.[12]

Thus fully understood, Brims's brief exchange with Fitch reveals a host of key characteristics of Creek society and warfare: political decentralization, the European impact on the peace-chief/war-chief relationship, Indian expectations of delegate immunity, and the passion behind blood revenge and the resultant endemic state of war. Indeed, one Creek town lost five dead or prisoners to Cherokee raids during Fitch's short visit.[13] The decision to send Chipacasi and his fellow warriors against the Yamasees lays bare two further issues, important both to Fitch and to an understanding of Native American restraints on war. The first was whether the Yamasees would be forewarned; Fitch had repeatedly noticed Yamasees visiting or living with the Creeks even as he tried to persuade the Creeks to go to war against them. Second, there was the need to mobilize and then ritually prepare sufficient warriors. Both requirements greatly affected the destructive capacity of Native American warfare.

First there was the problem of the Yamasee visitors. Not only did Fitch complain about the Yamasees having relatives among the Creeks, but he literally pointed to Yamasees attending the councils while he was speaking. Intermarriage and adoption or incorporation of peoples was a common practice among Native Americans and helped create links between peoples. Historian Steven Hahn argues that Yamasee-Creek links by this time were so extensive that they considered themselves one people, at least metaphorically.[14] Even when not related by blood or adoption, however, it was common for members from many nations to live among, or pay long visits to, another people for trading or political reasons.[15]

This widely shared cultural practice of "resident aliens," or long-term visitors, dramatically limited the probability of surprise and its inherent lethality. For example, the presence of visiting Creeks in a Yamasee town made it less likely that the Creeks would try to attack another Yamasee town, for fear of what would happen to the visiting Creeks. More systemically, resident aliens maintained interests in both camps, and frequently served as conduits of information between groups, both formally and informally.[16] Fearing the total destruction of either group in which they now had a vested interest, they warned of impending attacks and thus prevented the horrifying casualty rates of a successful surprise, even if they did not derail an attack completely.[17] It seems clear, for example, that Chipacasi's war party was preceded by warnings of its arrival.[18]

Resident aliens could also step into the role of peace negotiator, and indeed the Creeks seem to have formalized a specific titled office for this purpose. In Creek society (and among many other southeastern peoples) a "Fanni Mingo" was an adoptee who bore responsibility for representing his adoptive people in the councils of his natal people.[19] Even outside this formal system, resident aliens generally were not tainted by blood debt; they were not a part of the people who owed blood, and so were a safe choice to send to the enemy to open negotiations. The Powhatans, for example, probably imagined that Chanco (or Chauco), an Indian who in 1622 had tried to warn the English in Jamestown about the coming Powhatan attack, would have been considered by the English as a kind of "resident alien." Such a status, and his attempt to warn the English, later rendered him a suitable emissary from the Powhatans proposing peace, although he also brought with him a released prisoner as proof of his intentions.[20] James Quannapaquait and Job Kattananit, both Christian Nipmucs, expected to fulfill this role from their position as resident aliens among the English during King Philip's War. They offered to go to King Philip to either persuade him to submit to the English or suggest "making peace if they found the enimy in a temper fit for it."[21] At first, the Europeans fit into this "resident alien" system relatively naturally. Traders, forts, and garrisons were desirable both as conduits for material goods and as symbols of good feelings between the two peoples.[22]

The second issue Chipacasi confronted prior to his departure was the need to both mobilize and ritually purify his fellow warriors. Before a war party could depart, the community, the leadership, and the warriors partook in a series of ritual activities designed to achieve the purity that war required and insure their success. Most southeastern Indians followed a fairly similar pattern. They retired to the war chief's winter house and remained there for three days and nights, fasting and drinking potions intended both to purify them and protect them from danger.[23] For Native Americans the spiritual and material world were in fact coterminous; spiritual power was immanent in the world, but ritual skill was required to tap it.[24] Such a powerful belief in the importance of ritual meant that an offensive could easily be derailed by bad omens, whether natural or deliberately manufactured. If the spirits seemed unwilling to support the attack, a people sensitive to casualties could not afford to ignore them and risk defeat. Little Carpenter, a Cherokee leader, declined to send a war party to aid the British in 1758 because the "Conjurers" had produced omens of a "Distemper" that would afflict them after the first two months of the expedition.[25] A raid by the Gaspé Peninsula Indians in 1661 saw half of its participants depart when one of their number "just now recalled" the command of a dying relative (on the boundaries between the two worlds) not to participate in the raid.[26]

The demands of purity included a prohibition on sexual intercourse, both during ritual preparations and while on campaign, including the rape of enemy women. Although women were targets in other ways—for capture, adoption, and possible marriage—there appears to have been no equivalent in Native American societies to European soldiers' propensity to indulge in rape as a perquisite of war.[27] It is highly probable that this Native American expectation of war clashed with the Europeans'. There is not much direct evidence on the subject, but we have one example of a war started in part by English soldiers raping Cherokee women, and other examples that led to local retaliation.[28]

Finally, fearing the fundamental spiritual harm caused by the taking of life, some Native peoples expected returning warriors to undergo a period of purification prior to their full reentry into society. For southeastern Indians the process was much like that before they set out: three days of fasting under the guidance of the war chief, accompanied each night by the women singing songs outside the war chief's home. The Shawnees practiced a similar post-raid purification through isolation and fasting.[29] This need for purification after a raid put a basic limitation on the frequency of war, since it prevented warriors from simply resupplying and returning to the attack.

Where the needs of the spirit world provided a kind of halting and unpredictable restriction on warfare, the people's loudly voiced expectations of their leadership created a more immediate and abiding limit on their society's capacity to raise large forces.[30] Lacking coercive power, Indians leaders could neither raise large armies nor maintain their leadership position after a defeat. Furthermore, the division of authority between a peace and a war chief restrained the usually more aggressive tendencies of the war chief.[31] In military terms this divided structure of leadership dependent on group consensus imposed two significant limitations. The first was the inability to coerce warriors to go to war—the mobilization problem. Grieving relatives might persuade others, usually young men in search of status, to join them in a blood feud. But even raising troops for more "political" warfare required a similar and equally fragile process. Warrior mobilization and purification were intertwined. James Adair, a trader and traveler in the eighteenth-century southeast, described the process of mobilization:

> In the first commencement of a war, a party of the injured tribe turns out first, to revenge the innocent crying blood of their own bone and flesh, as they term it. When the leader begins to beat up for volunteers, he goes three times round his dark winter-house, contrary to the course of the sun, sounding the war whoop, singing the war song, and beating the drum. Then he speaks to the listening crowd with very rapid language, short pauses, and an awful commanding voice. . . . Persuad[ing] his kindred warriors and others, who are not afraid of the enemies bullets and arrows, to come and join him with manly cheerful hearts. . . . By his eloquence, but chiefly by their own greedy thirst of revenge, and intense love of martial glory . . . a number soon join him in his winter house [and commence the three day fast].[32]

The emphasis here is on volunteers, motivated by a desire for revenge and status, and presumably putting stock in their war leader's reputation. By its nature, this kind of recruitment proceeded town by town—requiring extensive negotiations, travel, and even gift giving to convince other townsmen to join in.[33] A war-party leader focused his mobilization efforts on the most closely confederated towns, but the towns' autonomy gave them options— sometimes at inopportune moments. Bound for the Yamasees, Chipacasi's army unexpectedly lost seventy warriors when a message arrived from the French that persuaded some to abandon the effort.[34] Despite a generalized martial enthusiasm among the young men, such mobilization techniques in a small-scale society usually restrained the size of war parties. In turn, a smaller war party had a more limited capacity to inflict damage.

Furthermore, the war chief's fragile authority limited the risks he could take while on campaign. Too many deaths could be a demographic disaster and would generally mark a campaign as unsuccessful regardless of whether the enemy had suffered more. The Iroquois believed that "a victory bought with blood is no victory."[35] The blame in such a case fell squarely on the war leader. Adair noted that "they reckon the leader's impurity to be the chief occasion of bad success; and if he lose several of his warriors by the enemy . . . he is degraded by taking from him his drum, war-whistle, and martial titles, and debasing him to his boy's name."[36] The Illinois apparently systematized the process: two failures simply ended a war leader's career.[37] Serious disasters could put the war leader's very life in jeopardy. After the English destruction of the Pequot village of Mystic in 1637, the Pequots nearly killed their leader Sassacus.[38] Such punishments encouraged the war leader to calculate a strategy that avoided risk, further limiting the destructive potential of any given raid.

In practice, avoiding risk meant pursuing a limited repertoire of tactical techniques. Consider the events of Chipacasi's march on the Yamasees. As the Creeks moved into the Yamasees' home territory, they were disappointed to find the enemy alert and inside their fortifications. The Creeks then went

> to a Fort in a Town Where we thought the Yamases were, and we fired at the Said Fort, Which alarmed ten Men that was Place[d] To Discover us which we [had] past [passed] when they were asleep. Our fireing awaked them and they Ran round us and gave Notice to the Yamasees Who was Removed from this town Nigher the Sea and had there Build a new fort which we found and Attacked but with litle Success [though] it happen'd the Huspaw Kings Family was not all got in the fort and we took three of them and fired Several Shott at the Huspaw king and are in hopes have killed him. There Came out a party of the Yamases who fought us and we took the Capt. We waited three days about there Fort, Expecting to get ane oppertunity to take some More but to no purpose. We then Came away and the Yamases pursued us.

Turning on their pursuers, the Creeks drove the Yamasees into a pond and were about to press them further when they were distracted by the arrival of a Spanish force. When the Creeks resumed their march home, the Yamasees pursued and attacked again, and it was in this "Batle in which they did us [the Creeks] the most Damnadge." All told, the Creeks killed eight Yamasees, bringing home those scalps, nine prisoners, and some plunder. The Creeks lost five killed and six wounded.[39]

There was much about this raid that was typical: Chipacasi and his men attempted surprise, but were foiled by fortifications and the belated alertness of a defending group of ambushers (who were clearly waiting for Chipacasi's arrival). The Creeks then attacked the individuals they could find outside the fort, before offering battle to the defenders. The Yamasees, perhaps to maintain prestige (see the discussion of "battle" below) accepted the offer but lost their "captain" in the ensuing fight. The Creeks then lurked around the fort hoping to snipe at the occasional exposed person, and after three days headed for home. The Yamasees immediately pursued, hoping to take advantage of a party spread out on the march. A series of running battles damaged the Creeks and sped them on their way.

This pattern was repeated by Native American war parties throughout the colonial period, and its contours were determined by several key factors. The first was the preference for surprise, largely to reduce the attackers' casualties. If achieved, either by ambushing a smaller party or by attacking a sleeping village, then the relative casualty count for the defenders could be quite high. There was no reluctance to kill large numbers of the enemy. As the Cherokees explained to George Chicken in 1715, "It was not plunder they wanted from them but to go to war with them [the Creeks] and cut them of[f]"; that is, to kill them.[40] To successfully kill very many, however, depended on surprise, and that was difficult to achieve. An alert enemy, or one warned by resident aliens or through other connections between peoples, as in fact the Yamasees had been, would be prepared. Their preparedness left the attacking war party with the options of returning home, offering open battle, or, if the defenders remained behind a palisade, laying siege. Each of these choices involves some interpretive controversy, so it is time to leave Fitch and Chipacasi behind and consider the problems one by one in a more general way.

The Cutting-off Way of War, Prisoners, and Peace

Chapter 5 suggested that Native American warfare prior to contact with the Europeans struck a relative technological balance between the offense and defense, and that combat comprised a mix of open battles, ambushes and raids, and sieges. The arrival of European technology upset some of those balances and rearranged some of the patterns. Indians quickly adopted firearms; they learned (somewhat more slowly) to avoid becoming trapped in their fortifications; and they largely abandoned the open battle in favor of more reliance on the ambush or raid. The process happened at different speeds around the continent, depending on the movement of steel, firearms, cannon, and settler colonies, but by the early eighteenth century, Indians across the eastern seaboard shared a tactical and operational style of warfare that I call the "cutting-off way of war."[41]

A hypothetical war party of some size—say four to six hundred warriors, and therefore with more options—had a number of tactical choices. Hoping to avoid friendly casualties, but still wanting to exact revenge, gain prestige, and perhaps administer a political lesson, they sought to "cut off" enemy villages, small groups, or even individuals through surprise and ambush. Repeated success at surprising a village could render it uninhabitable, in essence cutting it off from its cluster of related villages. Even short of complete success, repeated raids could finally force the enemy to abandon a region. It was not a Western-style "take and hold" campaign, but it nevertheless focused on territory in a kind of "conquest by harassment."[42] The cutting-off way of war used the same basic techniques whether the war party was large or small. The initial target for such a large group would probably be an entire enemy village, as it was for Chipacasi. Unfortunately for him, the Yamasees were forewarned. But if the enemy village remained unaware of their approach, as the Hurons of St. Ignace were unaware of the Iroquois approach in 1649, the attackers could sneak in before dawn and inflict considerable casualties. Out of four hundred people in the village of St. Ignace, 397 were killed or taken prisoner.[43] As in the Yamasee case, however, it was more common for the threatened village to learn of an enemy approach, to send for help, to ambush the approaching warriors by "waylay[ing] the pathe," or to gather behind a palisade wall.[44]

On some occasions, if the defenders felt confident enough in their numbers and preparedness, they could offer open battle, lining up their warriors to oppose the approaching enemy. Such battles were documented on several occasions by early European explorers. Samuel de Champlain participated in one against the Iroquois in 1609.[45] The Powhatans enacted a mock battle for the benefit of the early Virginia colonists, and John Smith described one war party as marching in a "square order."[46] Jacques Le Moyne de Morgues provided a drawing and description of a Florida tribe drawn up for battle in a deep, massed formation.[47] Even later witnesses in New England continued to describe Indians as occasionally lining up for battle. The Yamasees attacked by Chipacasi seem to have adopted this option. A very few casualties, and possibly even the first letting of blood, sufficed to end the battle and send both sides home.[48]

Rather than viewing such a battle as the main object of the expedition, however, it is probably more accurate to consider these encounters as moments in which the expedition *had already failed*, having lost the benefit of surprise. The battle served only to uphold the warriors' collective prestige. It was, in short, a face-saving measure, if perhaps also a kind of test of strength that could have decisive results in the right circumstances. An example of this was the Battle of Sachem's Field, fought between the Narragansetts and Mohegans in Connecticut in the summer of 1643. Miantonomi led a force of some nine hundred to a thousand Narragansetts into Mohegan territory (apparently still armed mainly with bows and hatchets), almost certainly hoping to surprise Uncas, the Mohegan leader. Uncas learned of their approach from his scouts, and he sent for help from his tributary villages. Gathering together about six hundred warriors, he declined to be shut up in his fortified town (Shantok), which at sixty square meters probably could not accommodate six hundred men anyway. He instead moved to meet Miantonomi on an open field, where the two sides approached to within bowshot. A parley was proposed, and Uncas offered to settle their dispute in a single duel. Miantonomi, with the greater numbers, declined, and Uncas gave the signal to begin the attack immediately. Miantonomi was wearing a European mail shirt and was so slowed by its weight that Uncas was able to seize him. Some Narragansetts, perhaps thirty, were killed. The vast majority fled successfully.[49]

If the defender chose instead to remain behind walls, the attacker had three basic choices: assault the walls, blockade the fort, or go home. Because of the technological balance between offense and defense, the first option was relatively rare, particularly during the precontact era. There was no easy way to overcome the defenders without absorbing significant casualties. Fire was one option to speed the assault, and was certainly used on occasion, even prior to the arrival of Europeans.[50] In general, however, attackers seemed to prefer the blockade, supplemented by sniping at the walls from the cover of the woods and trying to cut off any individuals who strayed outside as messengers or water carriers. The besiegers could keep this up as long as significant reinforcements from other villages of the defenders' people did not arrive. The blockade avoided casualties while offering the possibility of taking isolated prisoners or scalps and gaining the associated air of victory and prestige.

In short, the technological and tactical balance of the offense and the defense both in siege warfare and in open battle meant that warfare prior to the arrival of Europeans was usually a relatively mild affair. Successful surprise, however, could immediately produce high per capita casualty ratios.

There is no reason to suppose that Europeans introduced lethality to the Indian culture or calculations of war.[51] What they did do was increase the lethal capacity in Indian warfare. In 1649 the Iroquois, equipped with European metal axes, were able to cut through the palisade walls of the Hurons at St. Louis, even after surprise had failed.[52] In the face of the inability to dodge bullets or heal bullet wounds, the ritual battle, probably never the true centerpiece of Native American warfare anyway, rapidly disappeared.[53] It still happened, of course, that large war parties could encounter each other while far from their villages, neither having the benefit of surprise. Battle might follow, but not the linear open battle described by the early sources. As war's lethality increased and the cultural value of limiting casualties persisted, that way of war had become too costly. Instead, individual warriors "took to the trees," firing from cover and endeavoring to outflank their enemy with a "half moon" formation, rendering the enemy's use of a single tree as cover useless.[54] A seventeenth-century dictionary of the Narragansett language contains only a few phrases with tactical implications, but they highlight the most important issues in Native American warfare: "They fly from us/Let us pursue"; "They lie in the way"; "They fortifie"; "An house fired"; "An Halfe Moone in war"; and probably most important of all, "Keep Watch."[55]

In 1637 the Pequots failed to "keep watch" and soon found their palisade at Mystic surrounded and under assault by English settlers aided by Mohegans and Narragansetts. Prior to the campaign the Narragansetts had sent a message through Roger Williams expressing their desire that the killing of women and children be avoided.[56] The English claimed to be sympathetic to this desire, but when the Pequots frustrated the initial assault, the English, like their forebears at Rathlin, seem to have concluded that the fort and all its inhabitants were now at their mercy. They fired the village and shot or stabbed all those fleeing the flames, killing hundreds of all ages and sexes. One English commander reported the dismay of his Narragansett allies, who "came to us, and rejoiced at our victories, . . . but cried Mach it, mach it; that is, it is naught, it is naught, because it is too furious, and slays too many men."[57]

This protest by the Narragansetts has become quite famous among historians of Anglo-Indian warfare. It is often used to support the idea that Europeans in fact did introduce a culture of lethality into Indian warfare. Alternatively, the Narragansetts' complaint simply registers their desire for prisoners, not a cultural horror at lethality. The English style of war was only "too lethal" in the sense that it left no prisoners (especially in the wake of a storm) and thus denied their Indian allies the opportunity for prestige, incorporation, or adoption. And in fact the Mohegan and Narragansett sachems shortly fell to arguing over who should control the "rest of the Pequits."[58] It is in this attitude toward prisoners that one paradoxically finds the basis of both the most and least restraint in the overall frightfulness of Native American warfare.

Prisoners served four overlapping purposes. Bringing prisoners back to the home village certified the overall success of the mission, and warriors gained individual glory by pointing to particular prisoners as "theirs." Scalps were also a mark of success and prestige, but were considered a poor second best to a live prisoner. The preference for prisoners over scalps arose from prisoners' other three functions. A prisoner, particularly an adult male, became the target for their captors' rage and grief at their other losses. Elaborate and extended rituals of torture until death existed in many Indian societies. Scholars continue to struggle to understand the exact meaning of these rituals, but it is clear that at the center of the process was a tremendous outpouring of violent grief, in which the whole town—men, women, and children—participated. Native prisoners were well aware of what awaited them and sought to remain impassive in the face of excruciatingly inventive torture. Not crying out certified their personal bravery, and some captor tribes sought to partake of that bravery through a ritual cannibalism in which they absorbed the courage and spirit of their prisoner.[59]

The third role of the prisoner was that of adoptee. Adoption of prisoners may have been less universal than torture, but some incorporation into the captor group, either as kin or as a kind of servant, remained extremely common. For the most part it proved easier to incorporate women and children, but men could also be adopted. For the Iroquois in particular, the taking of prisoners in war both assuaged grief and restocked their own population. In this "mourning war" complex, Iroquois families adopted prisoners to replace dead kinspeople, even as they vented their grief by torturing those not selected. Daniel Richter and Jose Brandão have both traced the root of most if not all of the Iroquois wars of the seventeenth century to the ultimately vain pursuit of sufficient prisoners to replace their losses from war and disease.[60] Prisoner adoption was by no means exclusive to the Iroquois; the process is simply most clearly understood for them.[61]

The final role of the prisoner was as a source of ransom. Native Americans quickly became aware of the lengths to which Europeans would go to retrieve family members, and the imperial wars between France and England formalized the process of redemption through ransom. James Axtell has argued that the whole prisoner-taking complex of Indian warfare shifted in response to this economic possibility.[62] Prisoners could still bring honor, but now they brought material reward as well—but only if returned. Adoption or incorporation never disappeared, but over time it was substantially replaced by ransom.

This overdetermined desire for prisoners limited the numbers killed in Indian war.[63] Taking a prisoner alive in battle was no small achievement, and if a warrior succeeded, he was unlikely to return to the fray to take another, leaving the first prisoner either unguarded or subject to competitive confiscation by one of his fellows.[64] While some of the prisoners would be tortured to death, others would survive as adoptees (or ransomees), limiting the death toll of the conflict. Furthermore, the most likely adoptees were women and children, and so the goal of prisoner adoption tended to limit the killing of the defenseless. Finally, although this can only be a speculation, the extremity of torture practiced on individual prisoners may have played a role in limiting the need for further warfare. Since much conflict originated in a desire for revenge, the lengthy and elaborate rituals of torture practiced on a few symbolically "responsible" individuals may have limited the need to kill great numbers of the enemy.[65]

The logic of taking prisoners, and its apparent limitation on quantitative frightfulness, had a flip side in terms of qualitative frightfulness—at least as far as Europeans were concerned. The targeting of women and children as preferential prisoners for adoption violated European norms of war that nominally exempted women and children from the theater of conflict at all. The torture of prisoners, particularly communal public torture, also violated European norms.[66] In John Lawson's otherwise sympathetic description of Southern Native American life in 1709, he called the Indians' use of torture the one thing "they are seemingly guilty of an error in."[67] Furthermore, the long and difficult return marches after a successful raid, often under pursuit, meant that the captors ruthlessly weeded out the unfit or incapable. In their minds this was a mercy killing, to be preferred to slow starvation.[68] The logic of war for prisoners, however, extended beyond this simple calculation of fitness. While in theory a successful surprise of an enemy village could result in hundreds of prisoners, as after the Iroquois capture of St. Ignace in 1649, logistically there was almost no way to guard that many prisoners for the long trip home. Fearing pursuit, and not wanting to release their enemies, the Iroquois on that occasion summarily killed many of their prisoners.[69]

Native Americans were explicit about the contextual contingency of preserving prisoners, and they argued with their European allies against the European practice of exchanging or paroling prisoners. In their world view, to return prisoners to the enemy made no sense. They lost the chance both to grieve and to exult, and their enemies lived to fight them again. When some Creeks and Cherokees, assisting in an English attack on St. Augustine, witnessed such a prisoner exchange, they accused the English of conducting a sham fight.[70] In the incident made famous by James Fenimore Cooper's *Last of the Mohicans*, the French-allied Indians complained about the negotiated surrender of Ft. William Henry in 1757 and refused to abide by the terms of the surrender. They attacked and "massacred" the retreating British troops. Even in this incident, however, one can see the restraining effects of fighting war for prisoners. As soon as the Indians had dealt their blow to the British at Ft. William Henry and gained their prisoners and plunder, they promptly abandoned the French offensive.[71]

The end of an offensive, however, did not mean the cessation of all hostilities, and the resumption of peaceful relations required complex diplomacy. The full range of diplomatic activity and techniques lies beyond the scope of this book, but some aspects of the process of peacemaking are worth mentioning here.[72] First, Native Americans typically extended protections over enemy embassies to facilitate the beginning of negotiations. The Cherokees' violation of this code in 1715 brought them the lasting enmity of the Creeks.[73] This fundamental need to allow for the safe passage of negotiators was the sine qua non of limiting warfare among Europeans, and its role was perhaps even more important in an environment of revenge-based warfare. In the 1640s, for example, the French-allied Indians released an Iroquois prisoner to return to his people with a message about the possibility of peace. In this case, the Iroquois replied by sending back three negotiators and also released a French prisoner as their own gesture of good faith. The three negotiators were then treated as eminent guests during the ensuing negotiations.[74] Similarly suggestive of Indian diplomatic processes was the Pequots' approach to Lion Gardiner's fort on the Connecticut River in 1637, "calling to us [the English] to speak with us." Gardiner and an interpreter went out to negotiate, and the Pequots asked if they had "fought enough." Although no peace resulted from this attempt, both sides obeyed the dictates of the negotiating truce.[75]

The approach of one side asking for peace did not mean that the other had to accept it, but the divide between civil and military authority within most peoples helped the process of choosing peace. Civil chiefs, especially in those groups where their authority passed through hereditary succession, did not have their personal prestige or power vested in successful war.[76] While the war chiefs, supported by eager young men, might prefer to continue pursuing opportunities for victory, or to overcome recent reverses, such pressures did not apply to the civil chiefs in the same way. As discussed earlier, however, the

European diplomatic preference for dealing with one designated person, and the habit of distributing gifts through that person, tended to undermine the "behind the scenes" peace chiefs in favor of the war chiefs. Over time the decline of the peace chiefs' power made it more difficult to move from a state of war to one of peace.

In a similar way, Europeans affected the ability of women within Native societies to restrain warfare. The political power of women varied widely among Native American societies, but some of the most prominent, including the Iroquois, the Shawnees, and the Cherokees, reserved a substantial political role for women. In many instances their desires for vengeance, to grieve, or to replace their lost kin could start a war, but in other instances they could prove a driving force to end one.[77] In some ways women could shift more easily from war to peace, since they, like peace chiefs, did not derive their social status or authority from success in war. European failure to understand women's roles, and their preference for dealing with war chiefs in general, helped to undermine the potentially (although not universally) ameliorative role of women.[78] Like European technology, European diplomacy upset certain balances within Native societies that had helped to contain endemic warfare. Historian Claudio Saunt's comment about the Creeks is more generally applicable: "Though Europeans observed only inconstancy and disorder [in Creek social structures], Creeks saw a healthy tension between female and male, old and young, and peace and war."[79]

If making peace was difficult, and required a minimal level of trust, keeping the peace was perhaps even harder. Rather than as a permanent, preferred state of being, peace was conceptualized more as a temporary lack of hostile feelings.[80] The elders of the council realized that they could not fully control their young men's desires to seek glory—and perhaps revenge—and therefore their creation of a peace had to make such adventuring both less likely, and less likely to be successful.[81] Intermarriage and "resident aliens" helped to ease this process of sustaining peace. Much like European sovereigns, Native leaders used marriage as a diplomatic tool to secure peace. Most famous was the marriage of Pocahontas to John Rolfe. When the English sought another of Powhatan's daughters for that purpose, Powhatan refused, pointing out that he only had so many daughters and that he dealt with many nations.[82] For the English this sort of marriage normally occurred only at the highest political levels, between peoples or states, and implied a new relationship between those states. For Indians, intermarriage had a more pervasive impact on war, creating links between peoples at multiple social levels and making highly destructive surprise attacks less likely.

Another possible guarantor of peace was the exchange of "hostages," although this too was an institution with only partially shared meanings between Englishmen and Indians. Even to use the word "hostage" is to imply forced residence among an enemy people, and that definition does not always fit the Native concept. Powhatan, for example, deliberately took in and developed personal attachments to several young Virginia colonists.[83] To the English they were captives or runaways who needed to be restored to their own people. To Powhatan the exchange of people should have been both more lasting and a conduit of information between the two peoples. Powhatan regretted the flight of the English boy Thomas Savage, saying that he was "my child . . . in lieu of one of my subjects Namontacke, who I purposely sent to King James his land, to see him and his country, and to returne me the true report thereof."[84] For Powhatan, Englishmen living among his people served both as hostages at risk in the event of an English attack and as sources of information warning Powhatan of such an attack.[85]

Escalation and the Failure to "Mesh" Restraints

Structures that kept the peace naturally overlapped with those that helped to reduce the frequency of war. Thus we have come full circle. But it is important to note that restraint was not necessarily the dominant characteristic of Native American war. Peace chiefs (male or female) might play a role in limiting the resort to war or bringing a war to a halt, but the more persistent and durable cultural drive was blood revenge reinforced by the desire for personal status. That was the true circle of Indian war, where one act led to retaliation, and that retaliation demanded yet another response, and so on. Breaking the cycle of vengeance proved extremely difficult in a society ill suited to top-down coercion. It was in this regard that the Iroquois sachems quoted at the beginning of this chapter warned the governor of New York of their declining ability to control their young men. Nevertheless, if war was endemic, it was usually conducted on a small scale, in a context of technological parity prior to European contact, and usually without the more destructive goal of outright elimination of an enemy people. Where European warfare of the era was persistent, thorough, and—when believed necessary—all-consuming, Native American warfare was usually episodic and personal, with easily satisfied goals, although it might be no less fatal than its European counterpart.[86]

Here I return to the question of whether Native Americans began to shed their own restraints in response to the Europeans' way of war. Recently scholars have argued that a "collision of military cultures" occurred in which each people acted in ways that violated the other's expectations of war, leading to an overall escalation in violence. Examples include the Indian attitude toward rape and the European response to the capture of women or the communal torture of prisoners. Such violations of norms naturally led to angry retaliation, and war spiraled out of control.[87] Another explanation points to changed conditions rather than conflicting beliefs: the introduction of European diseases, technology, and materialist systems of exchange wrenched Native American warfare out of its comfortable, restrained path into something more terrifying and destructive.[88] There is much in both of these arguments, but the full story is even more complex.

Escalation in contact-era warfare was not just the result of simple anger at the violation of codes and expectations, nor can it be entirely attributed to the Native experience of dramatic demographic, technological, or economic change. A deeper problem was that neither Natives nor Europeans understood the systems of restraint imbedded in the other side's culture of war. As suggested in the introduction, the contours of war are shaped as much by social and cultural pressures restraining violence as they are by the calculations and fears that escalate it. In the meeting of different cultures of war, the conventional practices that tended to limit the destruction of war fail to work without the participation, or at least the understanding, of the enemy. The upward pressures of escalation overwhelm the downward pressures of restraint, and war becomes more frightful. As Lane's choices in Roanoke suggest, each side had a "grammar" of war, in which actions carried certain meanings that were readily understood by someone working inside the same cultural framework, but were often unintelligible to outsiders. This was not simply "cultural collision" producing escalation; it was also a case of *missed* messages, or missed opportunities, leading to a failure of restraint and the onset of a cycle of escalation.[89]

For example, Europeans understood hostages and diplomatic marriages. Ralph Lane took hostages and John Rolfe married Pocahontas. But the English did not really understand the cautionary or peacekeeping function of resident aliens; frequently they could not even tolerate their existence.[90] In an early dramatic instance, when an English emissary to Powhatan demanded the return of William Parker, who had been living with the Powhatans for some time, Powhatan complained "you have one of my daughters [Pocahontas] with you, and I am therewith well content, but you can no sooner see or know of any English mans being with me, but you must have him away, or else break peace and friendship."[91] This proved to be a constant pattern among

English emissaries, who persistently sought to retrieve whites they found adopted into an Indian people. Tobias Fitch himself sought the return of a white woman he learned was living with the Creeks.[92] Many English people did in fact become fully integrated members of Indian societies, but mainstream English society found the thought intolerable, and the culture crossers' ability to move back and forth from white to Indian society was correspondingly limited.[93] A few exceptional individuals, like Robert Beverly, a man who could be described as the quintessential eighteenth-century Enlightenment Southern gentleman, differed from their contemporaries by favoring intermarriage as a likely path to peace. But even he, while acknowledging the common humanity of Indians, expressed his hopes for their incorporation into white society through education and intermarriage in part because he believed the process would make it easier to acquire their land.[94] Most English people rejected intermarriage, distrusted whites among Indians, and proved even less willing to trust Indians, even Christian Indians, among themselves.[95] Most infamously, in 1675 the Massachusetts government ignored the Wampanoag John Sassamon's warning of King Philip's impending plans for war.[96] In an exception perhaps proving the rule, the Plymouth settlers allowed Squanto and Hobomock to live among them, where they seem to have served much as resident aliens. In at least one instance Hobomock and his wife played an active role in maintaining peace between the Pilgrims and Massasoit, the Wampanoag sachem and grandfather of King Philip.[97] In general, the English rapidly learned to use allied Indians as aids in war, but only rarely as aids to peace.

The particular inability to understand or relate to the resident alien as an aid to peace represented a more general failure of Englishmen and Americans to truly seek the incorporation of Indians as subjects or citizens on a level with themselves.[98] As with the Irish, it proved difficult to truly imagine the barbarian as a brother. The early Puritan settlers' desire for a program of Christianization, Beverly's Enlightenment-inspired admissions of a shared humanity, or even President George Washington's speeches about the justice due to Indians and their eventual citizenship rarely resulted in major changes in attitude or policy.[99] Much of the scholarship on white-Indian interaction in the last twenty years has adopted Richard White's idea of the "middle ground," an arena in which neither culture could dominate the other, leading to a blending of traditions, practices, and even people. But White carefully predicated his concept of the middle ground on the absence of coercive power, an absence that progressively disappeared in the eighteenth century in the face of white population increase and expansion.[100] Even before that balance of power tipped, however, the dominant English paradigm was to refuse to live with Indians, instead separating them into their own towns or reservations. The "Praying Indians" of Massachusetts, for example, arguably the most promising model in the English colonies for incorporating Indians as fellow subjects, were already segregated into their own towns before they were rounded up and confined on an island in Boston harbor during King Philip's War in 1675.[101] The tendency to segregate persisted despite repeated nominal hopes for and legal claims of sovereignty over Indian peoples.[102] In practice, English sovereignty in North America demanded separation. On the frontier, cultural practices blended, peoples peacefully traded, but tolerance remained fragile. The most vivid emblem of that fragility was American militias' repeated massacre of peaceful, converted Indian groups at the outbreak of a conflict, as they were apparently unable or unwilling to distinguish between friendly and enemy peoples.[103]

The inability to imagine "barbarian" Indians within an inclusive subjecthood resulted in part from hardening negative judgments of Indian culture. Chapter 8 will return to the problem of racism, but for now, consider the example of how one Indian structure of restraint was misunderstood by Englishmen. European witnesses to Native American war dances misread this ritual preparation as emblematic of a savage nature; such a judgment further fed the white racial prejudice used to justify violence against Indians. Although the ritual was designed to secure success, and any reasonably intelligent visitor would immediately recognize it as a process of whipping up an aggressive spirit, what European onlookers did not know was that participants were required to be ritually pure. Among other things, that requirement served to protect women from rape. Furthermore, the dance's public quality, combined with the delay it imposed, gave resident aliens (or others) time to warn the intended targets.[104] The English had their own forms of ritual preparation: sermons, military parades, speeches from political authorities, and the like. But those cultural signals were designed to reinforce coercive structures of mobilization and to cement social authority with appeals to the past, to the justice of their war, to necessity, and to God. Unlike Native preparations, they imposed no structural restraint on campaigning time or frequency, although they occasionally included rhetorical appeals to restrain unnecessary violence.[105]

Neither did the English expectation of social control translate for the Indians. The European method of restraining the violent behavior of soldiers, described in chapter 2, depended on the imposition of a code of discipline that had grown up in conjunction with the need for disciplined behavior on the battlefield, and was reinforced by the social prejudice of the elite leadership. When one side's soldiers violated truce terms or attacked civilians, it might be treated as an atrocity by the other side, but if the offenders' officers had attempted to restrain the soldiery, then retaliation was much less likely. The elite leaders within the European system understood that social control and discipline could fail, but making the attempt was key. Native Americans had no such coercive structure or vision. Their repeated claims that attacks had been conducted by young men on their own had little purchase with the English. When Fitch said to Gogel Eyes, "You Should have prevented thir Rogush proceedings," what he meant was that, at a minimum, Gogel Eyes should have been seen to try. From the Indian point of view, the statement that young men had acted on their own was a message that that act had not been done by the people as a group, and that the incident therefore did not demand full-scale war. That was their grammar of war, and the English rarely understood the message.

In fact, one of the most fundamental restraints within Native American warfare was their social structure that limited the power of Native leaders to coerce their followers. The inability to mobilize large numbers of warriors, and their individual urges to return home after an initial success, dramatically restricted their societies' overall capacity to wage destructive war. This fundamental sociopolitical characteristic of Native society underwent a number of shifts after the English arrived. Englishmen preferred to designate a small number of individuals as representatives of a Native people, who then received the diplomatic gifts. This process was not necessarily a "clash" of cultures. It was instead one in which European efforts to manage trade and diplomacy progressively altered Native American societies toward greater aggression and political unity, which in turn enhanced those societies' capacity to wage destructive war. In other circumstances, when English hopes for Indian allies stumbled on the Indians' non-coercive mobilization system, the colonists had to invent alternatives. One alternative was to persuade individual warriors to fight on behalf of the English through a system of economic rewards, especially scalp bounties. In this way the English deliberately encouraged a behavior that they normally cited as emblematic of Indian savagery, while simultaneously creating a situation where it was easier for their allies to take scalps indiscriminately, perhaps even from other allies, sowing the seeds for future conflict.[106]

Different cultural systems governing the treatment of prisoners also represented a way in which the two systems of restraint failed to mesh with or complement each other.[107] The problem was most visible when European systems of prisoner exchange and parole clashed with Native American ideas of the role of prisoners. There is no question of "better" or "worse" here. By the seventeenth and eighteenth centuries the European system preserved many prisoners' lives, but within the context of a much more lethal style of war. The Native American system typically killed fewer outright, but they then regarded prisoners as a prize of war. Ian Steele has pointed out that by the mid-eighteenth century, British imperial negotiators had begun insisting on the return of all prisoners as a condition of peace.[108] Such a demand usually proved impossible for Indian leaders to enforce, and it only prolonged conflicts. When Indian leaders did turn over captives, it broke up nascent resident-alien linkages between the two societies that might have helped defuse future wars.

Finally, in one of the most important examples of dissonance between the two cultural systems of war, Indians and Englishmen held very different notions of what being "at war" meant. Native Americans had a highly graduated or calibrated vision of war—the grand war, the little war, and even the "not quite war" (or better, the war that could be recast as young men out of control). They expected the targets of such violence to understand the meanings of war waged at different levels. Anthropologist Frederic Gleach has interpreted the early conflicts between the Jamestown colonists and Powhatan as the latter's effort to use violence to bring the colonists' behavior into line with his perception of their subordinate status. He had no desire to exterminate them, only to reform their behavior. In 1622, his brother Opechancanough, who had taken over the chiefdom, tried and failed to persuade the English to help him in one of his own wars to the west. He then changed his name, often a signal of coming war, and prepared to administer what he thought would be a decisive "lesson" in the proper subordination of the English to his control. He would strike at the English, punish them for their transgressions, and await the restabilizing of the relationship, with both parties in their proper roles.[109] It was this kind of thinking about war that motivated Indian leaders to drop comments about uncontrollable young men, like in the epigram opening this chapter. Seeking to reform their trade relationship with the English, the Iroquois did not threaten to go to war. They merely mentioned the possibility, even the probability, that their young men might commence their own independent raids—raids that the leadership would later disavow.

In contrast, much of the European intellectual approach to war over the preceding four hundred years had sought to define war as a distinct category of activity. A major driving force behind the effort to define the *jus ad bellum* had been to separate private war from public war and to eliminate the former.[110] When Opechancanough attacked, the English did not see it as a moderated corrective "punishment." They saw it much the way Americans understood Pearl Harbor in 1941—as an outrageous breach of norms that merited unrestrained and, if necessary, total war. In 1622, as Gleach has pointed out, the English never noticed that the Powhatans did not continue the attack for a second day. Powhatan had lashed out severely, but then stopped, waiting to see if an adjustment would be made.

This dissonance in views of "war" meant that the English could never fit within the natural limits on the revenge cycle of Indian war. When an Indian was killed by an Englishman, perhaps in a private dispute, the English some-times learned to "cover" such deaths with gifts, but they were rarely prepared to receive such gifts when an Indian killed an Englishman. For them an attack was an attack, meriting full-scale mobilization and retaliation. The English knew about retaliation from their own traditions, but not in the tit-for-tat way of the Indian blood feud. For settlers on the frontier, the only option, in their minds, was "war." For colonial officials, European-style state-based restraints sometimes suggested an initial round of diplomacy; they would demand, for example, that a chief hand over the murderers. Native American social structures rendered that sort of demand nearly impossible to accept or enforce, although long experience with European expectations and an acute sense of self-preservation sometimes could lead a group to surrender a culprit.[111] More often, however, a diplomatic impasse generated further conflict.

I suggested in the introduction that war is intended to communicate something to an enemy. Violent acts convey intention and meaning according to an internally consistent logic or grammar. For systems of restraint to func-tion, both sides must share a minimally similar understanding of those mean-ings.[112] In North America, cultural difference undermined that understanding with fatal and frightful consequences. In a study of Puritan New Englanders during King Philip's War, Jill Lepore argues that their religious convictions led them to consign Indian motives to the obscure intentions of a "distant but reproachful God," instead of recognizing their actions as a "loud shout from extremely disgruntled but very nearby neighbors, communicating a complex set of ideas about why they were waging war."[113] Failing to understand the grammar of Indian warfare led the New Englanders, and other Englishmen in other wars, to previously unimaginable levels of violence.

Over time in North America, the English and Indians learned about each other and reached a certain level of understanding. Diplomacy became possible;

some intermarriage took place; wars ended. But in a deeper sense neither side ever fully came to terms with the other culture's overall vision of war and thus never succeeded in meshing their systems of restraint. In the absence of such a shared understanding, and greatly aggravated by hardening English racism coupled with an insatiable desire for land, war became more extreme and more destructive. The English not only failed to discern the trees from the forest, they increasingly turned to clearing the land.

Conclusion

This chapter has been concerned almost entirely with war from a Native American perspective. As a result, it has only begun to hint at the reasons why the wars between Anglo-Americans and Indians became as violent as they did. Rather than focus on those aspects that have been described by so many others—English racial prejudice, greed for land, or even the impact of disease and economic disruption on Indian societies—this chapter has instead looked at the structures and ideas that normally limited Indian war and how they interacted with the European systems examined in the preceding chapters. This is not to say that racism, land greed, disease, and economic shifts did not matter; they reappear in greater detail in chapter 8. It was necessary to move beyond the standard explanations, however, especially those relying on blind racial prejudice, in part because even the very earliest contact, as in the story of Lane and Wingina, could produce frightful violence.[114]

Violence escalated so quickly from the outset of contact because men at war with "barbarians" (and both the English and the Indians saw each other as barbarians) began with certain negative assumptions about the other and quickly found that the normal grammar that defined the meaning of wartime violence did not work. This story was about more than prejudice. It was about a confluence of greed unregulated by social authority across the wide Atlantic, strategic calculations of necessity, a mismatch of tactical styles, systems of restraints that proved escalatory rather than complementary, and a loosening of individual morality created by all these processes together. Europeans arriving in the new world after the publicizing of the initial Spanish experience believed in certain basic differences: the Indians were uncivilized; they did not use or improve all the available land; they were not Christians, and therefore the centuries of rules designed for Christian peoples at war did not apply to them.[115] Even if we limit our vision of the ideology of arriving Englishmen

and their descendants to this short list of assumptions, they sufficed to generate frightful violence when left unchecked by other beliefs or structures of
restraint. In colonial North America, what started out as a basic belief in
difference quickly combined with an experience of violence that seemed to
violate the normal system of meanings in war. In time the combination produced a distancing and a categorization of the enemy that allowed sometimes
truly horrific violence.

So horrific, in fact, that the same men who in 1778 would carefully observe an
official and unofficial code of conduct in their war with the British would in
1779 skin the legs of dead Iroquois Indians and tan them into leggings as a
gift for their officers.

Part IV

GENTILITY AND ATROCITY

The Continental Army and the American Revolution

Few stories from the American Revolution are retold as often as that of the sufferings of George Washington's Continental soldiers, starving in their miserable huts in the winter encampment at Valley Forge in the winter of 1777–78. Nevertheless, the deeper one goes into the story, the more interesting it becomes. In the campaign that led Washington's forces to Valley Forge, and in their efforts to live through the winter and rebuild their army in the spring, officers and soldiers struggled to define and control wartime violence. They believed in the virtues of restraint; they understood the strategic necessity of cultivating the loyalty of the countryside; and the officers, especially Washington, committed themselves to making the Continental army into a regular European-style conventional army that would reflect the maturing military culture of discipline. The American militias, on the other hand, were notoriously less committed to this kind of discipline, and their war with Loyalists could be savage. Even more revealing, however, were the choices and actions made by many of the same Continental regiments, officers, and men from the Valley Forge winter of 1777–78 who in 1779 marched into Iroquois country. There they fought an entirely different kind of war. There could be no clearer contrast between wars among brothers and those with "barbarians."

7

"One Bold Stroke"

WASHINGTON AND THE BRITISH IN
PENNSYLVANIA, 1777–78

[The officers] are to consider that military movements are like the working of a Clock, and will go equally regular and easy if every Officer does his duty. . . . Neglect in any one part, like the stop[p]ing of a Wheel disorders the whole. The General [Washington] therefore expects, that every Officer will duly consider the importance of the observation: Their own honour, Reputation, and the duty they owe their Country claims it of them.

Continental army orders, October 10, 1777

A Conventional Way of War?

By the summer of 1777, the British and Americans had been trading blows for two years. They were beginning to understand each other's habits, strengths, weaknesses, and foibles. The modern myth tells us that Americans were adept at guerrilla war and hid behind trees to shoot at their too-conventional enemies marching in bright red coats. Myths are rarely entirely false, and Captain Johann Ewald, commanding a Hessian company of *Jäger* light troops, experienced a version of this myth on a sweltering September day in tidewater Maryland. He led a small party of men down the road in advance of the British army, recently landed at the head of Chesapeake Bay and now on the march for Philadelphia. American light infantry had been operating in the area, and a few of the *Jäger* had been wounded or killed in tiny skirmishes around the edges of the British army. Commanding the advance guard, Ewald was wary but reassured by being only a hundred yards from the army's lead elements. He moved forward with just six men to verify their choice of road. A small American party suddenly opened fire from the hedges surrounding them, and his six troopers immediately went down. Wrestling his horse back under control, Ewald yelled for the rest of his force to rush forward. The British army soon pushed through, but Ewald was lucky to escape; his horse died that evening of a wound in the belly.[1]

This kind of ambush was hardly unique to an American way of war, however, and would have been perfectly familiar to an Englishman in Ireland in 1599 or even in England itself in 1644. But there were also other ways to imagine an ambush. In almost identical circumstances, another Hessian officer encountered a completely different American conception of war the following summer. A similar small cavalry patrol, similarly surrounded by hedges in the New Jersey countryside (shortly after the Battle of Monmouth), was taken aback when a Continental colonel rose from behind the hedges and "very civilly" waved at them to "come to him." Here indeed was an unlikely vision of honorable war—springing an ambush without firing a shot! The surprise, or a confusion of language, brought a quick end to the calm scene. After a flickering moment of disbelief, the Hessians tried to wheel and flee, the American colonel again commanded them to yield, and only as they fled did the colonel order his men to rise and fire.[2]

John McCasland, a young Pennsylvanian drafted into the militia in January 1778, and serving as part of Washington's command based at Valley Forge, made a similar decision about honorable war. The men in and around Valley Forge were not idly starving. Many of them patrolled into the hinterland that supplied the British army based in Philadelphia. The British were as dependent as Washington on local supplies of food and forage, although they were better supplied with hard cash to pay the local farmers. While some of Washington's patrols were simply in search of food, others had a more directly military mission. Not unlike the garrison patrols of the English civil war, they were designed to make the British miserable—to deny them food and deny them peace. Washington started the patrols almost as soon as he moved the army into Valley Forge, and they persisted with varying degrees of intensity throughout the winter.[3] Men on these missions had a surprising amount of control over their choices. Desertion, plunder, and uncontrolled violence became options, at least when they could pressure their officer to go along—or give him the slip. European armies had come to recognize these small patrols as the source of much of the abuse suffered by civilians and had developed semiformal codes to try and govern them. Humphrey Bland's highly influential mid-eighteenth-century manual for British army officers declared that no detachments should be sent out with fewer than eighteen men, plus a sergeant. This eighteen-man minimum, Bland wrote, was designed "to prevent a small Number from being detach'd, who can only be sent to pilfer and steal, which is look'd upon, by all Sides, as an ungenerous way of making War."[4] McCasland's was a militia party of sixteen men, but other Continental foraging detachments from Valley Forge generally seem to have followed Bland's prescription. Private Joseph Plumb Martin's foraging party that same winter consisted of a lieutenant, a sergeant, a corporal, and eighteen privates.[5]

McCasland later remembered his mission as being to "prevent the Hessians from plundering and destroying property." On one occasion he and his party successfully crept up on a group of Hessians plundering an empty mansion. Seeing a sentinel outside, the group identified their two best marksmen and then cast lots to decide which of them would shoot him. McCasland drew the short straw. He recalled: "I did not like to shoot a man down in cold blood. [But] the company present knew I was a good marksman," and they would not believe a complete miss. McCasland "concluded to break his thigh." He did, and as they ran up to the house, the Hessians surrendered, one of them waving a bottle of rum "as a flag of truce." The Americans took them all prisoners and delivered them to camp.[6]

John McCasland and the unnamed Continental colonel, with his strangely honorable "ambush," were men struggling with what war meant. Violence had meaning for them, and they sought to use it in a way they could square with their personal vision of right conduct. The colonel clearly had a powerful notion of what "honor" in war meant and the risks that honor required. McCasland, notwithstanding the American legend of hidden riflemen, was appalled at facelessly killing a man from a distance and outside the context of open battle. At the same time, both men felt a burden of necessity and a desire to survive and to succeed—in the end the colonel's men did fire and McCasland did shoot the Hessian sentinel in the leg. Their dilemmas were a microcosm of their commander in chief's. General George Washington felt similar pressures—both to restrain the frightfulness of war and to somehow defeat the British. Crucially for the level of violence in Washington's war, he saw restraint as part of the path to victory.

This chapter and the next explore how George Washington and the officers and men of the Continental army coped with those pressures and modulated the violence they waged accordingly. This chapter focuses on the brothers' war of the Continentals versus the British in 1777 to 1778, but it will also deal briefly with the separate war fought between the patriots and the Loyalists. The focus here is primarily on American decisions and American violence, but some account is also made for the choices and actions of the British. Chapter 8 examines the actions of the same Continental units and commanders when they waged war against Indian enemies in 1779. There we will return to the problem of what happens when the other side's conception of war seems incomprehensible.

Rebelling Against a Distant King

We will not delve here into the complex causes of the American Revolution, but there are several factors to make clear. First, this was indeed a brothers' war. Few Britons were more enthused about being members of the British Empire than the American colonists at the end of the Seven Years' War in 1763. For many that enthusiasm and sense of belonging persisted well into 1776, a full year into actual warfare. For some the attachment never died, and they fought for the British as Loyalists, adding another layer to the brothers' war.

To be sure, not all the players in the war would have thought of themselves as "brothers": there were American colonists of diverse national origins; there were soldiers from Spain, France, Scotland, and various German kingdoms; and, not least, there were active Native American and formerly enslaved participants. Nevertheless, the political system and the political discourse within the colonies were firmly embedded in the larger "British" polity. The key question boiled down to this: exactly what was the subject relationship of the colonists to the crown? In the controversies of the 1760s and early 1770s, the nature of that relationship generated resistance and eventually rebellion. By the mid-eighteenth century power and authority in Britain flowed outwards from the king and Parliament, or, in the technical language of the time, the "king *in* Parliament." Male subjects in Britain were represented in that arrangement by their selection of representatives to the House of Commons—however much that franchise was in fact restricted or corrupted. The colonists lacked that direct connection to ultimate political authority. There were many other reasons why the colonists became first dissatisfied with, and then rebellious about, their relationship with Britain, but they framed their rebellion primarily as a debate over the nature of subjecthood. When the British responded with force, they too framed the conflict as a rebellion of subjects.

The British conception of the war as a rebellion carried with it the same possibility of unrestrained war seen in earlier chapters. Rebels merited no protections under international norms, although recent legal theories had begun to accept that when a rebellion acquired enough force to oppose the sovereign successfully, then the formal codes of war took effect.[7] British officers in America struggled with this problem and never achieved a uniform solution. Many of the senior officers acknowledged the subjecthood of the colonists, sought to return them to loyalty, and worked to win their "hearts and minds." Unfortunately, many of the junior officers (and a few of the senior ones) could not get past seeing the Americans as rebels, the only appropriate treatment for whom was fire and the sword. From the Americans' point of view, this appeared as vacillation and unpredictability, and produced a resentment perhaps worse than either policy consistently applied would have.[8]

For some months after the initial shots at Lexington and Concord in the spring of 1775, the British generals continued to imagine that the political controversy might be resolved without a major military victory. In reality the American leadership was long past reconciliation, but there was a caution and restraint to British strategy in 1775 that was not just a result of their limited resources at the outbreak of war. When General William Howe assumed command in October 1775, he initially hoped for "a decisive Action, than which nothing is more to be desired or sought for by us as the most effectual Means to terminate this expensive War."[9] The British remained in Boston under siege until the spring of 1776, all the while casting about for a strategy to reassert their control elsewhere. Finally, in the summer of 1776, having abandoned Boston, Howe launched a massive amphibious attack on New York City. Washington felt compelled to defend the colonies' most important port, but a series of mistakes and miscalculations in the face of greater British numbers led him to retreat to New Jersey and back across the Delaware into Pennsylvania by December of that year. Washington's famous surprise attacks at Trenton and Princeton then stopped the British advance. Washington and the newly inspired militia followed up with a persistent foraging war against the British in New Jersey for the duration of the winter and until the end of June, when the British army finally pulled back to Staten Island.[10]

The British leadership was divided over their military strategy for the 1777 campaign season. The ministry in Britain supported General John Burgoyne's audacious plan to launch a three-pronged attack along the Hudson River axis that would isolate New England (the presumed heart of the rebellion) from the other colonies. General Howe was less enthusiastic about this plan, and only half-heartedly cooperated with Burgoyne's movements that eventually led to the massive British defeat at Saratoga. Howe instead believed that he needed to destroy Washington's army by luring the general into a decisive battle. To that end he loaded up the majority of his army aboard ship, sailed south and then up the Chesapeake Bay (incidentally leaving Washington in a quandary, unsure which way to direct his forces). He landed at Head of Elk, Maryland, from where he advanced on the American capital at Philadelphia. Here was another target that Washington would feel obliged to defend, and Howe hoped to catch and destroy him.[11]

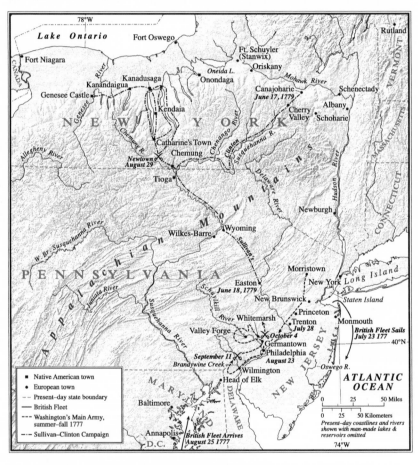

THE BRANDYWINE AND SULLIVAN CAMPAIGNS. Map by Justin Morrill, *the-m-factory. com*.

Brandywine

The British and Hessian troops had suffered during an extended stay aboard ship. Most of their horses had died, and the men could barely stumble ashore. Howe allowed them some time to find their feet before they began their march on Philadelphia. Initially the soldiers were surprised and even spooked by the eerie emptiness of the landscape.[12] Some advance American units of light infantry attempted to slow their march while Washington rushed the rest of his army south to confront the British advance. It was the advance units that killed the men in Johann Ewald's patrol. Few civilians turned out to witness the martial spectacle. Few even remained in their homes. The British soldiers readily interpreted such abandonment as proof of rebellious subjects and promptly appropriated whatever movables they could find.

Not all civilians chose flight, however. As the British army approached Washington's defenses on the Brandywine Creek west of Philadelphia, some residents believed the British would soon bring "peace and tranquility." Others were simply overcome with curiosity. Joseph Townsend, twenty-one years old and eager for new experiences, followed the rumors of the army on the move. He and some friends, somewhat awestruck, found themselves admitted into British lines (having told a sentry that they "wished to see the army, &c., if there was no objection"). Some British officers asked them about the Americans' whereabouts, and even about Washington's character. As Joseph headed for home, he found himself briefly dragooned into tearing down a fence to simplify the British advance. Quaker that he was, Joseph began to wonder at himself. Who was he to participate in preparing the field for battle? We too might wonder at his temerity, wandering an incipient battlefield and freely crossing army lines—he was once told to get beyond the pickets soon, because after dark he would have to stay inside the lines. Joseph was acting in accordance with the eighteenth-century belief that an army could march and fight without being a constant threat to the civilians around it. Earlier in the day, he and others had reassured some nervous women that the British army would hardly "murder all before them." The women's concerns may have been exaggerated, but they did have some basis in reality. The British army had torn up northern New Jersey, and stories of their atrocities were legion. Joseph and his friends had found the British army less threatening. Joseph's pacifist convictions led him to sneak away from the job of tearing down the fence, but even as he left, the sounds of battle echoed across the rolling Pennsylvania countryside. Riveted, he found a spot on a hill with some other locals and watched the battle unfold.[13]

Brandywine was a conventional battle in almost every sense of the word. It represented exactly the kind of contest that Washington was hoping his army could learn to win. For Howe it was another in a series of elusive opportunities to catch and crush the Continental army. Political pressure left Washington little choice but to try and defend Philadelphia, and so he moved his army into a defensible position between the British and the unfortified city.

Since his defeat at New York City Washington had seemed to adopt a "Fabian" strategy, fighting only when certain of not being caught and destroyed, and fighting to keep his army alive as a central symbol of continuing American resistance. It was a risky game, and it required an eye more for lines of retreat than attack. Colonel Alexander Hamilton described it earlier that summer: "We should not play a desperate game . . . or put it upon the issue of a single cast of the die. The loss of one general engagement may effectually ruin us. . . . Our business then is to avoid a General engagement and waste the enemy away by constantly goading their sides."[14] But at Brandywine Washington seems to have sought for a decisive victory, which would bring success and independence "in one bold stroke."[15] His army had rebounded to over sixteen thousand men, most of whom were on the battlefield, stretched out to cover the crossings of Brandywine Creek as Howe's thirteen thousand men approached.

The Culture of Discipline

Battlefield tactics had become simultaneously simpler and more psychically demanding since the English civil war—especially in North America, where cavalry forces were smaller than had become normal in Europe. The slow-firing matchlock musket had been replaced by the flintlock. Although still limited to three shots per minute, it was nevertheless handier and more reliable than its predecessor. Pikemen had been rendered obsolete both by this increased rate of fire, and by the invention of the socket bayonet that made the musketeer into his own pikeman. The fabled American rifle (and its less famous European version) existed, and was on occasion very effective. For the most part, however, battles were believed to turn on the ability of each side to deliver disciplined, rapid, linear volleys of musket fire, followed by a final and usually decisive bayonet charge.

European generals of the era increasingly believed that the only way to achieve this result was by imposing an iron discipline. Frederick the Great of

Prussia was the most famous practitioner of this kind of warfare, and his army became legendary for the precision of its battlefield formations and evolutions, as well as for the soldiers' rapidity of fire.[16] Of equal importance was the ability to stand and absorb this kind of punishment, and then to hold their ground when the enemy launched the charge. Sergeants stood behind the ranks of their men pressing a long halberd against their backs, keeping them in line, geometrically and psychologically.[17] This tactical summary naturally presents a simplified picture. European military writers regularly debated the optimal number of ranks in a formation, whether to fire by platoons or companies, how long to hold for the first fire, and more. But the essential quality was always a kind of mechanistic behavior that the soldiers must (so the officers believed) be savagely disciplined to accept.[18]

This tactical development was also important for its continued and enhanced impact on overall army discipline. The temporary orders of the sixteenth century had evolved into the printed versions of army regulations of the English civil war and subsequently, with continued reference to practices in other European armies, into the far more formal and legalistic articles of war created for the British army in the eighteenth century. Those articles continued to focus on the daily details of army life, sanitation in camp, alertness on sentry duty, and obedience and subordination to officers. They also retained the clauses that sought to limit the damage soldiers inflicted on civilians in much the same terms outlined in previous chapters—always allowing for the exception that they could destroy when ordered to do so. Perhaps most important, the problem and cost of desertion from a painfully acquired and trained standing army had led the British, along with other European armies, to rework their system of camp discipline to prevent soldiers from leaving without permission. Where even the 1686 articles merely forbade soldiers to "depart above a Mile from the Camp, or out of the Army without Licence," or to use any but the "Common Way" in going to and from camp, they lacked other specific measures against desertion.[19] In contrast, the 1765 British articles laid down specific rules designed to enhance accountability and added a section on desertion, supplemented by newly official mechanisms requiring men to be in their quarters at the beating of the retreat.[20] Measures that kept soldiers from deserting also kept them from unauthorized plundering in the countryside, at least to a point. Desertion remained rampant, as did plundering. The mechanisms of control, however, were clearly becoming more thorough and sophisticated.[21]

One shift in the mechanisms of control, for all these reasons, was a narrowing of the ratio of officers and noncommissioned officers to private

LINEAR WARFARE IN THE EIGHTEENTH CENTURY. A marvelous depiction of linear warfare, this was a large print originally produced in France but reproduced in England in 1771 for young officers. *The March of an army supposed to have been attack'd by a party of the enemy, shewing the form of a battle in front: This view is engraved to exhibit to the public & to young officers in particular the order of an army preparing to march* (1771) (Antoine Benoist, London: Francis Vivares). From the Robert Charles Lawrence Fergusson Collection. Reproduced by permission of the Society of the Cincinnati, Washington, D.C.

soldiers. A sixteenth-century English company had ranged from 1:25 to 1:100 officers to soldiers, with an unpredictable but small number of "sergeants." During the English civil war the ratio had improved. A company ranged from 1:33 to 1:50, plus five sergeants and corporals, as well as seven to nine additional officers at the relatively new regimental level. In the American Continental army of 1778, largely copied from the British model, a company of fifty-three privates was supervised by a captain, a lieutenant, an ensign, and six sergeants and corporals (a ratio of 1:18). The regimental staff included nine to twelve additional officers. Of course, this was only the hypothetical structure; Continental companies were notoriously understrength in privates.[22] Officers in the

eighteenth century continued to complain they could not prevent their men from deserting or plundering, but in this comparative light there is little doubt they had more supervisory capacity than their predecessors.

At Brandywine the American army showed that it had not yet fully absorbed all the structures and implications of the European system, but it was improving. A local Loyalist informed Howe that Brandywine Creek could be crossed several miles to the north and that the ford was still unguarded. Howe divided his army, left five thousand men to occupy Washington's attention at Chadd's Ford, and rapidly marched the bulk of his army on a wide swing around Washington's flank. Several American scouting reports suggested that the British were moving to Washington's flank, but the reports were contradictory, and Washington hesitated to rearrange his forces. Only when Howe's forces appeared in full view lining up on Osborne's hill, threatening not only the flank of the American army but also the line of retreat, did Washington fully comprehend the threat. He began a rapid reshuffling of his men. The soldiers rushed to a blocking position, but their inexperience and that of their officers hampered the reestablishment of the line. The British caught them before they were fully formed and pushed them back. Nevertheless, the Americans held long enough to allow Washington to pull his army out of the trap and retreat toward Philadelphia.[23]

In the Wake of Battle: Prisoners, Plunder, Provisions, and Property

The ratio of killed to captured in the losing army (roughly 1:2) suggests something about the restraint normally practiced on the battlefield at this point in history. The British suffered eighty-nine dead and nearly five hundred wounded. American casualties were never precisely determined, but a variety of estimates hover around two hundred killed, five hundred wounded, and four hundred taken prisoner. In previous eras the status of prisoners (especially common soldiers) had fluctuated. In sixteenth-century Ireland, thinking of the enemy as barbarians *and* brothers (or at least as potential subjects) had produced a bewildering combination of executing and pardoning prisoners. In the seventeenth century in England, in what was much more clearly a brothers' war, prisoners could easily be recruited into one's own army. Their value as recruits, and the mutual effort to adhere to international codes of war, preserved the lives of the vast majority of men who threw down their arms and surrendered. There were brutal and gruesome exceptions, especially when Englishmen fought Scotsmen, Irishmen, Welshmen, or even Cornishmen, but the precedent was clear.

During the American war for independence, the British and American regular armies regularly took each other prisoner, at a rate that saw many more taken prisoner than killed. Casualty figures from this era are always sketchy, but in the thirty combat incidents in the northern theater in 1777 that involved British regulars on one side, and which resulted in a draw or an American defeat, the Americans suffered a total of 907 dead and 1,709 captured (1:1.9). Removing the major battles of Brandywine, Paoli, and Germantown from these totals (as they represent the most extreme incidents) the ratio becomes even more "prisoner friendly" at 255 killed to 885 captured (1:3.4).[24] Looking at British defeats over the whole war, in engagements where they suffered more than a hundred casualties (and for which reports distinguish killed from prisoners), and even discounting the two major surrenders at Saratoga and Yorktown, the British suffered a ratio of 1,198 killed to 8,225 taken prisoner (1:6.9).[25] There were exceptions. In the most notorious incident, Colonel Banastre Tarleton's legion ran down the surrendering Virginia Continentals at Waxhaws, South Carolina in 1780. They killed 113 and took fifty-three prisoners—reversing the usual ratio.[26] In another infamous moment, carefully analyzed by historian Armstrong Starkey, the British launched a surprise attack on the Americans at Paoli, shortly after Brandywine, killing roughly two hundred Americans while capturing seventy one.[27] In general, however, these ratios, especially for British casualties, reflect a remarkable willingness to accept battlefield surrenders and preserve life.

To understand the restraint implied by these numbers, we must consider the fact that, unlike in the English civil war, only rarely were prisoners recruited into the opposing army. Washington and the Continental Congress both officially opposed recruiting from British or German prisoners (or deserters).[28] Some movement back and forth almost certainly occurred upon capture, but it never approached the wholesale recruitment that we saw in Part 2.[29] Prisoners instead were a burden; their care was the responsibility of their captor. Prisoners still had "value," as propaganda tools, as labor, and perhaps most importantly, as potential bargaining chips in the prisoner-exchange system.[30]

Dealing with prisoners, however, was not the first priority for the British army after Brandywine. Washington's army had escaped; Philadelphia seemed open to the British, but the army needed food. As the army spread out into the countryside foraging, a small group of British soldiers arrived at the home of Mary Worrall Frazer, provoking an incident whose details speak volumes for the hopes and fears of an eighteenth-century society at war with itself.[31] In the immediate aftermath of Brandywine, Major Persifor Frazer, an American officer present at the battle, returned to his home in Chester County, Pennsylvania. He found two Loyalists living in his house. They quickly fled, but Frazer still faced a problem. His allegiances and his military service were no secret, and, as the British steadily moved through the area, he was certain they would eventually target the house of a known rebel officer. He left home, but his wife, Mary, four children, a servant, and three slaves remained behind. Mary wisely began hiding the family's valuables in nearby woods and gardens. Within days a British foraging party arrived. The whole family, save Mary and a young slave girl, fled into the woods. Mary bluntly traded insults with one British officer, and the troops began to ransack the house even as neighbors gathered to watch. Eventually a second officer, Captain De West, entered the house, controlled or removed the most drunken and violent troops, and then personally searched the house. Although De West promised Mary that nothing belonging to her would be taken, she followed him from cupboard to cupboard and repeatedly protested the propriety of his searches and confiscations. De West soon discovered that an American unit had stored clothes and equipment in the house, and the British legitimately confiscated them. They also took salt, liquor, horses, and food. Eventually the British left, and a dismayed Mary realized that she no longer had enough food for her household. Worse yet, possibly on the same day, her husband had been captured and would spend the next six months in British custody.

Major Frazer, like the earl of Chesterfield in 1642, and like the Irish peasants in Shillelagh in 1569, believed his wife safer alone at home than on the road, even knowing that enemy troops would eventually come to his house. The fact that this was a brothers' war guaranteed that they would arrive—Loyalist neighbors saw to that. The residents of Shillelagh had discovered they were wrong—their families were killed. Chesterfield and Frazer were proved correct. In Chesterfield's case, however, the officers of 1642 threw up their hands in the face of the soldiers' culture, claiming they could only restrain the soldiers if Lady Chesterfield paid them. When she refused, the men plundered the house. In Pennsylvania in 1777, De West could control his men, albeit not perfectly, and not without effort, but Mary safely contested virtually every move he made. Save for two small glass creamers, Mary seemed unable to complain of them taking anything beyond what could legitimately be called military supplies.

Even the capture of her husband speaks to changes in the system and brings us back to the problem of prisoners. Surviving the transition from battlefield to captivity was one thing; treatment over the long term was another. Frazer was captured and held for six months. His value as a prisoner, like others, was as an exchangeable asset, and until that exchange could be negotiated, he was held. An officer in previous centuries would have been held until ransomed. A common soldier might have been killed, recruited, pardoned (in Ireland), or held only briefly and then released on his oath not to fight again. Most of those options (aside from ransom or pardon) reappeared in some form during the Revolution, but not on the scale of previous centuries. In the context of the American Revolution, prisoners had different "values," and their treatment varied according to that value. The normal pattern now was confinement, parole, and exchange, in assorted combinations. Initial confinement was almost inevitable, but it could stretch into years. In general the British record on the treatment of prisoners in confinement was poor, primarily because they chose to hold many common soldiers in unhealthy hulks anchored in New York harbor. The Americans did somewhat better in treating their large prisoner hauls at Saratoga and a few other instances, but they were hampered by an inability to feed their own soldiers, much less prisoners. At least one of the captured German mercenaries was surprised by the generally friendly manner of his treatment and noted, "Each day we received one pound of bread and meat, and we lay quiet." To be sure, this peaceable state was only reached after Washington had published a broadside describing the captured men as innocents forced into the war.[32]

Frazer was an officer, but even he experienced bouts of confinement alternating with the relative freedom of being on "parole": the freedom, guaranteed by his oath not to try to escape or to fight again, to live within defined boundaries wherever he could find (and pay for) lodgings. Frazer was captured in mid-September, and he initially complained only of "abusive Language" from passing British troops. At the end of September he signed a parole and was allowed to stay in homes in the city. On October 7 the British returned him to close confinement in the Philadelphia State House, where he remained until early January. Frazer complained of short provisions and that sentries posted in the rooms stole from the prisoners while they slept (incidentally attesting to the fact that they had not been stripped bare when first captured). Lieutenant James Morris was held prisoner there at about the same time, and he remembered the rooms being "destitute of every thing but cold Stone Walls and bare floors," with sixteen officers or thirty two common soldiers to a room. Many died of disease. Frazer was again briefly allowed his parole within Philadelphia, and was again confined in February, confinement from which he finally escaped. Under confinement his oath did not apply, and he was able to attempt escape honorably.[33]

Frazer's complaints in letters to his wife about conditions in Philadelphia may have played a role in reopening the then-stalled process of exchanging prisoners. Normally, prisoner exchanges took place under the terms of a formal agreement, or "cartel," defining the conditions, processes, and rank equivalencies. Both serious and trivial obstacles prevented a formal cartel during the Revolution until the very end of the war. Most importantly, such a treaty required the British government to recognize the legitimacy of the American state, something the Continental Congress persistently demanded. Washington also sometimes proved unwilling to exchange private soldiers, since the Americans' terms of enlistment were short compared to those of British privates. Returning prisoners enhanced British strength, while returned American soldiers were likely to go home. Nevertheless, more informal cartels were common, although frequently interrupted by specific disagreements. For example, when the British captured General Charles Lee in December 1776 and threatened to try him, a former British officer, as a deserter, Washington suspended exchanges. Such interruptions generally hurt common soldiers more than officers, but even Lieutenant Morris, taken in October 1777, was not exchanged until January 1781. Frazer's descendants, who compiled his papers, claimed that his complaints led to a renewal of the exchanges interrupted by Lee's capture. As a general rule, exchange proved to be conditional on cold calculations of political, military, and personal advantage. The intervening periods of confinement killed hundreds and probably thousands of common soldiers and sailors, but it had definite advantages over previous "systems."[34]

None of this is to argue that eighteenth-century warfare was gentle and kind, but several things had changed since the days of the English civil war. At the end of the seventeenth century and through the eighteenth century, cultural values stressing restraint had been joined to a new bureaucratic capacity to wage war according to those values. Restraint was not an inviolable rule. There were a variety of situations in which those values and capabilities did not converge, and in those we can see the unleashing of violence on a scale familiar from the worst periods of the previous two centuries—one such example is the centerpiece of the next chapter. But the incident with Mary Frazer represents, if not a perfect "ideal," then the best that might be hoped for in practice. Armies still had to be fed, and centralized logistical systems were far from perfected. Yet the model was clear, and Washington aspired to it, both in its structures and its ideals.

A major structure in that model was a shift in the supply system preferred by most European states in the eighteenth century. Two reasons are usually cited for this change, one ideological and the other bureaucratic. The excesses of the religious wars were in many ways driven by the natural needs of unsupplied and often unpaid soldiers to feed (and pay) themselves. This lesson was not lost on observers. As shown in Part 2, English and Scottish officers lamented these excesses even as they allowed them. Revulsion helped inspire a value-based effort to supply armies in a way less ruinous to the population. Furthermore, as William McNeill has argued, "war became, as never before, the sport of kings. Since the sport had to be paid for by taxation, it seemed wise to leave the productive, taxpaying classes undisturbed. . . . For soldiers to interfere with their activities was to endanger the goose that laid the golden eggs."[35] This value shift is discussed in more detail below, but with the desire to change logistics came an increased bureaucratic capacity to actually do so. France led the way in providing a logistical infrastructure designed to ease the burden of an army on its own people, which had corollary benefits for invaded populations.

The basic logistical change was twofold. First, there was an increased formalization and regularization of the contribution system rather than ad hoc collection or even outright plundering of a town for its provisions. The second aspect was a greater reliance on a system of depots or "magazines." Part 2 outlined how contributions emerged during the religious wars as an improvement over unregulated pillage as a means of supply. That system was nevertheless little better than extortion and was all too dependent on the whim of the local field commander. In France Louis XIV and his ministers hoped to further regularize the collection of contributions. In one instance they even set limits to their collection from enemy territory, and they generally prohibited the firing of villages as a means of extorting the payments.[36] Magazines, prepared ahead of time, allowed an army to move from point to point without (in theory) having to rely on the countryside for food (fodder for horses was another matter). Magazines had their limits, and an army laying siege to a city or on an extended march into enemy territory quickly became dependent upon the countryside for supplies (provided riverine supply was unavailable). There seems little doubt, however, that in general the proliferation of this kind of logistical system ameliorated some of the worst of the "pillage as supply" pattern of the previous era.[37]

Hand in hand with this structural shift was an intensified elaboration of the codes governing the conduct of war, a process that Geoffrey Best has described as the creation of an "Enlightenment consensus."[38] This consensus recognized that military necessity existed and could be used to justify almost any form of violence, but, in the words of the preeminent eighteenth-century codifier Emmer de Vattel, "The law of nature . . . does not allow us to multiply the evils of war beyond all bounds."[39] This balancing of necessity and humanity appeared in the work of all the codifiers, back into the Middle Ages, but the earlier writers had frankly put necessity first. The eighteenth-century writers never abandoned the principle of necessity, but they placed much greater rhetorical emphasis on the restrictions demanded by humanity. Enlightenment writers identified several categories of behavior that were of particular concern: plunder, requisitions and contributions, devastation (as a strategy), and the treatment of civilians as a result of siege. All these issues had been partially "regulated" in earlier eras, but those customs gained salience and clarity during the Enlightenment, in part through the recording of those rules and their linkage to new ideas of a pervasive and enforceable natural law.

In some ways, the process of codification merely elaborated upon the customs of the past, but the era also featured a new perspective on individual rights within natural law, especially regarding the rights to property. In essence, the definition of a "subject" was changing, and with it, the view of civilian rights to property. Two of the most important contemporary authors on laws of war, Hugo Grotius and Emmer de Vattel, highlight the change. Grotius was a Dutchman, born during the Eighty Years War against the Spanish and writing during the Thirty Years War. He produced a variety of treatises on the emerging concept of international law, the most famous being *The Law of War and Peace*, published in Latin in 1625 and translated into English in 1654.[40] Grotius drew on an extensive and violent set of classical precedents to argue that since being at war gave the right to kill, it also necessarily conveyed the right to "destroy and plunder the property of enemies," although he sought to limit such a right to situations where it was "necessary" to deny resources to the enemy. Grotius further argued that property, including land seized in war, had its ownership legitimately transferred, including even property "taken from people subject to the enemy," which can be "regarded as taken from the enemy." In this vision a "civilian" was merely a subject, and if the sovereign was an enemy, then so was the subject (although he did encourage leaving those "guiltless of the war . . . all articles which we can go without more easily than they can").[41]

Vattel's attitude toward property transfer, appearing in *The Law of Nations* in French in 1758 and in English in 1760, was more restrictive, and implied a new attitude toward the treatment of civilians. Vattel, like Grotius, admitted the right of a warring state to seize the property of its enemy, and also to lay it waste if necessary for military purposes. But he more carefully differentiated between seized lands that were "conquests" and moveable property or booty. Both, when taken, belonged to the sovereign, but tradition allowed for some redistribution of movable property to the soldiers, and they could usually keep the property of enemy soldiers taken on the battlefield or while on detachment. Vattel noted, however, that the custom of "pillaging the open country and defenceless places" had been replaced by contributions (preferably "proportion[ed] to the abilities of those on whom they are imposed"), which was now the legitimate way for a general to make the enemy's country "contribute to the support of his army." As for land (and this was key), Vattel argued that although sovereignty might pass through conquest (provided it was made stable and secure, as Grotius required), "private individuals are permitted to retain their [property]."[42] In short, Vattel admitted the same basic right to transfer property in war, whether land or booty, but he hedged the principle more carefully. The definition of booty should be confined to state property or the personal goods of enemy troops, and private land should remain in private hands, even if the owner acquired a new sovereign in the process of war.

In Vattel's work the concept emerges of a civilian as something more than just a noncombatant. In the simplest terms, a noncombatant was someone who deserved protection for reasons of humanity; a civilian, however, was a legal category of person who had rights under international law. For Grotius, a subject in legal terms had been little more than a "surety" in dealings between sovereigns. If one sovereign had a just claim against another and went to war to recover it, he was justified in taking it from his enemy's subjects, just as one might make a claim on collateral in the case of an unpaid debt. A hundred years later Vattel deliberately differentiated himself from Grotius on this topic. He pointed out that Grotius found only necessity as a reason for sparing peasants (that is, they might be useful). Vattel argued more strictly that "at present war is carried on by regular troops: the people, the peasants, the citizens take no part in it, and generally have nothing to fear from the sword of the enemy." He clarified that such protection depended on their submission to "him who is master of the country," and that they pay the contributions and refrain from active hostilities. As noncombatants became civilians, and, simultaneously, as subjects became citizens, they acquired more theoretical protections under international law.

In many ways, Grotius, Vattel, and other early modern jurists merely codified customs created in previous centuries, but this process of codification created a kind of feedback loop in which custom and written premise (or "law") reinforced each other.[43] Even as these attitudes were shifting, they were also penetrating into North America. In one example, an American colonial author, trying to justify using French prisoners as surety against an attack by French-allied Indians, referred specifically to Grotius's and Charles Molloy's 1676 commentary on "reprisals," and he quoted extensively from Molloy.[44] Similarly, Lieutenant Governor William Bull of South Carolina wrote to British army colonel Archibald Montgomery in 1760 in a manner that deserves a moment's consideration. Bull had just learned that Alexander Miln had seized some Cherokees who had been invited to dinner under truce. Troubled, Bull wrote:

> On this occasion turned to my Grotius, where I see that faith is the surest Bond of Human things, The reputation of faith is sacred between Enemies. Faith and Justice are to be strictly observed in War and are to be kept even against Interest. A confidence in public Faith procures Truces even among Enemies, or War could never be ended, but by compleat Victories. Faith is to be kept with Slaves, and even with those that are perfidious. . . .

But of this enough, As I doubt not but you are perfectly acquainted with the rights of War and Peace.[45]

Bull's admittedly halting attempt to apply the writings of international law to the rough and tumble of the colonial frontier nevertheless speaks to its salience in the mental landscape of the North American colonists.

Defining Victory

Before returning to the ways in which this background would influence Washington, one must examine how eighteenth-century generals imagined achieving victory. What were their calculations of how to win? One key element was the increasing expense involved in the national standing armies and the infrastructure developed to support them. By the mid-seventeenth century most European states, led by France, had begun to shift from the "aggregate-contract" army, composed of small cores of household or royal troops fleshed out by much larger numbers of contracted mercenary formations, to the "state commission" army. The new armies relied on soldiers recruited for long periods of service and then intensively trained. They were expensive investments.[46] An army had become a precious component of state power, and using it invariably entailed loss, whether it was victorious or not. In one example, Louis XIV waged a "successful" war in the Spanish Netherlands between 1672 and 1678. By the end of the war, although the French military establishment had increased in size, it had suffered as many casualties (roughly 150,000) as there had been soldiers in the initial invasion force.[47] Such risk demanded careful calculation, and a ruler could not lightly commit himself to war. Once at war, leaders preferred the course of least risk. Because of the close-range musketry confined in space and limited in time, battles could be horribly destructive of an army. Generals thus learned to prefer maneuvers that would force an enemy's defeat or retreat without entailing battle. Battles were *believed* to be decisive and so were both sought and feared. The result was a kind of operational tug of war between the hope of winning a decisive battle and fear of its consequences.[48]

Furthermore, most of the wars in western Europe from the 1660s through the middle of the eighteenth century, and the ones most central to Washington's mental horizons, had been fought for relatively limited goals. In David Kaiser's words, western European warfare took on a "pattern based on frequent wars, generally cautious military tactics, rapid diplomatic changes of front, subsidies from richer to poorer powers, almost continual peace negotiations, and a general willingness to compromise issues." John Lynn has described this as "war as process" (in contrast to "war as event"), in which each military move, gain, or loss was associated with a parallel set of diplomatic maneuvers that might solidify a gain and end the fighting. Battles and sieges were indecisive, operations unfolded slowly, war fed war, attrition dominated calculation, and ongoing diplomacy remained constantly in the background. Limited territorial exchange was also a part of the pattern, but these exchanges were qualitatively different from extensive efforts at conquest. There were European wars of conquest during this period, notably under Sweden's Charles XII, but they had less impact on the Anglo-American vision of war than the wars of Louis XIV and the other western European dynastic conflicts, which, among other things, had generated colonial equivalents in North America.[49]

Given this combination of limited goals, constrained means, technological parity, and fighting often within a heavily fortified theater of war, calculations of "how to win" usually focused on pressuring the enemy state through the financial burden of war. One way was to occupy stretches of enemy territory that could then be placed under contribution to make war pay for itself. Alternatively, one could target the expensive military arm of the other state. In this approach the war was still a contest of resources, but the financial and military "center of gravity" became one and the same: the enemy army and the enemy fortification system. In terms of violence, this calculation put the enemy's identifiable military resources at the center of the conflict and ostensibly freed the civilian population from the fear of devastation. The fear of losing one's own expensive army could produce endless maneuvering and a preference for the geometrically predictable siege, both of which could seriously affect the nearby civilian population, but it nevertheless did not deliberately target civilians. To be sure, one could attack the enemy army "indirectly" by devastating its food supplies while avoiding battle, or even to scorch the earth defensively in an invader's path. These had been the English tactics in Ireland in the sixteenth century, and they had been common enough elsewhere well into the seventeenth century. But such devastation had become the exception in western Europe due to the ideological shifts associated with the Enlightenment and embodied by codifiers like Vattel, as well as the new mental focus on attacking the enemy army itself as a prime way to pressure an enemy state. There were notorious exceptions, notably Louis XIV's deliberate devastation of the Palatinate in 1689, but that campaign was most frequently referred to by contemporary authors as the exception that proved the general rule. Vattel made a particular point of condemning it.[50]

As for Washington, he held the same fears and hopes for decisive battle, based in his concerns for his own standing army. He too saw destroying an army as the core of victory or defeat. He needed to preserve his own force and damage the British, and, in Hamilton's phrase, he feared "put[ing] it upon the issue of a single cast of the die." He was tempted by the opportunity to inflict major damage on the British at Brandywine, but as events there turned against him, he immediately reverted to the greater need to preserve the Continental army and continue the "process"—to draw out the war and its cost on the British. As for the indirect approach of devastation, the British relied on resources from across the Atlantic and on food procured from a countryside that Washington considered to be American. Others had scorched their own earth, and Washington, as will be seen, considered doing so against the British, but he quickly ruled out that option. In a brothers' war that choice had little appeal. There were too many people whose opinion mattered to the outcome, and Washington was actively worried about the people in the countryside, not merely as subjects, or even as civilians, but now as *citizens* in the new republic. But Washington's choices and his thinking about how to achieve victory are complicated, perhaps even contradictory, so let us consider them in more detail in the wake of his defeat at Brandywine.

The Continental Army and Valley Forge

In appearance, in the conventions he followed, and even in his definition of success, Washington sought to wage war in a typical eighteenth-century fashion. Not all his contemporaries shared his cultural viewpoint; many of them passionately rejected the standing army as an institution. But once Washington was appointed commander, he controlled the shape of the Continental army, although he had much less influence on the state militias. He wanted a mechanistically disciplined eighteenth-century army capable of fighting on the same field and in the same linear style as the British army. To achieve that vision, he needed long-term recruits who could be trained to a British standard. And it was over the issue of recruiting that he almost immediately clashed with the Continental Congress's republican fear of a standing army. Washington and Congress continued to debate the issue until the fall of 1776, when Congress authorized him to recruit men for three years or the duration of the war.[51] The men recruited under those conditions were the ones who began to show promise of standing up to the British during the fight at Brandywine and who would famously be trained the following spring at Valley Forge by Baron Friedrich von Steuben in European-style discipline.

Washington's understanding of conventional war included both the Enlightenment ideal of limiting war's ravages and the aristocratic conception of the role of honor in determining right conduct in war.[52] Lacking hard statistical data, it is difficult to make hard and fast conclusions, but as a whole the Continental army proved relatively restrained in its application of violence. Its poor logistical infrastructure and general lack of pay meant that the soldiers all too often relied on rough-and-ready foraging, especially for firewood, but the Continentals committed few violent crimes, and even their foraging practices were comparatively restrained.[53]

These two broad generalizations are supported by Washington's actions in the campaign that followed Brandywine and especially by his decisions related to the winter encampment in Valley Forge. Following Brandywine the British occupied east-central Pennsylvania and threatened the surrounding area with devastation, both deliberately and indirectly through the process of supplying themselves. Their soldiers were not always as well controlled as those with Captain De West at Mary Frazer's house. Joseph Townsend, the young man who had visited the British camp, described the countryside after the British passed as "a scene of destruction and waste." The inhabitants "had their stock of cattle destroyed for the use of the army—their houses taken away, and their household furniture, bedding, &c., wantonly wasted and burned." Joseph added that the British used furniture for firewood rather than timber or even readily available fences, since "being in an enemy's country, inhabited by rebels, there was no restraint on the soldiery and rabble who accompanied them."[54]

If the British officers failed to restrain their men (and some were clearly not interested in doing so), Washington hoped to restrain them militarily. Having failed in conventional battle at Brandywine, and despite the British occupation of Philadelphia on September 26, Washington gambled again with a surprise large-scale attack at Germantown on October 4. He openly hoped to "destroy the Enemy's Grand Army" if given a favorable opportunity to attack it.[55] The gamble almost paid off, but a too complicated plan, poor weather, and a lack of battlefield command and control allowed the British to recover and push Washington's army away from Philadelphia.[56] Washington then tried and failed to isolate British-controlled Philadelphia via a series of forts on the Delaware River. With winter approaching, he reassessed his strategy for the remainder of the year.

Three basic principles shaped Washington's strategic thinking: to keep an army intact and visible as the physical manifestation of the rebellion; to maintain at least the tenuous support of the countryside; and, crucially, the belief that both those objects were best achieved by a conventional force. At the same time, however, he had to find a way to pressure the British. Their only accessible "center of gravity" was their army, and through it, British public opinion.

Keeping the army alive essentially meant avoiding a crushing defeat. This proved a crucial factor in Washington's decisions after Germantown, not just in Pennsylvania but for much of the rest of the war. He increasingly avoided defending fixed positions unless he felt he had an overwhelming advantage, and so he avoided battle, or more accurately, began to seek battle only on his own very restrictive terms. These considerations became explicit at the councils of war Washington held in late fall 1777. One of his brigadiers, William Maxwell, argued, "If we throw the Armey away we have, without some good appearance of success we are much more likely not to get another one nor support the Credit of our money." Maxwell favored "harassing" the enemy "by every means in our power." In short, Maxwell concluded, "If they cannot meet [catch] us in the field they will make very slow work in conquering the Country."[57]

Washington hoped that an evasive war of attrition would pressure the British militarily, but he had to reconcile that course with his second imperative: maintaining the loyalty of the countryside. The preceding chapters repeatedly showed the burden armies placed on the countryside, whether deliberately or simply because of their vast needs. European armies in the eighteenth century were now better paid than ever before and were more often able to depend on centralized supply, but they nevertheless still depended heavily on locals for provisions and especially fodder for animals. Washington's army depended entirely on local supplies, and worse, most of the time he could not pay his troops. But, in this brothers' war, in which the loyalties of the locals might prove crucial to the outcome, Washington had to avoid alienating them. The motives of utility and morality are difficult to separate in these circumstances. General Horatio Gates once concluded that the irregularities of previous foraging parties had made it impossible for him to buy supplies from the countryside.[58] But Washington's concerns and the concerns of his officers were more all-encompassing than merely utilitarian. Although Washington certainly believed, in his words, that "no plundering Army was ever a successful one," he and much of the Continental leadership also realized that "restraint towards civilians . . . was essential to having the war effort serve as a mainspring of national legitimacy and resultant nationhood."[59]

Obedient to this strategic calculation to cultivate the loyalty of the countryside, Washington tried to ease the impact of his army as much as possible. First, he and the Congress refused to devastate the countryside to make a desert for the British. From our perspective today, it might seem obvious that avoiding such destruction was the only real choice, but devastation was not an unknown strategy in the eighteenth century. General Greene strongly encouraged burning New York City during the American retreat in 1776, and Washington expressed ambivalence on the subject until Congress specifically resolved to preserve the city.[60] The willingness to consider burning New York was partly tied to the perception of a heavy Loyalist element in the city, in addition, however, burning it would have deprived the British army of shelter. In the countryside around Philadelphia, Washington had to steer a more cautious course designed to preserve revolutionary loyalty, and he did so openly and consciously. At one point during the Valley Forge winter, for example, Washington rejected General Lachlan McIntosh's proposal to depopulate an entire district between the lines so as to cut off residents' illicit trading with the British.[61] Even though Washington could not create a desert, he could deprive the British of certain key resources, destroying, for example, the Pennypack flour mills outside Philadelphia.[62] There was a fine line between an army's natural process of foraging and a deliberate decision to devastate, and so some British observers were convinced that Washington *had* decided to devastate the area around Philadelphia.[63] As chapter 8 will make clear, however, a deliberate decision to devastate the countryside has immediate quantitative and qualitative effects that foraging achieved only when it was carried on for long periods of time in the same area.

Perhaps of even greater importance was Washington's attitude toward martial law as applied to civilians. Sutlers supplying the army, and women accompanying the army as wives and domestic laborers, were easily folded into the embrace of army regulations—these were civilians to whom "military law" was easily and relatively painlessly applied. In contrast, "martial law" or the use of speedy military courts, tribunals, or even preemptory decisions to "try" civilians in the area of army operations, was a much messier business, and, as discussed in Part 1, fraught with the potential to escalate the frightfulness of war. Eighteenth-century articles of war, including those passed by the Continental Congress, allowed the use of martial law against persons caught acting as spies. The "necessity of war" also readily justified the use of martial power to confiscate food and forage for the use of the army, and Congress explicitly authorized as much in December 1776. It stretched the point only slightly by also authorizing Washington to arrest those who refused such requests, but Congress specified that such persons be turned over to the states for a civil trial. Then, in the emergency of the winter of 1777–78, Congress authorized Washington to impose a much more comprehensive version of martial law on the regions around Philadelphia, declaring all persons who acted on behalf of the British, even if merely bringing them supplies, to be traitors to the states and subject to trial by a court-martial.[64]

This posed a real dilemma for Washington. The need to cultivate loyalty in a brothers' war demanded restraint. Simultaneously, however, the rhetoric and passion of "treason" justified the means and mechanisms of martial law, which, as in Ireland, could easily escalate the violence of the war. Washington used the powers of martial law but remained aware of its limits. Orders and accounts from that winter make it clear that people caught passing goods into Philadelphia could be severely punished, including with lashes and even death. Washington authorized what historian Wayne Bodle calls an "informal roadside triage," in which patrol leaders confiscated goods intended for the British, turned some loose, flogged others, and brought the worst offenders in for trial.[65] The number of people actually tried, however, appears small, and when one court-martial confiscated the land of an offender, Washington intervened, pointing out that martial law did not allow for the confiscation of real property.[66] He probably based this conclusion on the English tradition of martial law discussed in chapter 1, in which land could only be confiscated by the act of a civil court or through a declaration of attainder by parliament. Although in Ireland martial law had nevertheless been turned to the purposes of property confiscation, first de facto and then slowly becoming de jure, Washington, per Vattel's new emphasis on the inviolability of private land in wartime, sought to confine land confiscation to the civil courts. Congress did seek such confiscation; it resolved in November 1777 that the states should "confiscate and make sale of all the real and personal estate" of active Loyalists.[67] The state legislatures in turn passed their own versions that law, confiscating Loyalists' land on a fairly large scale (as had Parliament in the 1640s) and using their militias to enforce the seizure. Washington wanted his army kept out of it.[68]

In addition to destroying key local resources and trying to prevent the movement of civilian resources into Philadelphia, Washington hoped to use his army directly against the British ability to move or supply themselves. In deciding to encamp at Valley Forge, so close to Philadelphia, he sought to limit the British ability to ravage the countryside. Washington explained his choice of dangerous and unsheltered winter quarters as an effort "effectually to prevent distress [of the inhabitants] & to give the most extensive security" to the countryside.[69] Throughout the fall and winter he kept some of his forces moving around the countryside, not only to feed his own army but also to interfere with British efforts to do the same. It was in those circumstances that James McCasland encountered the party of Hessians and shot one in the thigh.

Meanwhile, winter set in, and in eighteenth-century rural Pennsylvania it could do so with a vengeance. The so-called "little ice age" that lasted from 1500 to roughly 1850 was in full swing; rivers and harbors, famously including the Thames near London, and even New York harbor, might freeze solid.[70] For soldiers on the move, or for men in ragged canvas tents, ill supplied by a nation not yet able to mobilize and move the necessary resources, it was even more fierce. One Continental army surgeon confessed his despair at yet another river crossing in early December: "Sun Set—We were order'd to march over the River-It snows-I'm Sick-eat nothing-No Whiskey-No Forage-Lord-Lord-Lord."[71] Another soldier complained that "we had no tents nor anithing to Cook our Provisions in and that was Prity Poor; for beef was very leen and no salt nor any way to Cook it but to throw it on the Coles and Brile it[,] and the water we had to Drink and to mix our flower with was out of a brook that run along by the Camps and so many a dippin and washin it which maid it very Dirty and muddy."[72] The winter only grew worse, and Valley Forge became a byword for the sufferings of Washington's army. More trying than the weather, however, was the fact that Washington was subsisting his army on a countryside strained to the limit.

Even the relatively small armies operating in North America during the Revolutionary War were huge compared to the largest nearby cities. The comparison to cities is apt, because armies, like cities, required food to be brought to them in order to survive. Cities benefited from long-developed infrastructure and a market system to transport that food. Armies had to invent their own infrastructure, and the supplies had to come to them wherever they happened to be. Washington estimated that an army of fifteen thousand men required 100,000 barrels of flour and twenty million pounds of meat in one year.[73] In contrast to that scale of demand, Commissary General Joseph Trumbull estimated that Philadelphia, sitting at the center of America's largest flour-production industry, could only supply twenty thousand barrels of flour. Armies also required large numbers of horses and carts. One estimate from 1778 suggested that the American army alone needed ten thousand horses, all of which had to be fed.[74] None of these estimates account for the simultaneous demand created by the British. Nor do they convey the seemingly insoluble problems of actually moving all that food to the army.[75] General Nathanael Greene, appointed quartermaster general during the Valley Forge winter, was so desperate for forage for the army's animals that he ordered his subordinates "to forage the country naked." Anticipating that locals would complain that they had no forage left for their own animals, Greene coldly recommended solving their problem by taking "all their Cattle, Sheep and Horses for the use of the Army."[76]

It was in this context of need and organizational weakness that Washington's army marched into Valley Forge in the winter of 1777–78. There capacity again confronted control as we saw in Part 1 and Part 2: unsupplied and unpaid soldiers only too naturally turned to the countryside to fill their needs, and their officers were unable or unwilling to control them. The difference now was that although Washington could barely muster the resources to feed his army, he was determined, as a matter of strategic calculation, to control his soldiers' behavior. In this new era, and in this context of a brothers' war, the Americans were surprisingly successful at containing the violence of their unpaid, ill-fed, and ill-clothed army. The culture of discipline developed over the preceding century, though far from perfect, provided Washington the tools he needed to control his men, who otherwise were experiencing a kind of wartime life more familiar to soldiers of the sixteenth or seventeenth century in terms of supply and pay. The great irony of Washington's position was that he had greater strategic need to protect the countryside from his own troops, but he had to do so without the one tool that most European states had finally begun reliably to have: pay.

Controlling the soldiers started with regulating how the army gathered supplies. Washington began the war determined to avoid the outright impressment of supplies, and historian Don Higginbotham has argued that "even when impressing, the supply officers under Washington's immediate control made every effort to obey state laws on the subject."[77] In the wake of Brandywine, Washington worried that forcibly confiscating private arms from the citizenry might create resentment, although at the same time he authorized the impressment of clothes, carefully specifying that they be paid for or receipts provided.[78] In general the Continental leadership observed this latter requirement, and although Continental money grew worthless, they usually did provide receipts or certificates.[79] Continental officers also helped prevent abuses by their more consistent practice of sending quartering officers ahead of troops. The warnings they provided allowed the town sufficient time to gather food.[80] The opportunity to prepare and the tendering of receipts may seem small comfort, but most inhabitants recognized and appreciated the difference between soldiers "carving for themselves" and the regulated process of purchase or impressment.[81]

As always, setting the rules for gathering food was one thing; actually controlling the troops was another. From the outset the Continental army promulgated articles of war closely modeled on the British articles of 1765. Unlike the British, the American articles initially limited the number of lashes to the biblically approved thirty-nine, but Washington soon believed that number insufficient and convinced Congress to raise the limit to one hundred. As the war wound on, the number of lashes increased along with the general severity of the articles—especially their severity against plundering and other crimes.[82] In an intriguing lapse, the original articles omitted one of the classic antidevastation clauses that had appeared in almost every preceding set of English articles, but it was reinstated in the 1776 version, dictating that:

> All officers and soldiers are to behave themselves orderly in quarters, and on their march; and. . . shall [not] commit any waste or spoil, . . . or. . . maliciously destroy any property whatsoever belonging to the good people of the United States, unless by order of the then commander in chief.[83]

Regulations notwithstanding, there is no doubt that casual plundering was sometimes a problem, and Washington determined to stamp it out. Continental army order books are quite literally filled with warnings to the soldiers not to plunder, accompanied by dire threats of punishment.[84] In one of Washington's more famous imprecations against plunder he thundered:

> Why did we assemble in Arms, was it not to defend the property of our Countrymen? Shall we then to our Eternal shame & Reproach be the first to Pillage & Destroy it. . . . The Commander in Chief most solemnly assures all, that He will have no mercy on Offenders against these orders—their Lives shall pay the forfeit of their crimes.[85]

Those same order books record the regular infliction of such punishment, including flogging and execution. Unlike the dearth of records from the English civil war, for the Continental army court-martial records are exceedingly common, and they show that punishments were severe.[86] The Continentals were never perfect, but one cannot doubt that Washington's disciplinary efforts greatly reduced the damage they inflicted on the countryside.

In general, as with the articles of war, Washington took his model for this kind of discipline from European precedent and sometimes even from the specific advice of the foreign officers in his service. In the month after Germantown one of his German officers recommended, mangling the French term, creating a "Core of Marishosy." A maréchaussée corps, he suggested, was used by "every power in Europe" as a kind of special mounted military police. Their first role was to control the "Damage done by the Soldiers of the Army in Stealing or Destroying the publick Effects," but they were also to take up stragglers, control the sutlers, and help maintain march discipline in general.[87] Washington duly created the corps, and he enacted a host of other institutional measures designed to enhance his control over the army.

Many if not most of those measures were directly related to his belief that control and discipline were necessary to a successful army, but there was an element of the process that reflected more than mere calculation. The third major principle of Washington's strategy for the war, after maintaining an army and cultivating the loyalty of the countryside, was to fight the war with an army that *looked* right.[88] For Washington all these things went together. Discipline in the camp or on the march meant discipline on the battlefield. It was to him, as he said in the quotation that opened this chapter, like the "working of a Clock," which was "easily disorder'd" by "neglect in any one part, like the stoping of a Wheel disorders the whole." Synchronized collective discipline indeed.

Much of the Valley Forge winter was spent simply surviving, but as the weather grew warm in the spring, Washington's mind turned toward the next year's campaign. As part of his preparations he embarked on a new and ambitious plan to make his army's "clock" run more smoothly. He sought to make it even more conventional and, in his mind, thereby more effective. This was the setting for Valley Forge's other famous story. The pretended and infamously profane "baron," Friedrich von Steuben, arrived at the camp on February 23, 1778, and soon convinced Washington to let him retrain the army. He standardized and systematized the marching and firing procedures of the army, beginning with a model company who in turn went on to train others. Steuben's system made some concessions to the troops' rawness and to their resistance to the full panoply of European discipline, but there is little doubt about the overall result—in effectiveness and in appearance.[89] The seventeenth-century emphasis on synchronized collective discipline for infantry armies firing volleys of muskets in line was born of perceived necessity—a response to changing technology and the expansion of the army to include the socially inferior. But by the third quarter of the eighteenth century, it had become accepted as the way an army should *look*. Washington sought international respectability, and to achieve it he had to win, but the army also had to look right.[90] It was this kind of expectation that led Washington to banish the women (and the camp kettles) from the march column as the men paraded through Philadelphia prior to Brandywine.[91] He would even make sure that his men shaved and powdered their hair with flour before marching into the siege lines around Yorktown in 1781.[92] More recent veterans will recognize a parallel in the modern assumption of a connection between battlefield performance and highly polished boots. As historian John Shy put it, "Washington and other native American leaders stressed a regular army . . . because they felt a need to be seen as cultivated, honorable, respectable men, not savages leading other savages in a howling wilderness."[93] The victory at Saratoga proved key to gaining the French alliance, but the French army would not fight the war on its own; it needed a recognizable and respectable American equivalent to fight alongside.

Washington's preference for a conventional army, and his personal commitment to the Enlightenment/aristocratic culture of war, profoundly affected the nature of violence in his war with the British. Adopting the forms of the eighteenth-century European army brought in a host of structures that tended to limit violence. All the mechanisms established by eighteenth-century armies to control the rate of desertion also served to keep soldiers' movements more carefully controlled. Camp gate guards, passes, patrols to scoop up stragglers, and the like cut down on the number of armed men out seeking food or plunder. And, as suggested by John McCasland's party of sixteen men, Washington even sought to structure his foraging parties in accordance with an established norm designed to limit random destruction or plunder.

Washington also followed British practice in his effort to limit the number of women authorized to accompany the army. It is important to remember that there are always two kinds of plundering of private property: official, for supply purposes or as part of a deliberate campaign of devastation, and unofficial. As Joseph Townsend had complained, much of the disorderly, unofficial plunder associated with the British army was conducted by the "soldiery *and rabble* who accompanied them."[94] This was an old complaint. Women and other camp followers (usually servants) were often described as the worst plunderers; in Part 2 we looked at the role of women in the "pillage economy" of the seventeenth century. The horseboys in Ireland had a similar reputation. Seemingly in reaction to this kind of problem, eighteenth-century European governments reduced the number of women they allowed to accompany their armies, and they also brought the women more fully under military discipline—something made easier by the institution of regular pay. The British army arrived at a usual ratio of four to six women per hundred men.[95] Even this small number of women, when supplemented by Loyalist women trailing their units, and (worse, from the white American point of view) by slaves following the army in hopes of eventual freedom, generated loud complaints about their behavior.[96] On the American side, Washington spent a surprising amount of time on the role of women in his army. He did not like the additional logistical load they entailed, or their impact on the martial appearance and the march readiness of his army. He regularly reminded his officers to keep the women out of the line of march and off the wagons, and he banished them from the public army parade through Philadelphia. Eventually Washington also tried to fix a ratio of authorized women to men, arriving at one for every fifteen men, very similar to the upper end of the official British ratio of four to six women to one hundred men.[97]

Washington's reliance on British army practices included adhering to the "customs and usages" of war discussed in previous chapters, now even more carefully elaborated and more broadly applied. Those codes mandated a minimal respect for prisoners, allowed for the paroling of captured officers, set the norms for the negotiated surrender of a besieged town, and demanded respect for flags and messengers of truce. Sometimes the formalities and punctilios could become obstacles, as when the British and Americans repeatedly failed to reach a standard agreement on the exchange of prisoners. On the other hand, communication between the two sides was regular and reliable. During the Valley Forge encampment alone, Washington sent at least six "flags," official catch-all deputations, to British headquarters, and even briefly established Germantown as a truce town where the two sides could meet to converse.[98] This kind of communication was crucial to containing escalatory retaliation. When Washington or the British commander felt that the other side had violated a basic humanitarian principle, they used their channels of communication to threaten an equivalent retaliation, usually leading the other side to back down. Perhaps the single most important factor in containing the frightfulness of the war, as conducted by the regular armies, was this ability to effectively *threaten* retaliation, without having to resort to it.[99]

The Honor of Soldiers

For the most part this chapter has examined Washington's personal preference for, and experience with, the eighteenth-century British practice of war, and in particular his sense that following that practice had strategic value in winning international respectability and domestic support. There is little doubt about the extent to which Washington put his stamp on the character of the army. Furthermore, his strategic choices (made in council with his senior officers, often under pressure from the Congress) set the boundaries for the violence his army waged. He avoided deliberate devastation around Philadelphia, and he applied martial law in a very limited way. On the other hand, as we have seen in previous chapters, the nature of war almost always meant that junior leaders and common soldiers retained a tremendous freedom of action through which they could drastically reshape the frightfulness of war. The European culture of discipline increasingly aimed to limit their opportunities to do so unless ordered, but the daily process of moving and foraging created countless opportunities for soldiers and junior officers to make their own decisions outside the moderating force of the army hierarchy.

Many of the junior officers found themselves in an unusual and unexpected situation—they were authorities in a standing army, supposedly one of the most reviled institutions in the colonial American ideological universe. Furthermore, they were fighting in a republican cause as part of an institution that normally awarded officership based on social status. In truth, the officers of Washington's army behaved in ways that explode old assumptions about either their republicanism or their higher social status. They readily accepted their commissions; they greedily pursued promotion; they were jealous of any slight to their status, and they were touchy about their positions relative to each other. Many of the men came from artisanal or middling backgrounds, to the point of inspiring British ridicule. One sarcastic observer remarked that the "Col. of New York Regt. is a Butcher & the Major a Chimney Sweeper. Quelle Canaille!"[100] One lieutenant was even disdained by soldiers in his own unit, who claimed he was "not worthy of wearing rufled shirts [and] that he was a Servant before he had his Commission."[101]

This sort of criticism made the officer corps sensitive to slights to their honor.[102] That sensitivity, however, had a positive effect on their desire to enforce the codes of war. The whole complex of quarter, paroles, prisoner exchanges, and so forth had arisen from the notions of mutual self-interest in an international aristocratic officer class. The Continental army as a whole observed those rules in part from calculation; observing their forms helped them gain international respectability and, hopefully, an ally. But the codes had also acquired moral force regardless of calculations, through the medium of personal honor. To violate one of the codes was to stain one's honor, and Continental officers were particularly touchy about their honor. It was perhaps this sensitivity to honor combined with a certain naïveté that led the Continental colonel described in the opening of this chapter to spring his strangely honorable ambush.

Individual soldiers had their own sense of honor, restraint, and simple morality as well. McCasland's unwillingness to kill the Hessian sentry speaks for itself. The soldiers' attitude toward plunder and other forms of violence directed at civilians is harder to tease out and must have been highly variable. Nevertheless, a substantial body of evidence suggests that the Continentals, in this brothers' war and in a new era of collective discipline, worried a great deal more about the limits on their behavior than their brethren of an earlier century. Private Martin remembered one incident during his own personal foraging expedition of 1777, when he found himself among a flock of "geese, turkeys, ducks, and barn-door fowls." He recalled that he could "have taken as many as I pleased, but I took up one only."[103] Elijah Fisher, in what is generally a very laconic journal (six years of wartime service covered in twelve printed pages), dwells at length (almost two of the twelve pages) on four members of his unit who robbed a house while he was on leave. He detected their actions immediately by their change of clothes, and he said to one of them that he hoped the new clothes had not come from robbery. From Fisher's journal it is not at first clear if he was morally offended by their actions or merely afraid for his comrades' skins if they were caught. But he followed their fates carefully as they were caught and informed on each other; three were sentenced to die (two escaped), and one received one hundred lashes. What is quite clear from Fisher's reaction and interest is that such plundering was not the norm for him or for most of the men he knew.[104]

Sergeant Benjamin Gilbert's diary provides another perspective on what life and violence in the Continental army were like on a daily basis. Again, this is not to argue that the army was perfectly civilized, but consider the following: Gilbert made a diary entry nearly every day for two years and seven months between January 1778 and July 1780. It is a remarkably frank, if not loquacious, document, detailing the visits of women to his hutmates' beds, long nights of "keeping it up" drinking, getting "well Sprung," playing cards and games, and attending local dances. He kept meticulous track of his regular purchases of alcohol, and then made sly asides about the all-night consequences. One of his boon companions was finally caught sneaking his regular bedmate into camp. Both the man and the woman were court-martialed, and the woman was drummed out of camp. In many ways this was a manifestation of the libertine soldiers' culture of the aggregate-contract army of the sixteenth and early seventeenth centuries. But here it lurks within the supposedly disciplined army of the eighteenth century. The libertine qualities of a soldier's life had not died so much as gone underground. There, however, it inflicted much less damage on the countryside. Gilbert freely admitted to a full range of minor peccadilloes, and he kept careful track of punishments meted out to his and other nearby units. In all that time he admits to stealing only once—some peaches from a nearby farm. Furthermore, during one march, after a morning's hike of merely a mile, the officers searched the men's packs. A sergeant was reduced to the ranks for going into an orchard for apples, and two officers had their swords taken from them for "Gettin into a Turnip yard after Turnips."[105]

Beyond the everyday kind of morality or restraint we find in some soldiers' behavior, the men also developed their own compromises between the opposing pulls of an authoritarian culture of discipline and their participation in a supposedly republican cause. The men's humble social origins accustomed them to social discipline and undermined their ability to avoid the army's more corporal version thereof. Recruits were frequently those who lacked the political or financial influence to avoid service (or at least to limit their service to the less onerous militia). Where soldiers in the militia possessed some leverage over their officers—often local men whom they knew and dealt with regularly—Continental soldiers had no such influence. Most historians now agree that after 1777 many of the common soldiers came from the least powerful segments of society: escaped slaves, freedmen, vagrants, British deserters, Irish, and so on. Even the white men of English or Scots-Irish descent generally came from relatively poor backgrounds. Furthermore, many of them were substitutes or draftees, serving for monetary gain or because they could not afford to avoid the draft. The increasingly longer terms of service also created a more stable organization, in which the men were more likely to serve under the same officers for longer periods.[106]

On the other hand, those same long terms of service, with the physical suffering that they entailed, inspired a communal sense of righteousness among the soldiers of the army, who began to feel that they were the true carriers of the spirit of the Revolution. Admittedly, the army experienced a lot of turnover in soldiers, but it had existed as an institution since the summer of 1775 and included long-term recruits from late 1776. And the officers, generally more permanent presences, loudly connected cause to conduct by stressing that good behavior and success in battle would prove the virtue of the army and thus of the Revolution.[107]

While evidence for the ideological motivation of the common soldier is difficult to tease out, occasional comments or incidents reveal how soldiers may have come to see themselves as the heart of the Revolution, inspired by revolutionary principles.[108] One common soldier (later an officer) in the Massachusetts line composed an imaginary dialogue on the justness of the war, probably near the time of his first enlistment in 1777. He noted not only the atrocities of the British army but also the king's violation of his duty to his subjects: "Was not the king bound to us as much as we were to him to protect us in the free exercise of our Laws and liberties? [Instead he has] . . . entered our Country with fire and sword to reduce [us] to slavery[.] What then remains for us but manfully to take up the sword and defend ourselves and our injured country?"[109] On July 4, 1777, another Continental soldier wrote that "the eyes of all our countrymen are upon us, and we shall have their blessings and praises if we are the means of saving them. . . . Let us . . . shew the whole world, that freemen contending for liberty on their own ground are superiour to any slavish mercenary."[110] Such words from common soldiers, however, are rare. For most soldiers the evidence of their commitment lies in their remaining with the army at all. Many may have had few choices, and some may have stayed in fear of punishment, but the army's promises of food and pay were broken far more often than they were fulfilled.[111] Furthermore, even during the Continental mutinies of 1781 the soldiers spurned British advances, turned over British spies, and carefully restrained their violence toward civilians along the line of their march to protest to Congress.[112]

This is by no means an argument that Washington's men were perfect, either in their behavior or in their commitment to the cause. Although subject to the culture of discipline, and not lacking a belief in the revolution, they retained a freedom of action while struggling under appalling conditions. There is probably no more consistent complaint in Washington's orders than that his men were tearing up fences for firewood. Compared to Joseph Townsend's complaint about the British burning furniture and leaving the fences alone, this seems virtuous restraint indeed. Though to be sure, there were individual offenders. A dragoon private named Judah Gridley (or Gudley) made himself particularly notorious with his personal extortions around Washington's army in the fall of 1777, until he was finally trapped by three officers and convicted at a court-martial.[113]

Perhaps equally important for containing the frightfulness of the war, the Continental soldiers did not generally acquire a lasting desire for bloody revenge on British soldiers. They have been accused of massacring British soldiers trying to surrender at Cowpens in 1781 and, later that same year, a column of Loyalists in what became known as "Pyle's Hacking Match." There is little doubt that at Cowpens the Americans were indeed fired by anger at British commander Banastre Tarleton's recent killing of surrendering Continentals at Waxhaws. But despite the British surrendering in the midst of battle, six hundred men survived to become prisoners, while suffering one hundred killed and two hundred wounded. Some individual British units at Cowpens appear to have suffered a steeper killed-to-prisoner ratio, but Cowpens's overall ratio of 1:6 hardly deserves to be considered a massacre.[114] The destruction of Pyle's column can be interpreted in different ways, but it is not unreasonable to see it as an extremely successful and short-range ambush created by a confusion over uniforms. It was certainly not a massacre of men who were intent on surrendering.[115] Perhaps most tellingly, Armstrong Starkey has documented the remarkable restraint shown by the Continentals in their successful night bayonet assault on the British garrison at Stony Point, which he contrasts to the much bloodier version of the same kind of attack carried out by the British at Paoli.[116]

Conclusion

Stories of individual soldiers choosing restraint and contrasting stories of plunder and violence can be multiplied almost indefinitely. Weighing the two against each other is difficult because the usual sources are victims' recollections or complaints. Plunder is complained of; restraint rarely recorded. Only when looking at a sustained and apparently honest diary, like Elijah Fisher's or Sergeant Gilbert's, do we get a comparable sense of the *absence* of violence. That absence becomes even clearer when we contrast the events of this chapter with those narrated in the preceding chapters and the one that follows. The European disciplinary system had improved the ability of commanders to control the impact of an army, but only within limits, and, crucially, it was only true insofar as they *chose to try*. Deliberate devastation remained an option, and in this brothers' war the British all too often chose to treat the Americans as rebels. For them devastation was both a possible calculated choice and a cultural and legal temptation. For Washington, devastation was likely to harm his cause more than help it, and the ideology of the revolution demanded at least a veneer of virtue. For all their complaints about each other's behavior, the British and the American leadership communicated in a language of war with a shared grammar, and doing so restrained the frightfulness of the war. When the upward pressures of escalation threatened new levels of violence, the downward pressures of restraint usually emerged and were understood by both sides. In Washington's case, his finer sense of the political value of restraint kept the Continental army under better control than did the British army's more developed disciplinary mechanisms. For as many complaints as were lodged against the Continentals, there were also civilians who acknowledged their better discipline than either the British or the militias. One preacher in North Carolina reflected that the army was "first victorious over its enemies . . . and then over itself."[117] Starkey's comparison of the Continentals' restraint in their victory at Stony Point to the British massacre of sleeping or surrendering men at Paoli suggests the differences. An even better comparison is to the American militia. The militia's way of war, especially against the Loyalist militias, was decidedly less restrained, for a host of reasons that have absorbed the attention of a legion of historians.[118] But an even more striking comparison exists ready to hand: many of the same men and units who fought the British in Pennsylvania in 1777–78 marched in 1779 into the Iroquois country of northern New York. Notwithstanding their greater experience and more thoroughgoing discipline, during that campaign they wielded an entirely different kind of violence.

8

"Malice Enough in Our Hearts"

SULLIVAN AND THE IROQUOIS, 1779

*You defeated the savage army and conquered those barbarians who
had long been the dread of four frontiers. . . . Led by the consideration
of our just and complete conquest, of so fertile a part of the western
world, I will venture to look a few years into futurity. . . . I see all these
lands inhabited by the independent citizens of America.*

Thanksgiving sermon delivered to the army at
the end of the Iroquois campaign, October, 1779.

"Some Skinned by our Soldiers for Boots"

The men (and women) of the expedition had been literally and figuratively
struggling uphill for weeks.[1] For many if not most of them, the rugged, forested, and yet somehow also swampy uplands of Pennsylvania and New York
were a new world. They could see the potential in the land, and they could
even see where the people of the Iroquois confederacy had nurtured that
potential into abundance. Meadows, fields, and orchards proclaimed their
bounty, but the lurking Iroquois and their Loyalist allies also lent menace
to the landscape. Every river crossing might be contested. Every defile
defended. Every patrol, however minor or close to camp, ambushed. For the
most part, eighteenth-century campaign diaries are perfunctory lists of
dates, distances, camps, and food. In a striking contrast, the diaries from this
campaign into Indian territory vibrate with hope and fear. Hope for the
abundance of land they might add to the new nation. Fear of the "savage"
still abiding within it. One day they were "Incampt upon a Very Elegant
Piece of Land" or surrounded by "trees Exceedg large Some Measuring

14 foot 4 Inches thro," or in "Excellent pasture" in a place "pleasant & Well situated" on the river. The next they slept "on the Edge of the Hill by the Shades of Death" (an actual place name), or climbed a pass so "Steep that man & Beast was in danger of falling & Endangering his life or theirs." One day a dinner to celebrate the 4th of July; on another a soldier killed and scalped near the edge of the camp.[2]

No temporary militia, these men constituted four divisions of Washington's Continental army under the command of General John Sullivan (see p. 211). Many of them had fought in Pennsylvania in 1777 and 1778. Now in 1779, although marching into the wilderness, they brought with them all the military formalities so painfully learned in the preceding three years. Daily march orders were regular and precise. When marching into a civilian town, they cleaned themselves up for the parade. Men who deserted were rounded up, tried, and sentenced to death. Some were executed; some pardoned; some had to run the gauntlet, whipped along the way by their fellows. The officers persisted in their habits of squabbling over rank and honor. General Edward Hand believed that his suggested surprise raid was rejected from "jealousy" and was astonished that "men engaged in their country's cause would oppose salutary measures because the Honor of a brilliant action could not be immediately attributed [to] themselves." One of the chaplains used his time in camp to read (and enjoy) Jean-Jacques Burlamaqui's 1748 treatise *The Principles of Natural and Politic Law*, containing an extensive discussion of Grotius's work on the laws of war.[3] As always seemed to be the case for the Continentals, logistics were a problem, but here at least there was a river to help transport the supplies and no competing British regular army to contend for them. When they ran short near the beginning of the march, Sullivan simply camped and waited.

Sullivan's 2,500 Continentals began their long march in Easton, Pennsylvania, on June 18 (see map, p. 176). He paused for supplies for nearly five weeks in the Wyoming settlement, and paused again to await the arrival of a second column under Brig. Gen. James Clinton. Clinton had begun his march on the Mohawk River in New York with 1,500 more Continentals. The two columns converged south of the Finger Lakes, at the "door" to the center of Iroquois country. There they built a fort to secure their retreat and proceeded to march together into the Seneca and Cayuga country around the lakes named for those peoples, burning crops and destroying villages as they proceeded.

The Iroquois, spurred on by the Mohawk leader Joseph Brant, and the Tories commanded by John Butler, had long foreseen the army's approach.[4]

MAJOR GENERAL JOHN SULLIVAN. By Richard Morrell Staigg. By permission of Independence National Historical Park–Library.

Earlier that spring and well into the summer, they had launched raids intended to slow or divert the Continentals' approach. One of those raids devastated the settlement around Minisink in late July and then killed as many as forty pursuing militiamen.[5] The Indians and Tories were slower to confront Sullivan's or Clinton's much larger forces directly, but even so several men who strayed from the camp had been cut off and killed. The Iroquois also began to evacuate their vulnerable towns. Somehow this seemed to surprise the Continentals. Lieutenant William Barton of the New Jersey Brigade described Sullivan's attempt to make a surprise night march on the village of Chemung. After a long night of stumbling, occasionally moving only yards at a time, daylight found them short of their goal. They sprinted "a great part of the way," only to find "to our mortification the entire town evacuated." An hour later a small party of Indians sprung an ambush, killing six or seven soldiers and wounding more before melting back into the forest. The soldiers burned the town and the standing corn and returned to their camp to await Clinton's arrival.[6]

The united army set off along the Chemung River, prepared to continue their mission of destruction. On August 29, the first day of their resumed march, they encountered an Iroquois roadblock. The Indians, with the assistance of a very few British regulars and a somewhat larger group of Tory

JOSEPH BRANT. By Charles Willson Peale. By permission of Independence National Historical Park–Library.

rangers, perhaps six hundred men all together, lay astride the narrow road through the Indian village of Newtown, near the modern city of Elmira, New York. Approaching the fortifications blocking the road, Sullivan commenced a cannonade, while sending nearly half his troops on a flank march into the wooded hills above the enemy position. The combination of cannon fire and overwhelming numbers soon sent the Indians and their allies into a fighting retreat. Their retreat set the stage for a signature atrocity in the history of white-Indian warfare, an atrocity all the more marked by its apparent casualness.

As the Americans scoured the woods in the wake of the fleeing Indians, they found twelve dead and took two prisoners. They presumed that additional dead and wounded had been carried off by the Indians. The differing accounts of the dead, of the prisoners, and of their respective fates tell an interesting story of prejudice, blindness, and differential treatment. Virtually all accounts agree that the Americans took two prisoners. Most acknowledged that one prisoner was white (he apparently had feigned death), and the other a "negro." For some diarists the prisoners' race was irrelevant—they were merely "two tories." One noted that the white man was in fact a "Painted Tory." Had he dressed and fought like an Indian, or was he merely tarred by association? Under interrogation, both men delivered similar accounts of the size and composition of the now retreating Tory and Indian force.

The picture grows starker with the treatment of the Indian dead, but questions remain there too. One source specified eleven dead Indian men and one dead Indian woman. Another noted that the troops scalped eleven of the dead. One of the best witnesses, Lieutenant Barton, reported eleven or twelve Indian dead, all of whom, he said, the soldiers scalped. Sullivan's report merely listed twelve dead, one a woman. Were the Americans reluctant to scalp a woman? Or were they simply reluctant to admit to doing it? Or was it so common as not to be worth mentioning? Of thirty-one descriptions of the battle, only four mentioned that one of the dead was a woman (a detail surely not invented). Ten acknowledged the scalping of the Indian dead (claiming anywhere from five to twenty, but the real total was almost certainly twelve). By 1779 for Americans to scalp dead Indian warriors was no real surprise, but some men clearly had qualms about the scalping of women.

Others had fewer concerns and went much further. Ensign Daniel Gookin hinted darkly at worse matters in his journal, saying that the soldiers had taken "several Indian scalps &c &c &c." Three other diarists filled in those blanks, informing posterity that the soldiers had skinned the legs of two of the Indian dead, dressed the skin into leggings, and presented them to their officers. Lieutenant Rudolphus Van Hovenburgh of New York recorded the event, but decided to abbreviate it "Sm. Skn. by our S. fr. Bts." His diary is otherwise free of abbreviations. Was he ashamed? Embarrassed? Sgt. Thomas Roberts of New Jersey was more matter of fact. He described the incident without comment, while Lieutenant Barton seemed not to hesitate to accept the leggings as a gift from his men. One last sordid detail highlights the banality of these accounts. John Jones, a Continental deserter interviewed by the British in February, claimed that a major paid two soldiers a gallon of whiskey to make the leggings.[7]

After Newtown, the combined army proceeded deeper into the mountain and lake country, seeking their enemy, or more precisely, seeking Indian towns and fields to destroy. Although the landscape alternated between rugged and fertile as they approached the main Iroquois villages, the men were astonished at its productivity and potential as they came through the narrow defile after Newtown and emerged into the country around the lakes. At the sight of Seneca Lake, Lieutenant Van Hovenburgh called it "as pretty a Lake as ever I Beheld and most beautiful Land and it appeared to be very good Land on the other side of the Lake."[8] General Edward Hand called it the "most delightful I ever beheld." In this context, however, comments on fertility went hand in hand with assessments of destruction.

Hand went on: "The Quantity of corn & other vegetables destroyed Immense. The Towns Burn'd many & pretty." Hand immediately qualified that last judgment by elaborating, "considering the inhabitants."[9] Between early August and the end of September, Sullivan and Clinton destroyed thirty to forty Iroquois and allied towns, burned 160,000 bushels of corn, and girdled thousands of fruit trees. No significant groups of Iroquois were killed or captured—or even seen for that matter, other than at Newtown. They did find a couple of white captives, including Luke Swetland, taken by the Indians in their raid on the Wyoming Valley the year before. Throughout the expedition Sullivan lost approximately forty men to Indians, accident, and disease. While extraordinarily destructive of property, the expedition did not destroy the Iroquois' will to resist. Provisioned by the British at Fort Niagara, which Sullivan did not attack, the Indians vigorously resumed raiding the frontier in 1780, killing or capturing over three hundred Americans and destroying six forts and over seven hundred houses and barns.[10]

Its effectiveness aside, the Sullivan expedition was an entirely different kind of campaign than that waged by many of the same troops in the countryside around Philadelphia in 1777 and 1778. More frightful both in its level of destruction and its casual atrocity, it represented a different way of war. Prejudice, hatred, and a long history of frightful intercultural warfare can explain much of the violence of the Sullivan expedition. This was a "barbarian" war, and cultural assumptions played a weightier role than actual cultural knowledge in authorizing forms of violence against Indians that were qualitatively different than those considered appropriate for brothers (or even for more traditional "foreign" enemies). But other considerations mattered too. Logistics and conscious strategic calculation dramatically shaped the character of the campaign. Finally, the men's fixation on the productivity of the landscape hinted at the way long-term competition for land, without an honest vision for incorporation of the conquered, could generate violence without regard for postwar political reconciliation.

To make sense of all these details, from culture to logistics to strategy, would require a full history of English-Indian relations back to Ralph Lane and Wingina/Pemisapan in 1585–86. Part 3 covered much of that ground from an Indian perspective and hinted at the extent of possible cultural conflict. For much of the rest we can turn to the Anglo-American debate at the outbreak of the war over the use of Indians as allies; to the Indians' actions in 1778, which spurred Washington to order the Sullivan campaign; and to further details of Sullivan's expedition itself.

Indians as Allies

To Europeans, Indians had long been both enemies and allies. Their material needs and the traditional enmities between indigenous peoples had always allowed Europeans to convince one group to aid them against another, or even against other Europeans. It is not a stretch to say that every war fought in the colonial era included Indians on both sides. By the time of the French and Indian War of 1754–1763, however, the white demographic explosion in the British colonies had made the colonies much less dependent on Indian aid than they had been in the preceding decades. In contrast, the heavily outnumbered French had become increasingly dependent on their own Indian allies—a fact that stirred controversy even within the French ranks. As for the English colonies, the less reliant they became on Indian aid, the easier they found it to decry Indian tactics. And those tactics were devastatingly effective all along the frontiers of white settlement, especially in the northeast and stretching down into Virginia. Indian raids were terrifying, frequent, and unpredictable. Worse, local white militias tended to retaliate on the most vulnerable Indian populations—those trying to remain neutral, and even Christianized Indians living in mission towns near the frontier zone.[11]

At the close of that war, when the British government absorbed both the former French territory and the former French allies, it was difficult to persuade the colonists that they could all be subjects to the same king. To be sure, the Indians were not yet to be full subjects; they lacked the essential trappings of civility. But they were expected to be obedient. Many Indian groups had their own interpretation of what the defeat of the French meant, and were unprepared to accept the British understanding of its meaning. It is not even clear if the British themselves were entirely sure what status the Indians would have in the newly enlarged empire. Clashing interpretations among British administrators and Indians generated a new and equally devastating Indian war. Known as Pontiac's War, it encompassed many more groups than those directly inspired by the Ottawa leader Pontiac. The British government quickly identified colonist-Indian tension over land as a major cause of the war. Their efforts to pacify the frontier included establishing the 1763 Proclamation Line, nominally forbidding white settlement west of the Appalachian Mountains. Colonists across the social spectrum decried the line, from the poorer farmers already encroaching across the mountains in their hosts of small farms to the vastly ambitious and wealthy land speculators who had hoped the war would free up their title to hundreds of thousands of acres in the Ohio country.[12]

This built-up colonist-Indian hostility, rooted in years or even decades of mutually violent warfare, put the British government in the unusual position of being a relatively pro-Indian mediator. When the American war for independence broke out, both sides assumed that the British had the upper hand in the quest for Indian allies. But that quest was fraught with controversy. British generals looked for Indian allies from the very beginning. As early as September 1774 General Thomas Gage in Boston wrote to his counterpart in Quebec to ask if a force of Indians and French Canadians could be collected "for the Service in this Country, should matters come to extremities."[13] And come to extremities they did. Yet although some British officials recognized the inflammatory potential of appealing for Indian aid, others greatly valued its utility. There is little doubt that correspondence like Gage's, rumors of Cherokee attacks in the summer of 1775, and actual Cherokee raids in the summer of 1776 raised the ire of large segments of the colonial population and confirmed them as rebels.[14] Britain's efforts to rouse the Indians became one of the grievances listed in the Declaration of Independence. Despite the controversial nature of using Indians as allies, however, as the war gained steam both sides actively sought them out.

Some few Indian peoples chose to side with the Americans. The Stockbridge Indians of New England stalwartly provided a small company of men who served the Continentals as scouts. The Catawbas on the North Carolina/South Carolina border also fought on the rebel side.[15] Most significantly for the northeast region, the Six Nations Confederacy of the Iroquois split. At the outset of the war the confederacy struggled to remain neutral, but eventually the Senecas, Mohawks, Cayugas, and Onondagas sided with the British, while most of the Oneidas and the Tuscaroras took the part of the Americans.[16] This alliance did not stop the soldiers in Sullivan's army from voicing their contempt for their allies. The commander of the 2nd New Jersey Regiment expressed dismay at this, threatening to punish soldiers who had spoken "Contemptously and Ridicule[d] the Indians who have Come to join Us." On an earlier campaign General Hand had seen the possibilities for Indian aid destroyed by the indiscriminate violence of frontiersmen.[17] Such strong and open dislike for Indians as allies boded ill for those Indians that the men regarded as enemies. Their attitude in part derived from long conditioning to view all Indians as "savage," but it had been reinforced by what had happened at Cherry Valley and Wyoming the previous year. Or, at least, what was said to have happened.

Cherry Valley and Wyoming, 1778

The British were far more successful at gaining Indian allies.[18] As the British army was evacuating Philadelphia in the summer of 1778, their agents in the north successfully persuaded some of the Iroquois to join with Tory rangers led by Maj. John Butler, a Connecticut native, to raid the frontiers of New York and Pennsylvania. A key player in this recruitment was the Mohawk leader, Joseph Brant. Brant spoke excellent English, had attended Ebenezer Wheelock's school for Indians, had fought for the British during the French and Indian War, and had even traveled to Britain to attempt to secure land guarantees for his people. Brant's sister Molly had married the famous Sir William Johnson, the British Indian agent and adopted Mohawk, who died in 1774. Brant's loyalties were entirely with his people, but when it came to choosing whether the British or the Americans were more likely to help them, his choice was easy. His efforts, combined with intense British diplomacy, pulled four of the Iroquois nations into the war on the British side. Brant and Butler made a formidable team, and they began active operations on behalf of the British in 1777.

White settlers in New York responded to the Iroquois' open hostility with a series of uncoordinated but devastating plundering raids on Mohawk land and property. Much like Edward Waterhouse in 1622, who cited the Powhatans' attack as a justification for a full conquest of their territory, settlers in New York in 1777 acted as if they had been released from constraint. To make matters worse, to the southeast of the Iroquois heartland lay the Wyoming Valley of the Susquehanna River, the scene of twenty years of violent struggle among Connecticut settlers, white Pennsylvanians violently resisting the wave of New Englanders, and the original Delaware Indians. White-Indian and white-white struggle along the Susquehanna had left a bitter and ongoing legacy of conflict throughout the region, and the Revolution provided new opportunities to fight old battles. To the Iroquois the expanding white settlement along the river pointed like an arrow into the Iroquois heartland further upriver. The controversial British-Iroquois treaty line established in 1768 and further refined in 1773 seemed unlikely to contain white expansion, and the ongoing dispute between the Connecticut and Pennsylvania settlers provided a stimulus to some of the Loyalists who joined the Iroquois in 1778.[19]

One of the Connecticut settlers who arrived to swell the white population in Wyoming in 1776 was Luke Swetland. Swetland and his fellow Connecticut settlers in the Wyoming Valley were the targets of Butler's and Brant's planning for 1778.[20] Faced with provocations to the northeast and the southeast, and in the absence of a large British campaign, Butler and Brant determined to raid the frontier, hoping to "make [it] . . . a no-man's land and deprive the state and the Continental army of one of the richest granaries in the colonies."[21] The most notorious of their raids was the July 1778 attack led by Butler and supported by four hundred to eight hundred Indians, mostly Senecas, against settlements in the Wyoming Valley near what is now Wilkes-Barre, Pennsylvania. On July 3 the Tories and Indians soundly defeated a militia and Continental force in open battle and then killed many of the fleeing soldiers. Butler offered terms to the surviving troops and civilians in a nearby fort and then successfully protected their lives after the surrender. He was not able to protect their property. In the days after the battle the Indians and Tories harried the inhabitants out of the region. Butler reported that his force burned some one thousand homes in the wake of the surrender. American reports of the attack were a good deal more colorful, if mostly fabrications. The battle soon became known as a massacre, perpetrated by a merciless Butler leading indiscriminate Indians who killed all before them and tortured their prisoners. The reality was terrifying enough, but the propaganda and rumors of even greater frightfulness helped empty the valley.[22]

Luke Swetland experienced that reality, and later in life he produced a somewhat more honest account of his captivity. Taken in the weeks after the battle, he was treated as a captive in the customary Indian manner. He was bound, brought along on the return march to the Iroquois villages, beaten on several occasions as part of a process of grieving and exulting, and then finally adopted into a family as a new "grandson." From that point on he lived as well or ill as his new family did, until the approach of Sullivan's army offered him an opportunity to escape. That opportunity was made possible by the level of trust with which he was treated. Ironically, when he escaped to the army, even after some initial confusion over whether he was a freed Indian captive or a just-captured Tory, the army's officers and men considered him tainted by having been an Indian captive and held him a close prisoner for some time, even while he provided intelligence and guidance to the army.[23]

In September 1778 Brant led a raid on German Flats, destroying the whole settlement around Fort Herkimer and Fort Dayton—although virtually all of the residents successfully found protection inside the forts. Then in November Butler and Brant combined forces in an attack on Cherry Valley. Before the attack they persuaded the Indians in council to spare women and children, but the attack quickly spiraled out of control, and neither man could stop the killings that followed. The Indians afterwards claimed to be angry, both at having falsely been accused of atrocities at Wyoming and also at an American force that had surrendered, promising not to fight again, and then returned to the war. The Indians complained to Butler that "they would no more be falsely accused, or fight the Enemy twice."[24] Small American forces tried to retaliate during the ensuing months, but their retaliation succeeded only in bringing down yet more destructive raids by the Indians.

Reports of those raids, and particularly of the flight of inhabitants from productive fields needed by the Continental army, led Washington to address the problem.[25] Washington needed to stop the raids, restore stability to the region, and renew the locals' faith in his army's ability to defend them. He detached General John Sullivan and a significant slice of the Continental Army on a campaign against the Iroquois homeland. Of particular interest is that this army, unusually, received almost no local militia support. This greatly simplifies the question of "laying blame" for atrocities: these were virtually all regular troops.

The Clinton-Sullivan Expedition

The Sullivan campaign is best understood if we begin not with cultural prejudice but by analyzing the capacity of the American army to do damage and the calculations that determined their strategy. To begin with capacity, the level of possible destruction was much higher than it had been in the Philadelphia campaign (or in most other Continental operations) for two simple reasons. First, the enemy homeland and population was accessible. Second, the total population of the Iroquois relative even to the limited size of the Continental army theoretically put the Indians at serious risk of destruction. Washington's sources suggested that a maximum of 2,050 men (Indians, British, and Loyalists) would oppose the campaign, defending a total Iroquois population I estimate at 6,400.[26] To confront them, Washington initially proposed sending six thousand men, although he eventually dispatched only four thousand.[27] It is rare in European military history to see such a skewed ratio of invading army to invaded population. That ratio created the potential for very high levels of destruction, although it did not foreordain them.

Despite this army-to-population ratio, the army's capacity to inflict violence was limited by its inability to surprise a society that had long fought warfare based on surprise. Strategically, the Iroquois had plenty of warning of the expedition, and tactically, the Continentals repeatedly tried and failed to surprise populated villages. The failure at Chemung has already been described, and Continental surgeon Jabez Campfield described a similar scene when the army entered Conadasego: "The Indians had deserted the place some short time before our arrival. It seems we are not to see any more of these people. It was expected they would have made a great stand at this place. . . . It is difficult to account for the conduct of the Indians, who quit their towns, & suffer us to destroy them, their corn, their only certain stock of provisions, without offering to interrupt us."[28] Accounts of the expedition and the raids immediately prior to it document at least ten separate occasions when American troops made a concerted effort to surprise a village they believed was still occupied. In all but one of them they failed completely. Colonel Goose Van Schaick's attack on the Onondaga towns in April had succeeded. His men took thirty-three or thirty-four prisoners, mostly women, and killed twelve or thirteen.[29] This inability to achieve surprise drastically limited the total possible human damage that the Continentals could do and shaped their calculation of how to win.

Time and the army's size (despite its being nearly as large as the Iroquois population) also limited the Continentals' capacity to wreak human and property damage. Strategic priorities limited the time available; the British continued to be the main opponent, and Washington could ill afford to keep so many troops deep in the interior for too long. Furthermore, it was possible that the British might intervene or take advantage elsewhere once they realized just how much force Washington had committed to the Iroquois campaign.[30] Finally, once destroying crops became the goal of the operation, that also limited the optimal time for the campaign. General Nathanael Greene advised that the campaign begin in mid-June when the corn could be caught half-grown. As it was, Sullivan did not get into the Iroquois cornfields until August.[31]

Other strategic and logistical considerations further limited the size of the force and the time available. Advancing so deeply beyond the frontier required Sullivan to drop off detachments along the route to secure his communications, and only so many troops could be fed in such an environment.[32] As it was, the expedition severely strained Continental logistics, and the army ultimately had to live off the land for much of the march. Pre-campaign planning focused on how far upcountry the rivers could convey the column's logistics and artillery. Would the rivers retain enough water late in the summer? What time of year would the Indians' corn be most vulnerable?[33] After the pause to collect stores at Wyoming, Sullivan's column was accompanied by 120 boats, in addition to the packhorses and live cattle, which were regularly lost or damaged in the march up country.[34] Delays and losses meant that once the two columns united and prepared to move into the heart of the Senecas' country, the army had "only 27 days Provision & Scarcely Horses to carry it—our force or rather number of mouths is almost double what it should be," leading Hand to conclude that "our Campain must therefore soon end." General Hand hoped for an early confrontation with the Indians as the only way to prevent the campaign from becoming a laughing stock.[35] Quartermaster sergeant Moses Sproule recorded the preparation of special bags for the packhorses, the oversetting of wagons, and "roads extraordinary bad" that cost them horses. After the fight at Newtown, Sullivan put the army on half rations of beef and bread, and promised to take advantage of the ripening vegetables in the Indians' fields.[36] In the end, logistical considerations cut short the campaign and sent Sullivan and his troops back to Pennsylvania.[37]

What about Washington's calculations? Within the larger goal of protecting the army's bread basket, Washington's immediate goal was to stop the Iroquois raids.[38] Washingon's eventual orders to Sullivan, after extensive discussion and consultation with Generals Philip Schuyler, Greene, and Sullivan himself, were twofold: "The immediate objects [of the expedition] are the total destruction and devastation of their settlements and the capture of as many prisoners of every age and sex as possible. It will be essential to ruin their crops now in the ground and prevent their planting more." Washington went on to suppose that these actions would create a "disposition for peace," but he directed that overtures for peace be ignored until "the total ruin of their settlements is effected—It is likely enough their fears[,] if they are unable to oppose us, will compel them to offers of peace." Peace would then be maintained through hostages, which were "the only kind of security to be depended on."[39]

Where did this thinking come from? Why were devastation and hostages perceived as the way to win—the calculation of military necessity? Note that the problem is not entirely unlike the one Washington had faced in the Philadelphia campaign: preventing an enemy from devastating the countryside. In part this decision to devastate and imprison was a result of capacity. The Iroquois homeland was accessible and could be made a desert. This calculation lay at the heart of Schuyler's advice: "Destroy the Seneca towns and the Indians must fall back to Niagra . . . This is a long distance from the frontier. With no intermittent place to use as a supply base, no sizable body of Indians can raid the frontiers through the winter and into the spring."[40]

But the real assumption underlying this calculation was the now-old expectation that regular troops would not be able to surprise Indians and destroy them in battle. Schuyler and Washington reassured each other of this belief, although they continued to hope for hostages. Schuyler first wrote to Washington: "Much as I wish that the Onondagas & Cayugas should be surprised because I esteem the having their familys in our possession as an almost certain means of bringing the whole confederacy to proper terms, yet, I confess I have many doubts whether matters can be conducted so secretly when a large body of troops moves as to give reasonable hopes of success."[41] Washington concurred with that opinion, writing back just a few days later: "I should esteem it difficult to effect a surprize upon an enemy so vigilant and desultory as the Indians."[42] Such expectations proved well founded. Surprise was difficult, if not impossible, and some of the soldiers on the expedition even began to believe that starving the Iroquois was unfeasible. Major Jeremiah Fogg expressed his frustration at being unable to catch any of the enemy Indians or convince them to be friendly, and he went on to opine that "to starve them is equally impracticable for they feed on air and drink the morning dew."[43]

Major Fogg's doubts notwithstanding, by the eighteenth century the "feed fight," the deliberate destruction of crops and villages, was the Americans' presumptive strategy against Indians, not merely an expedient borne of frustration. One historian has identified this strategy of devastation as one component of an American way of "extirpative" war.[44] The violent consequences of such a strategy could be dramatic. Devastation as a deliberate choice was not only indiscriminate in its treatment of the land and the property within it, it also opened doors for individual soldiers to make more drastic choices about the people and property they encountered. As a technique it fit within a European tradition of war already seen in Ireland in the sixteenth century, and in some foreign wars on the continent up into the eighteenth century.[45] This was a strategy the English had avoided employing among themselves from 1642–46, but which they willingly turned on the Irish during that same period. If it was seen as necessary, the resources of the enemy could be considered legitimate targets, and all the codifiers of international law admitted the iron law of necessity, however much they then tried to hedge it. The strategic choice to devastate and take hostages opened a wider scope for a frightful style of war, but cultural considerations contributed to the resulting violence as well.

A crucial cultural variable was the desire for, and belief in, retaliation. The legitimacy of retaliation was widely accepted within American society: If an enemy act exceeded certain bounds, then retaliation was equally authorized to exceed those bounds. In European notions of war, the customary right of retaliation had been codified explicitly as the only means of enforcing expectations of behavior between enemies. Emmer de Vattel acknowledged in his treatment of the laws of war that retaliation should be avoided, but "when we are at war with a savage nation, who observe no rules, and never give quarter, we may punish them in the persons of any of their people whom we take."[46] Vattel thus acknowledged a "law of retaliation," a phrase frequently used by eighteenth-century combatants. In their minds retaliation was not only a human urge but a right, practically, if not actually, embedded in written English law. This legalistic view of retaliation blended with and reinforced the more popular ideology of an individual's right of self-redress—the right "to make oneself whole" in response to injury or affront.[47] This expansive and quasi-legal vision of retaliation played a crucial role in the minds of individual soldiers. It transformed acts usually seen as immoral into legitimate acts of war, clearing consciences and creating the freedom to be frightful. What was more, the popular vision of the right of retaliation was not confined to tit-for-tat equivalence. This popular vision had some standing in contemporary interpretations of international law. Writing at the end of the seventeenth century, jurist Samuel von Pufendorf argued that it was not "always unjust to return a greater evil for a less, for the objection made by some that retribution should be rendered in proportion to the injury, is true only of civil tribunals, where punishments are meted out by superiors. But the evils inflicted by right of war have properly no relation to punishments." Military practice, as exemplified by Washington in his dealings with the British high command, had shifted to favoring one-for-one equivalency in their threats of retaliation, but militiamen on both sides lived by the older code and regularly threatened to hang six or more for one.[48]

As suggested in Part III, there were so many ways in which Indian and English notions of war differed that it was almost inevitable that one side or the other would perceive the others' wartime acts as opening the door to a war of retaliation. For the Indians such a war was perfectly normal, but its built-in restraints normally prevented it from escalating in scale. For Europeans, to call a conflict a war of retaliation opened a frighteningly wide range of violent options. By this time Americans virtually assumed that a war with Indians would be waged as a war of retaliation. In the specific context of the Sullivan campaign, many of the participants believed they had good reason to wage such a war. Rightly or wrongly, the public perceived the Indian and Tory attack on Wyoming in July as a massacre, and the November 1778 attack at Cherry Valley, New York, had in fact been quite vicious.[49] Much of the army probably lacked a direct personal connection to those events, but there were regiments from New York and Pennsylvania who may have had connections to Wyoming or Cherry Valley, and a few were closely tied. Two small militia units, no larger than a company, came from Wyoming. Others may have been like Obadiah Gore, a lieutenant in the Continentals whose family had been among the Connecticut settlers in Wyoming. His three brothers and two brothers-in-law were killed in the Wyoming battle and all their farms, as well as Gore's, burned.[50] Even soldiers who lacked that intimate experience called the expedition "a just & speedy retaliation for British & savage barbarity."[51]

Furthermore, a number of incidents or reminders on the march cemented the Continental soldiers' urge to retaliate and defined the nature of the war they saw themselves fighting. When the army arrived at Wyoming, the site of the previous year's battle, they were shown around the massacre site, where they saw unburied bodies and scalped skulls. Lieutenant Daniel Clapp wrote that the place "might be called a Golgotha[;] we here found the Bones of Many whose flesh had [been] consumed above ground indeed there was scarcely a day but there was brought to our view scull bones and other bones."[52] In a more personal affront to the men of the army, a few days after the Newtown fight, the Iroquois caught, overwhelmed, and killed most of a Continental patrol of twenty-nine men, including two allied Indians, under the command of Lieutenant Thomas Boyd. The Indians took Boyd and Sergeant Michael Parker alive from the battle site back to another village; there they tortured and killed them. Their heavily mutilated bodies were left for the rest of the army to see. Almost every diarist carefully detailed the extent of the damage. Sergeant Moses Sproule lost track of his own sentences as he described how they had been "mangled in a most Inhuman & Barbarous Manner, having pulld their nails out by the root, tied them [to] trees & whipd them with prickly ash whilst tied to the trees, threw dart at them, Stabd them with spears, cut out their tongues cut off their heads & cut out their Tongues &c." To make matters even worse, after the Indian torture, dogs "had knawd their bodys till they were shocking to behold."[53]

There is little doubt that the men of the army reacted passionately to Boyd's death and the treatment of his and Parker's bodies. But Boyd was killed well after atrocities committed by Continental troops, most notably after the skinning of the legs of two Indian dead at Newtown. The legitimacy of direct retaliation, atrocity for atrocity, is an insufficient explanation for the frightfulness of the campaign as a whole, or even for the personal and individual violence meted out by Continental soldiers. We have to look deeper.

Essentially, by the middle of the eighteenth century at the latest, white American society had come to presumptively "authorize" a level of wartime violence against Indians that it simply did not sanction against European enemies. This social authorization generated an atmosphere of permissiveness that assured individual soldiers that they would not face censure from their peers for such violence. It did not *require* that they act violently, but it swept away the restraints that might otherwise govern individuals' choices about violence. This absence of restraint opened the door for acts of violence even when the leadership's calculations of necessity or notions of honor suggested mercy and forbearance.

The most obvious example of the authorization of atrocity was the Continentals' natural scalping of the Indian dead, so natural as to seemingly not be worth commenting on. The historical sources for this campaign, and for others, suggest no Loyalists were scalped. In the Sullivan campaign the soldiers did not get many opportunities to scalp Indian dead, but they did after Newtown. And Boyd himself encountered, killed, and scalped several Indians on his patrol immediately prior to his own capture and death.

That such violence would be assumed speaks to the power and thoroughness of American social authorization for violence against Indians. But perhaps even more telling are those incidents in which the standard was contested or seemed less clear. The potential squeamishness about scalping women has already been mentioned. It is intriguing that nearly every diarist could accurately describe the prisoners (one Tory and one Negro), but very few mentioned the woman among the dead at Newtown, and none clearly indicated that she had been scalped, although several implied it.

In another instance of contested standards of violence, on September 2, after the army destroyed Catherine's Town (Shechquago) south of Seneca Lake, they discovered one old woman hiding in the bushes on the edge of town, too feeble to flee. Most accounts mentioned the incident, both because she provided good intelligence on the Iroquois' and Butler's intentions, and also because of her apparently extreme age. The army's allied Indians built her a small hut, and Sullivan ordered that sufficient food be left for her. Stories conflict regarding what happened next, but one thing is clear: when the army returned to her hut on September 23 they found her still there, but nearby was the corpse of a younger woman, who apparently had returned to care for her. The diarists presumed that she had been shot by some soldiers serving as messengers as they passed through the ruined town.[54]

An even more striking example of contested standards for violence against Indians occurred when a detachment under the command of Lt. Col. Henry

Dearborn, a future United States secretary of war, was marching along the western shore of Cayuga Lake and encountered three women and a crippled boy hiding in a house. Under orders to take hostages, Dearborn seized two of the middle-aged women. The others were too old or crippled to move. Dearborn left their house standing amid the ruins of the town and ordered his men not to harm them. As the troops began to march away, two soldiers sneaked back, locked the woman and boy in the house, and set it on fire. When the marching column of men noticed the fire, some officers and soldiers tried to save the two.[55] Although Dearborn had not wanted them hurt, enough freedom of action existed, and enough social authorization outside the military hierarchy remained, to allow such an extreme act of cruelty. And note that while this act may not have reflected the army's overall willingness to inflict that kind of violence, it was precisely the kind of act that was most remembered by the Iroquois. Indeed, the Iroquois reported atrocities beyond those recorded in the soldiers' diaries. An Onondaga chief claimed that the Americans under Colonel Goose Van Schaick who attacked his town "put to death all the Women and Children, excepting some of the Young Women, whom they carried away for the use of their Soldiers & were afterwards put to death in a more shameful manner."[56] Atrocity created its own momentum.

This social authorization for violence against Indians rested on a long-established prejudice, itself grounded in a host of real and perceived differences between Indian and European culture. Over the last thirty years historians have documented the many ways in which English cultural standards seemed to conflict with those of the Indians. Part 3 of this book added the failure of Native and English systems of restraints on war to synchronize. Both cultures found some similarities in the other, and both found a basis in mutual self-interest to establish diplomatic and commercial relations, but they also found repeated, persistent, and fundamentally disturbing *difference*. Difference, combined with a long history of conflict over access to land, fostered ever more entrenched beliefs among Americans not only about the extent of difference but also in its connotations of barbarism. Indians were not just "savages," a word that originally simply implied men living without the trappings of "civilization" such as clothes, cities, or writing. They became barbarians, bent on cruelty, and indifferent and even resistant to the attractions of Christianity and a civil life as defined by the English. While the travel literature of the early explorers had preconditioned the first English colonists to expect to find both "good" and "bad" Indians, by the middle of the eighteenth century at the latest, for most American colonists, all Indians were simply "bad." Thus, while New England publicist William Wood, writing in

1634, was generally positive about the local Indians, his 1764 editor, Nathaniel Rogers, instead grumbled about "their immense sloth, their incapacity to consider abstract truth . . . and their perpetual wanderings," and commented that "The feroce manners of a native Indian can never be effaced."[57] The Indians had been seen from first contact as an "other," someone different, someone not "us," someone not automatically granted trust or credit, but now they had evolved into the *notorious* Other, the barbarian, someone against whom extreme violence was presumptively justified. General Hand, in March 1778, had led an expedition of militia against Indians near Fort Pitt, and he found, as he expected, that the soldiers were "well Disposed, savage like, to murder a defenseless unsuspecting Indian."[58] One wonders if General Hand recognized the irony in his turn of phrase.

Historians have debated when and how this shift in views of the Indians took place, frequently citing one war or another as the moment when all Indians became "bad Indians." Candidates have included the 1622 attack on Jamestown, the Pequot War of 1637, King Philip's War in 1675–76, and, most recently, historians have emphasized the French and Indian War (1754–63).[59] To point to one war as the turning point is probably fruitless. The change instead took place over a long process of accumulating and hardening prejudice, very much stimulated by the competition for land. That desire for land may explain why, despite more than 150 years of direct experience with Indians, some assumptions about cultural differences could persist in the face of overwhelming evidence to the contrary. Within Sullivan's campaign one can still find references to some of those old assumptions, including, of all things, the inaccurate belief that the Indians left the land vacant and unused. Major Jeremiah Fogg could wonder, after having marched through the Iroquois country burning countless fields of crops, "whether the God of nature ever designed that so noble a part of his creation should remain uncultivated, in consequence of an unprincipled and brutal part of it." He therefore thought it appropriate to "lay some effectual plan, either to civilize, or totally extirpate the race."[60]

Fogg's willful blindness left him open to genocidal war in a way that had nothing to do with calculations of military necessity. His and others' values and attitudes set the parameters for the frightfulness of the violence used within a campaign that was already designed to destroy. Even where calculations of necessity might seem to be the most important factor in choices about how to wage war, culture played a role in authorizing a particular style of violence. The social authorization that influenced Fogg's vision of the landscape, and that allowed soldiers automatically to engage in scalping, also

allowed the Continental army's leaders to choose tactics that would otherwise have been unacceptable. The decision to take hostages, for example, was
intended to force the Indians to return to neutrality. Continental leaders
sought a similar outcome in dealing with Loyalists, but they never considered
using hostage taking as a tactic against the vast majority of that population.[61]
The hostage strategy for the Iroquois was a conscious strategic calculation,
rendered acceptable by a specific cultural vision of the enemy as a barbarian.
Washington did not invent the hostage strategy; it had started to emerge
as an option against Indians during the eighteenth century and had been
specifically advocated by British major general Jeffery Amherst during the
French and Indian War, but only against Indians.[62] Ironically, in one sense
the hostage option represented a kind of restraint. The usual colonial practice
had been simply to kill Indian prisoners. The Iroquois themselves noticed
these two ways of war. In a conference in 1782 the Iroquois pointed out that
when the Americans took British prisoners, "the Rebels don't put them to
death; But we have no mercy to expect, if taken, as they will put us to death
immediately, and will not even spare our Women and Children."[63]

Even the seemingly prosaic destruction of fruit trees was an escalation that
some Continental officers perceived as excessive and unnecessary. Europeans'
expectations of war included the possible necessity of devastating the countryside, but even in those circumstances they found the notion of destroying
fruit trees egregious.[64] That thought also apparently disturbed some of the
officers on the Sullivan expedition, who "thought it a degradation of the army
to be employed in destroying apple and peach trees, when the very Indians in
their excursions spared them, and wished the general to retract his orders for
it." Sullivan rejected this argument, countering that "the Indians shall see,
that there is malice enough in our hearts to destroy everything that contrib-
utes to their support." Ignoring Sullivan, some of the officers "would see no
apple or peach tree; so that they were left to blossom and bear. . . . Thus did
gen. Hand and col. Durbin [Dearborn] do honor to their own characters."[65]

Thus, even in the midst of a campaign of destruction, there occurred decisive moments in which individuals or leaders deliberately chose to contain
the violence of the war. Culture is never uniform, and the level of social
authorization for unrestrained violence fluctuated. We have seen how some
care was taken for abandoned elderly women. A surgeon on the expedition
expressed guilt and remorse at the extent of the army's devastation: "I very
heartily wish these rusticks may be reduced to reason, by the aproach of this
army, without their suffering the extreems of war; there is something so cruel,
in destroying the habitations of any people . . . that I might say the prospect

hurts my feelings."[66] In another incident General Philip Schuyler successfully interceded on behalf of several captured Mohawk families who, he
argued, had obeyed agreements to remain peaceful and should not have been
taken.[67]

Such efforts at restraint had a long history. Practical calculation, fear of
retaliation, and even concern for the laws of war in general occasionally had led
Englishmen and Americans to obey truces, preserve prisoners, and observe
surrender terms with Indians. In the late seventeenth century, at least some
New Englanders were willing to admit that an Indian could "kill his enemie in
the feild in a souldier like way" and thus not merit execution once captured.[68]
John Barnwell, at the beginning of the eighteenth century, acknowledged that
there was a danger in Indians believing he would not keep his word—his allies
might desert him—and so he refused to violate the terms of a truce with the
Tuscaroras.[69] There are plenty of such examples, but in general these moments
of restraint in war, or in the wake of war, were exceptions. What is interesting,
however, is that when Europeans *wanted* to express restraint, they did so by
referring to the legal structures and practices with which they surrounded war.
Given how important the laws and customs of war had become, why were
moments of restraint in war with Indians so rare?

Conclusion

Their rarity stands in contrast to the trends of restraint in eighteenth-century
warfare laid out in chapter 7. By the 1770s a substantial intellectual and philosophical platform nominally encouraged this kind of humanitarian restraint.
Furthermore, generations of often friendly interaction with Indians, admittedly in parallel with hostilities, should also have provided some foundation for
sustaining these fragile attempts at restraint. Prejudice and cultural difference
were no doubt part of the problem. Military worries about the elusiveness and
stubbornness of Indians, like that of the Irish, played a role in escalating the
frightfulness of war. In the end, however, the key to escalating wartime violence
against Indians was the inability of most Americans to imagine a place for
Indians among the citizenry of the new republic. The distant government in
Britain, less affected by prejudice or the local hostilities experienced on the
frontier, had begun to make room for non-Englishmen as members of the
empire, especially in the wake of the Seven Years War. Highlanders, Catholic
Irish, German colonists in North America, free blacks, French Canadians,
South Asians, and even Native Americans gained purchase in the metropolitan

imagination as subjects. The English still considered Native Americans savages, but savagery was merely a stage on a natural progression to civility. Given time and education, they could become subjects like any other, and as subjects they would owe taxes and allegiance, and they would be owed protection under the law. General Jeffrey Amherst even acknowledged the legal transfer of the Indians' subjecthood from the French to the English in the French capitulation agreement of September 1760. He allowed that they should continue to enjoy the use of their own laws and lands, thus acknowledging the shift in international law discussed in chapter 7, which sought to preserve private property even during the transfer of sovereignty. This metropolitan vision for the subjecthood of Indians and other ethnicities did not always survive the end of war and the end of their usefulness as allies, and some of those groups would need many more years before full inclusion became a reality. Still, the American colonials' attitude toward Indians was a great deal more hostile. Amherst may have accepted the principle of the Indians' lands remaining inviolate; the colonists did not. So hostile were they that one historian has described the frontier experience from the 1760s to the 1790s as the "war of each against all."[70]

The key difference between the colonial and the metropolitan imaginations was that the colonists were in direct competition with the Indians for land. Some of the Continentals were clearly eyeing Iroquois country with more than a little acquisitive interest. Prior to the start of the campaign, six of Sullivan's senior officers cooperated to lay claim to the land that they were about to invade.[71] And other soldiers fighting on the frontier during the war were documented as "using their time in the service to claim land for later use." Officers occasionally "chastised those who 'presumed to mark trees in the wood with initial letters and their names at large, thereby giv[ing] great uneasiness to our good friends and allies, the Delaware nation.'"[72] Other officers, however, may have been on the lookout for themselves. In the midst of preparations for the 1779 campaign, General Hand was actively speculating in land on the Wyoming frontier. He worried about Connecticut's claim to the area; he calculated that the flight of white settlers had lowered the price on land, and he complained after the campaign that he had not had time to make all the land purchases he had hoped.[73] To be sure, Sullivan's campaign was not designed to wrest the Iroquois's land from them. But some soldiers were thinking along those lines, and who can blame them for thinking that way, given the long history of Anglo-Indian wars and land cession or seizure? This vision of *supplanting* the Indians required no *incorporation* of them as subjects or fellow citizens.[74] It required no postwar reconciliation of wartime violence as did the English civil war, or the rebel-Loyalist war within the American Revolution (although in fact many Loyalists would leave the new nation in fear of retribution).

Frightfulness in any war comes down to choices: choices of leaders and of soldiers. But choices are rooted in experience, shaped by cultural perception, given form by calculation, and confined by physical possibility. In Sullivan's expedition, calculation suggested a strategy of devastation was necessary, and cultural attitudes toward Indians authorized high levels of personal violence, even atrocity. Control of the soldiers was relatively strong, as strong as it had been in Pennsylvania in 1777, but soldiers retained the freedom of action to make their own choices, and against Indians those choices faced fewer social barriers to violence. The combined result of all those forces was a frightful way of war, waged by many of the same regiments who had struggled so hard around Philadelphia to restrain the violence of their war.

Conclusion

Limited War and Hard War in the American Civil War

IN THE BROTHERS' WAR OF 1642, Sir William Waller was not alone in hoping that war could be fought "without an enemie" and "without personal animositie." The ideal vision of war in America's eighteenth-century brothers' war included the even wider hope that an army could march and fight without (much) damage to or intrusion into the surrounding countryside. Neither war lived up to these ideals, but the vision of limiting war, and of limiting its impact on civilians, reemerged at the outset of America's largest and most destructive brothers' war. Ironically, it was William Tecumseh Sherman, after the First Battle of Bull Run in 1861, who sounded the call for restraint, fulminating at his troops, whom he called "goths or vandals" utterly lacking in "respect for the lives and property of friends and foes." Back in the 1640s, as Waller's war had gone on, he had hardened his vision, declaring that anyone pretending neutrality in England's brothers' war was "detestable" and would be taken for "nothing more than enemies to the State and men accordingly to be proceeded against." In the Revolution George Washington and his officers often bowed to necessity—leading Nathanael Greene in one instance to order the country to be foraged naked. Sherman underwent his own transition during the course of the American Civil War, famously concluding that "war and individual ruin" were necessarily "synonymous terms," and formulating his strategy accordingly.[1]

Each of these men—Waller, Washington, and Sherman—began their war within an existing international code of conduct that suggested a certain level of restraint in wars between states, a code they intended to apply to their own internal war. Each man also found additional reasons within their brothers' wars to reinforce their initial inclinations toward restraint: they imagined an eventual postwar reconciliation and they saw some strategic value in cultivating popular loyalty. The paths they took to greater extremes of violence, however, arose from different conjunctions of culture, capacity, calculation, and control.

The participants in the American Civil War, especially its leadership, began that conflict imbued with an eighteenth-century vision of war and its conduct. The scale and stakes of the French Revolutionary and Napoleonic Wars had generated new horrors within intra-European war but had not destroyed the Enlightenment ideal of constraint. In the relative peace following Napoleon's defeat, Europeans and Americans attempted to further elaborate and strengthen restraints on warfare, increasingly emphasizing explicit codification, including international treaties regulating wartime conduct. Americans had carefully observed the Napoleonic Wars, but those who had paid the closest attention were military professionals focused on tactics and strategy. By 1861, however, the context had changed. The American and French revolutions had revealed the vast potential military power of a population mobilized by a new vision of individual participation and investment in the state. Meanwhile, less spectacularly but no less influentially, the industrial revolution had quietly changed the nature of military power and the capacity of a state to wield it.

American observers in 1861, however, were unprepared for the impact of these political and economic changes. Their vision of regulated war, their sense of the modern discipline of armies, and even their strategic calculations of victory all suggested an increase in the value of restraint, and the social power devoted to it. In the aftermath of the Napoleonic Wars, all Western nations devoted an increasing amount of attention to the legal meanings and limits on conduct within war. International-law theorists never denied the iron law of necessity, but they continued to develop ever more elaborate and restrictive codes designed to limit the damage of war. The 1850s stand out as the crucial decade in the accelerating development of more formal efforts to define appropriate conduct in war. The Crimean War and the Second Italian War of Independence produced two crusaders who achieved international prominence—Florence Nightingale and Henry Dunant. Nightingale's experience in the Crimea led her to create a movement to improve medical care for common soldiers, while Dunant's horror at the suffering of soldiers at the Battle of Solferino in 1859 led him to found what became the International Red Cross. In 1856 the European great powers agreed in the Declaration of Paris to forswear privateering (essentially state-licensed piracy). Then, in 1864, the International Conference for the Neutralization of Army Medical Services in the Field laid the legal foundation for the modern Geneva Convention. At that conference twelve of the fifteen nations in attendance signed a convention governing the treatment of wounded soldiers and neutralizing medical services. More nations signed on thereafter. The idea was not original, but in the form of an international treaty it gained greater purchase, uniformity, and visibility.[2]

Military officers and writers in the United States followed these developments and in one respect led them. Regular army officer, lawyer, and future Union commander in the western theater Henry Wager Halleck published *International Law; or, Rules Regulating the Intercourse of States in Peace and War* in 1861. That work, and his subsequent correspondence with jurist Francis Lieber, laid the basis for the so-called Lieber Code, issued to the Union army in 1863 as General Order 100. Lieber built upon past articles of war and explicitly combined them with the natural-law tradition of the Enlightenment codifiers like Emer de Vattel. The result was a comprehensive and positive declaration of the rules, with a much-expanded concern for the treatment of the enemy. It was copied by the governments of Prussia, France, and Great Britain, and had a significant impact on the Geneva and Hague treaties that followed. Like the older codifiers, Lieber was fully cognizant of "necessity" and even of retaliation. His code had hard edges. It stated that "military necessity admits of all direct destruction of life or limb of armed enemies, and of other persons whose destruction is incidentally unavoidable in the armed contests of the war." It further admitted that "the laws of war can no more wholly dispense with retaliation than can the law of nations, of which it is a branch." Lieber even found utility and legality in starving enemies, armed and unarmed, as a legitimate way to shorten a war. Much like Vattel, however, Lieber insisted that such allowances did not "admit of cruelty—that is the infliction of suffering for the sake of suffering or for revenge," nor should retaliation be resorted to "as a measure of mere revenge, but only as a means of protective retribution, and moreover cautiously and unavoidably."[3] Lieber's code is often taken as the point from which to begin study of the modern international law of armed conflict, and it represented the culmination of a long European cultural movement to constrain the violence of war to carefully defined targets.

In terms of control, the nineteenth-century belief in military discipline was, if anything, even stronger than it had been in the eighteenth century. The need for precise linearity of the line of musketeers had eased in the face of new technologies and tactics, but the broad outlines of the culture of discipline remained. This was particularly true for the small professional American regular army in the interval between the War of 1812 and the Civil War. As this army was heavily composed of immigrants who lacked the social power to challenge their officers, and relegated to distant frontier posts, institutional discipline and control could be extreme.[4] Admittedly, wartime volunteer units, which comprised a heavy percentage of the U.S. Army during the war with Mexico, set a different pattern for officer-soldier interaction. Such units continued to elect their officers (or at least their junior officers), and they were unwilling to accept the kind of discipline practiced in the regular army. Even so, General Winfield Scott, who commanded the American forces advancing on Mexico City in 1847–48, consciously recalled Napoleon's problems with guerrilla resistance in Spain and was determined to control his troops. To a large extent he succeeded, and he did so over the volunteers' protests.[5] The massive call-up of volunteers in both armies during the Civil War initially leaned heavily on the looser precedent for discipline, but the presence and authority of veteran regular officers injected a strong element of the old culture of discipline, one that took even greater hold as the war dragged on. That vision of discipline no longer demanded the same kind of mechanistic battlefield precision, but it retained the expectation of holding soldiers responsible for their conduct.[6]

Especially crucial to hopes for restraint in 1861 were the calculations of Union officers at the outset of the war; they envisioned a short, limited conventional campaign capped by a decisive battle. They saw Napoleon's greatest victories in 1805 and 1809 against Austria and in 1807 against Prussia as the paradigmatic form of war. Cadets at the American military academy at West Point were taught that Napoleon had moved his armies rapidly along geometrically appropriate lines of approach and forced his opponent to fight a climactic battle at a disadvantage, usually followed by the occupation of a capital or major city and a negotiated submission. This pattern seemed confirmed by Winfield Scott's expedition to Mexico City in 1849.[7] The Northern public, and Union officers in particular, imagined the Civil War would follow a similar pattern, in which victory in battle would lead to the occupation of Richmond, followed by a negotiated readmission of the Southern states into the union.[8]

Popular opinion and military calculation within the Confederacy followed a similar pattern. Confederate leaders recognized the need to defend their territory as a whole, preserving the stability necessary for slavery, while also defending specific psychologically and industrially important targets. But General Robert E. Lee also understood the expanded role of national will in determining the outcome and the way that battlefield victory and defeat could shore up or erode that will. Like Napoleon, he was a master at crafting a battle so that it resolved in a singular moment of decision. He desperately sought to avoid grinding, multiple-day contests, though he fought several, even before Ulysses S. Grant's 1864 campaign. Strategically, however, Lee sought to batter Union armies and, by doing so, undermine Union public opinion and willingness to continue the fight. In that sense he foresaw the nature of the war's later years.[9]

Lee may have been prescient in his attention to civilian will, but there were other changes in context that he seemed less aware of, and the Union leadership had to learn hard lessons over two years of war before they were fully able to respond to those changes. The fundamental problem was a vast increase in demographic and material capacity, and that change in capacity alone almost proved sufficient to undermine the restraint suggested in cultural developments in the law of war, the culture of discipline, and even initial strategic calculations.

Washington and other American leaders during the American Revolution had worked with and tried to exploit the demographic power resulting from a wide commitment to a cause. They calculated much of their strategy around the need to cement local loyalties and exhaust British political will. Some historians have gone so far as to call it the first "people's war," focusing on the substantial mobilization of militias and their struggle for local political dominance.[10] The French Revolution, however, truly drove home the lesson. The state-organized mass mobilization of young men provided armies on a vast new scale scarcely imagined by the dynasts of the eighteenth century. In part this was a bureaucratic success; the revolutionary French state invented and successfully pursued means of tracking and tapping their population. More crucially, that populace imagined themselves as citizen-participants in the state, and proved willing to serve or suffer in the interests of the state in previously unheard-of numbers. Even setting aside whether patriotic fervor generated more violence by its ideological underpinning, it is clear that the simple expansion of capacity in sheer numerical terms generated the *possibility* of a vastly more frightful way of war.[11]

To gain a rough appreciation for the extent of this change, consider that the peak English effort in Ireland at the end of the sixteenth century approached twenty thousand men, which by all accounts represented an enormous strain on English resources. In the 1640s both sides put sixty thousand men into the field at the peak. The Americans in the 1770s rarely had more than twenty thousand men in the field at once, although the militia's role is harder to quantify; a rough estimate might be an upper limit of forty thousand men. Respectively, these mobilizations represent 0.5 percent, 2.3 percent (counting both sides), and 1.4 percent of the mobilizing state's population. None of these match the kind of numbers for the American Civil War. The

Confederacy's peak effort in June 1863 was an armed force of 307,464, representing 3.4 percent of the total population of the Confederacy. If slaves are not counted in the total population, the percentage becomes 5.5 percent. Even though the theater of war had enlarged, this change in demographic power represents a substantial increase in the capacity of an army to inflict damage on the countryside, as well as a marked increase in the per capita demands on the population for the support of an army.[12]

This level of popular mobilization necessarily required popular commitment. The state did not merely develop a capacity to enforce a service requirement; rather, the populace experienced a dramatic shift in their willingness to imagine themselves as members of the state, hopeful for its military success and fearful of the consequences of national defeat. A further consequence of that imagined commitment was the potential for irregular warfare waged by an aroused populace on the fringes of the army or on the home front in an effort to solidify loyalty.[13] Small "fringe" theaters of conflict during the Civil War generated an intense and vicious kind of localist warfare.[14] One of the motivations for creating the Lieber code had been to define Southern guerrillas who operated out of uniform as illegal combatants and to justify harsh measures against them. In this way codification created pressures for restraint but also defined the conditions in which restraint could "legitimately" give way to escalation.[15]

Slavery too played a role in the nature of the war's violence. White Southerners had spent generations, possibly reaching as far back as the late seventeenth century, developing a theologically, and eventually a "scientifically," justified ideology of black racial inferiority. Building that hard shell of racism had required instilling a fear and hatred of slaves rooted in their supposed natural tendency to brutality, especially sexual brutality. Firmly convinced of those "truths," white Southerners reacted passionately to the threat of freed slaves, or worse, freed armed slaves in Union uniforms. The Confederate government and military leaders regularly promised no quarter or mercy to black troops or their white officers, and they fulfilled those promises in some of the worst atrocities of the war. Although Northern white troops were not initially enthusiastic about the service of black troops, they nevertheless took notice of the atrocities committed against them. The blood and slaughter associated with those incidents caused Northern white troops to loosen the bonds of discipline or moral control among men already frustrated at a long war fought by men they considered rebels.[16]

These key cultural and demographic shifts set the stage for the Civil War, but there had also been a structural shift in the material aspect of war. By 1861 the processes of industrialization, especially the construction of railroads, which allowed the massive movement of troops and military supplies, had changed the relationship between the battlefield and the home front. It is important not to press this idea too far. Armies, especially in Virginia, where campaigns moved over relatively short distances, continued to forage and feed themselves from the local countryside, inflicting incalculable damage in the process. But, especially as Virginia was eaten bare, or farther west, where armies operated in areas of relatively low population, both sides depended on food and supplies brought in by rail. This ability to supply troops year round meant that although the Civil War was a good deal shorter than the English civil war, the American Revolution, or, of course, the Thirty Years War, it was a great deal more intense in terms of continuous operations.

Furthermore, this logistical connection of battlefront to homefront required the continued willingness of "civilians" far from the front to contribute both their sons and their produce. Even repeated defeats could be recovered from quickly, provided civilians remained committed. Since those civilians were far from the front, most of them did not experience the physical destruction and horror of the war, at least at the outset, and not compared to the "garrison war" of the English civil war or the localist "militias' war" of the American Revolution. Those forces had sought to wrest materials, taxes, and men from their immediate vicinity, sometimes transmitting the fruits to larger field armies farther away, but nevertheless *fighting* to make it happen. The American Civil War had a clearer sectional division, with armies advancing along fronts and armies at the fronts supplied from a relatively peaceful rear.

In its initial stages, therefore, the Civil War avoided the violent consequences of the garrison war, and its generals could keep their men concentrated in field armies, supplied by industrial production and transportation. At first they focused their attention on destroying each other's armies and capturing each other's capital. However, Ulysses S. Grant, William Tecumseh Sherman, and Philip Sheridan came to recognize the possible war-winning consequences of making those distant civilians into targets.

Arriving at this new strategic calculation took time. The Union defeat in the first attempt on Richmond in 1861, and repeated defeats in Virginia over the next two years, did little to dampen the Northern impression that the right approach and the right battlefield would achieve a clear victory, the capture of Richmond, and the readmission of the Confederacy into the Union. Meanwhile, in the western theater, Grant's and Sherman's experiences of trying to occupy Kentucky and Tennessee convinced them they needed a different approach to the war, one that Grant and Sherman would develop into the strategy of "hard war." Taking note of how much Southerners valued property, they began experimenting with a raiding strategy designed to bring the South back into the Union. Sherman realized that "all the South is in arms and deep in enmity," and concluded, therefore, "We cannot change the hearts of those people of the South, but we can make war so terrible that they will realize the fact that, however brave and gallant and devoted to their country, still they are mortal and should exhaust all peaceful means before they fly to war."[17] In 1864, having just completed his march from Atlanta to Savannah, Sherman reiterated this position in perhaps his second most famous quotation (after "War is hell"). Clearly articulating the changed role of civilians in sustaining the capacity of the state to make war, he argued,

> I attach more importance to these deep incisions into the enemy's country, because this war differs from European wars in this particular. We are not only fighting hostile armies, but a hostile people, and must make old and young, rich and poor, feel the hard hand of war, as well as their organized armies. I know that this recent movement of mine through Georgia has had a wonderful effect in this respect.[18]

In adopting this "hard war" strategy, Union armies under Grant and Sherman not only overcame the reluctance to target civilian property that marked the early years of the war, they made that property a key part of their strategy.[19] Nevertheless, even as civilian property became a target, the Union leadership successfully (for the most part) restrained violence against civilian persons. Southern mythology and even some recent revisionist attempts to emphasize atrocity, rape, and uncontrolled retaliatory war have been firmly challenged by seemingly irrefutable research.[20] There is little to no evidence of Union soldiers killing, assaulting, or raping white Southerners (slave women proved more vulnerable). Even the property destruction was targeted and restrained. Union generals and the men in the army alike believed in the necessity of punishing the Confederacy, but they nevertheless aimed that passion and frustration at the Confederate infrastructure, broadly defined.[21]

Notwithstanding these limits on violence during the American Civil War, restraint remained shaky. The context of a brothers' war, suffused as it was by the passions revolving around slavery, provided plenty of emotional impetus toward atrocity. The scale of the war, the number and variety of men involved, and even the diversity of opinions among politician and generals all created room for many small and some large actions that we now, and they then, would judge to be atrocities. Especially in the last years of the war, the generalized desire for vengeance, the adoption of a deliberate strategy of devastation, and rumors and reports of specific atrocities generated significant escalation, official and unofficial.[22] Perhaps the most instructive lesson to draw from this example of a brothers' war is, again, the importance of definitions of subjecthood—now citizenship—in determining the nature of wartime violence. Rebellion had always aroused the state into considering the harshest forms of violence, and now citizens felt a similar disgust at and commitment to suppressing rebellion. Union soldiers never tired of derogatorily calling their Southern counterparts "secesh" (secessionist). They nevertheless firmly anticipated an eventual reunion. The exact terms of this reunion, the extent of suffering that the South would have to endure, or how much would have to be destroyed in order to make that happen remained to be seen. But they believed in the fundamental unity of their society and hoped to restore it to a working order.

Thus the American Civil War reinforces the lessons from the earlier chapters, both in the complexity of the interaction of the capacity, control, calculation, and culture, and in the significance of visions of citizenship or subjecthood. The Englishmen of the 1640s similarly had preserved a level of restraint in their war with each other because they continued to see their enemy as fellow subjects, and because the elite leadership was invested in a set of codes designed for international war that touched their personal honor and their belief in God. The mutual accusation of "treason" carried emotive and legal potential for violence, but it only rarely erupted. On the other hand, seventeenth-century Englishmen were only just beginning to exert a comprehensive form of discipline over their soldiers, and the contending sides' administrations proved insufficient to the challenges of feeding and paying troops from the state's divided resources. The wars of the three kingdoms turned truly frightful when the defeated royalists returned to war and thereby undermined the hope for reconciliation as fellow subjects implied by their initial surrender, or in the fighting with the Scots and Irish, who had less claim on being fellow subjects (in English eyes) in the first place.

For George Washington the power of the codes of war and the respectability they could bring him and the new nation was even stronger, and therefore, despite the British losing their status as brothers (in the sense that he was no longer interested in reconciliation), he fought his war as if with legitimate international combatants. Unlike most European states in the eighteenth century, however, Washington's new nation lacked the resources to pay and feed its soldiers, but the maturing culture of discipline and the soldiers' sense of their own membership in the nation helped contain the worst of the violence normally associated with armed unpaid men. Here too "treason" as an escalatory accusation threatened restraint. For that reason the rebel and Loyalist militias were less clear about their willingness to reconcile, and many Loyalists did flee the new United States at the end of the war—although the actual extent of postwar retribution they faced proved limited.

This logic of imagined subjecthood or citizenship partly explains the qualitatively uglier violence wielded by Southern troops during the American Civil War. They were not invested in postwar reconciliation. Worse, the Southern culture of masculinity discouraged restraint and obedience, while the arrival of uniformed black troops provided a particularly enticing target for Southern rage.[23] Indeed, for Southerners the presence of black troops shifted their fight in some ways into a war against barbarians, a mode defined by their unwillingness to imagine former slaves as citizens.

In contrast to the hopes for eventual reconciliation in their brothers' wars, English and Americans fighting against the Irish and the Indians struggled with the prospective subjecthood of their enemies. To be sure, the English and Americans did not go to war simply because they saw those enemies as barbarians; conflict arose fundamentally from the competition for land and resources. But defining them as barbarians profoundly shaped the nature of the war that followed. Care is needed here, however. Prejudice and racism do not suffice to explain the violence of these wars. Englishmen and Americans began those conflicts from an entirely ethnocentric point of view, demeaning, deriding, and misunderstanding their enemy, but nevertheless were prepared to absorb them if they changed their ways. In their conception of incorporating the barbarians, they assumed that submission would produce subjects who would swear allegiance to the English monarch and gain the protections and obligations of English law.[24] English wartime strategies in Ireland were calculated to produce submission; peacetime policy was designed to impose English law and enrich the English crown. The Irish, however, repeatedly declined the "civility" the English expected to accompany submission, used submission to their own ends, and regularly and effectively resisted efforts at reform. The more the Irish resisted becoming "good" subjects, the less willing the English became to consider them as such. That did not stop the English from maintaining their legal claim of dominion over the island and its resources. In time, strategies of submission became strategies of starvation and expropriation, with the war's violence further propelled by differences in ways of war and failures in the state's ability to pay, feed, or control its troops.

Indians faced similar pressures to become good subjects, and they too resisted. Despite massive social trauma from diseases and dislocations, they resisted as effectively as they could. They resisted so well, in fact, that they engendered in white society a level of presumptive hatred that rendered all wars against them horrifically violent. But even here, within deeply impassioned white-Indian wars of retaliation and ethnic or racial prejudice, one must account for other factors. Calculation, for example, was central. To an extent even greater than in Ireland, Englishmen fighting Indians struggled to come to grips with their enemy and felt compelled to resort to indiscriminate methods of destruction and even starvation. With the Indians in particular, there was a marked dissonance in the two sides' grammars of war, especially in the normal restraints that both sides assumed to apply in war. Ironically, war practices that normally restrained violence in this cross-cultural context instead unleashed it—the upwards pressure to escalate, present in all wars, in this case "missed" or simply did not encounter the normal downwards pressures of cultural and structural restraint. As Englishmen became Americans, and as subjects became citizens, the problem of incorporation worsened. The Enlightenment construction of citizenship defined it as a participatory process—to be a citizen meant to actively participate in the governing of the state. In theory there was no fundamental obstacle to Native American citizenship (and indeed all Native Americans became U.S. citizens in 1924), but for a very long time such citizenship proved a prohibitive imaginative leap for white Americans, and without the hope for future reconciliation, war grew increasingly frightful.

In the end, in Ireland and in North America, it was the combination of frustration, ethnic or racial prejudice, and an overwhelming desire for land that prevented the incorporation of the barbarians as "us." If "they" could not become "us," then the only alternative was to exterminate them or, more simply, drive them elsewhere, while remaining willfully blind to the fatal consequences thereof.

Although unfamiliar to many Americans, the wars in Ireland profoundly influenced the way early colonists imagined and then fought against the Indians. The English civil war then profoundly influenced American political development and reinforced the colonists' already existing prejudices against armies. The political and emotive influence of the Anglo-Indian wars, the American Revolution, and the American Civil War are generally taken for granted. What is not often noticed, however, is that these five sets of conflicts, formative as they were, were all fought against either barbarians or brothers, and this simple truth has had a pervasive impact on the American culture of war.[25] Each demanded intense consideration of the nature of the imagined community, or "nation." Who should be allowed in, and who must be thrust out? Furthermore, they all seemed to demand absolute solutions—no mere diplomatic exchange of slices of territory would suffice. The preservation of Protestantism, the supremacy of Parliament, survival on a new continent, independence, and preserving the Union were all ends about which there could be no compromise. And all of those wars were fought in circumstances in which the presumptively normal conventions of combat and diplomacy did not apply, or at least not in immediately obvious ways. Because they were wars with barbarians or brothers, they were *extraordinary*. As a result Americans have tended to think of all wars as extraordinary and even absolute.

That tendency to consider war in extraordinary and absolute terms has not always been productive—especially given the modern capacity to wage vastly destructive war. The American Civil War marked the last brothers' war in the Anglo-American experience. Americans would go on to fight "barbarian" wars against the Indians in the "old west," Filipino guerrillas in the expanding American empire, and the Japanese in World War II.[26] But the main reason to conclude with the American Civil War is because it marked such a dramatic shift in the demographic and industrial capacity to destroy, at the same time that Europeans and Americans increasingly sought to draw boundaries around acceptable forms of violence. European states in this period claimed increasingly exclusive rights to declare war at all, and they simultaneously placed more restrictions on its conduct.[27] This seeming divergence of capacity to destroy and the cultural desire to restrain was mediated by the calculations of generals, who in two world wars tended to choose escalation, even when that may not have been their initial predilection. Americans in World War I and especially in World War II entered those wars focused on their enemies' armed forces, but they also brought with them extreme assumptions about war. President Woodrow Wilson in 1917 sought to wage a war to end all wars. Americans in World War II would tolerate nothing but unconditional surrender, and when achieving that goal became more difficult, they escalated their means until they included functionally indiscriminate bombing and even atomic weapons.

How had American military leaders gotten to the point where they could reconcile this capacity to destroy with the increasing codification and regulation of wartime violence? And what does this continued disconnect between capacity and restraint promise for the future? Again, the nature of citizenship proves central. The trend of greatest importance has been the gradual ideological expansion of the category of "noncombatants" into the more inclusive "civilians," and the more fraught corollary of "citizens." The oldest humanitarian and theological principles suggested and encouraged the protection of women, children, and the aged as noncombatants, although their property, which supported their sovereign's power, had no standing. Ideological shifts visible in the seventeenth century, and especially by the eighteenth, suggested that a "civilian" had an independent legal existence other than being the subject of a sovereign, and civilian property therefore also deserved some protection from the ravages of war. In the revolutionary age, however, civilians became citizens, participants in the state, and thereby implicated by its actions and perhaps even urging the state on to greater extremities. War had become a struggle of nations—nations of citizens.[28]

At the outset of the Franco-Prussian War in 1870, for example, the king of Prussia, Wilhelm I, explicitly promised protection to civilian property. But after the defeat of the French armies and the emergence of irregular resistance, his army chief of staff Helmuth von Moltke explicitly blamed the French citizenry and thereby justified reprisal operations against them. To be moderate with them, he argued, merely awoke their "incorrigible self deceptiveness and arrogance"; it was necessary to "strenuously fight this nation of liars, and finally thrash them."[29] Similarly, in the opening months of World War I, an Austro-Hungarian general, commanding a force invading Serbia, announced to his men that "the war is taking us into a country inhabited by a population inspired with fanatical hatred towards us, into a country where murder, as the catastrophe of Sarajevo has proved, is recognised even by the upper classes, who glorify it as heroism. Towards such a population all humanity and all kindness of heart are out of place."[30]

Beyond their participatory responsibility, even when not in uniform, industrialization made citizens into actual producers of military power, and thus they became targets again. Ironically, when states attempted to use violence in accord with international norms that at least seemed not to target civilians, but which did permit attacking the enemy's infrastructure, civilians became subject to the most indiscriminate forms of violence: strategic bombing, blockade, and, in the most modern formation, economic sanctions.[31] In this sense, what made the German atrocities of World War II so repulsive was their deliberately selective—indeed, discriminate—targeting of civilians at the individual level.

Since World War II, partly in the face of an almost limitless capacity to destroy, and partly in response to demands by an ever more informed and interconnected international political public, the restraints both on going to war and on its conduct have gained greater purchase among Americans. Lapses in restraint, and even atrocities, however much instances of them are held up for public scrutiny and condemnation, are now very much the exception in the British or American conduct of war.

Nevertheless, Americans continue to struggle with the legacy of an absolute vision of war and with the contradictions in the now highly developed sense of international law and human rights vis-à-vis the technological capacity to destroy. Americans have imagined that they can resolve this contradiction through a reliance on technology: Precision weapons will solve the problem of indiscriminate violence; they will attack only what needs to be attacked to destroy a ruling regime, whose citizens are increasingly described as fundamentally innocent. This kind of thinking shaped American operations against Iraq from 1990 to 2003, in Bosnia in 1995, in Kosovo in 1999, and in Afghanistan in 2001. But when the opposition in Iraq and Afghanistan turned into at least semi-popular insurgencies, Americans again confronted the problem of identifying and distinguishing the innocent from the guilty, and technology proved ill-suited to the problem. Furthermore, the terrorist attack in 2001 activated an old fear of barbarians and, with it, a public demand for an absolute solution without regard to its cost or the probability of absolute success. In these circumstances the old, persistent, and powerful Enlightenment ideal of restraint and discriminate violence has proved harder to sustain, although the slippages in America's modern conflicts pale beside those described in this book. The ideal of restraint and the abhorrence of atrocity are stronger than ever, even in a world that has an unimaginable capacity to destroy. The worry remains: in a great emergency, will cultural preferences for humanity and restraint continue to shape the calculations of a society and a military that has access to such capacity?

Abbreviations

BL   British Library, London

CCM   J. S. Brewer and William Bullen, eds., *Calendar of the Carew Manuscripts, Preserved in the Archi-episcopal Library at Lambeth, 1515–1624*, 6 vols. (London: Longman, 1867–73)

CJ   Great Britain, Parliament, House of Commons, *Journals of the House of Commons*. Vol. I (1547/1628–) [London: H.M.S.O., 1628–]

CSPI   *Calendar of State Papers, Ireland*

CSPI(NS)   Mary O'Dowd, ed., *Calendar of State Papers, Ireland, Tudor Period 1571–1575*, rev. ed. (Kew, UK: Public Record Office, 2000)

CSPD   *Calendar of State Papers, Domestic*

DNB   H. C. G. Matthew and Brian Harrison, eds., *Oxford Dictionary of National Biography* (Oxford: Oxford University Press, 2004), online edition, http://www.oxforddnb.com/

DRIA   William L. McDowell Jr., ed., *Documents Relating to Indian Affairs, 1754–62* (Columbia, SC: Colonial Records of South Carolina, Series 2, 1970)

HMC   Historical Manuscripts Commission

HNAI   Bruce G. Trigger, ed., *Handbook of North American Indians*, vol. 15, *Northeast* (Washington, DC: Smithsonian Institution, 1978)

JCC   *Journals of the Continental Congress, 1774–1789*, 34 vols. (Washington, DC: GPO, 1904–37).

JR   Edna Kenton, ed., *The Jesuit Relations and Allied Documents: Travels and Explorations of the Jesuit Missionaries in North America, 1610–1791*, 73 vols. (Toronto: McClelland & Stewart, 1925)

LJ	Great Britain, Parliament, House of Lords, *Journals of the House of Lords*. Vol. 1 (1509–) [London: H.M.S.O., 1767/1770–]
PGW-RW	*The Papers of George Washington: Revolutionary War Series*, 19 vols. (Charlottesville: University Press of Virginia, 1985–).
PNG	*The Papers of General Nathanael Greene*, ed. Richard K. Showman, 13 vols. (Chapel Hill: University of North Carolina Press, 1976–)
RV	David B. Quinn, ed., *The Roanoke Voyages, 1584–1590: Documents to Illustrate the English Voyages to North America under the Patent Granted to Walter Raleigh in 1584*, 2 vols. (London: Hakluyt Society, 1955).
SP	State Papers, The National Archives, Kew, UK
SPH	*State Papers Published Under the Authority of His Majesty's Commission, King Henry the Eighth* (London: HMSO, 1830–34)
STC, Wing, Thomason	Books published in English between 1473 and 1700 have been cataloged in standard bibliographies using these three abbreviations (plus one more not used here). Notes to books from this era include this catalog name and a catalog entry number to clarify which edition has been cited.
TCD	Trinity College Dublin
WGW	*The Writings of George Washington from the Original Manuscript Sources, 1745–1799*, ed. John C. Fitzpatrick, 39 vols. (Washington, DC: GPO, 1931–44).

Notes

Introduction

1. Some of these stories are told in later chapters. For the others, see William Farmer and C. Litton Falkiner, "William Farmer's Chronicles of Ireland from 1594 to 1613," *English Historical Review* 22 (1907): 127–28; Thomas Stafford, *Pacata Hibernia; or, A History of the Wars in Ireland during the Reign of Queen Elizabeth, Especially within the Province of Munster under the Government of Sir George Carew, and Compiled by his Direction and Appointment*, ed. Standish O'Grady (London: Downey, 1896), 2:156–211; Michael Leroy Oberg, *Dominion and Civility: English Imperialism and Native America, 1585–1685* (Ithaca, NY: Cornell University Press, 1999), 199; Carl P. Borick, *A Gallant Defense: The Siege of Charleston, 1780* (Columbia: University of South Carolina Press, 2003), 206–22.

2. Carl von Clausewitz, *On War*, ed. and trans. Michael Howard and Peter Paret (Princeton, NJ: Princeton University Press, 1976), passim, esp. 77, 80, 90–3.

3. Europeans did not introduce scalping to North American Indians, although they greatly aggravated and escalated its use by awarding bounties for scalps. James Axtell, "The Unkindest Cut, or Who Invented Scalping?: A Case Study," in *The European and the Indian: Essays in the Ethnohistory of Colonial North America* (New York: Oxford University Press, 1981), 16–38.

4. David Edwards, "Some Days Two Heads and Some Days Four," *History Ireland*, January/February 2009, 18–21.

5. A few historians have dealt with the "limits of the possible" in their consideration of wartime violence. On the broadest scale, John Landers examines the limits imposed by the "organic" economy of the preindustrial world. *The Field and the Forge: Population, Production, and Power in the Pre-Industrial West* (Oxford: Oxford University Press, 2003). See also Frank Tallett, *War and Society in Early-Modern Europe, 1495–1715* (London: Routledge, 1992), 50–68; Geoffrey Parker, *The*

Military Revolution: Military Innovation and the Rise of the West, 1500–1800, 2nd ed. (New York: Cambridge University Press, 1996), 45–81; Rhoads Murphey, *Ottoman Warfare, 1500–1700* (New Brunswick, NJ: Rutgers University Press, 1999).

6. Azar Gat, *War in Human Civilization* (Oxford: Oxford University Press, 2006), 494.

7. This is a paraphrase and slight modification of Clifford Geertz, "Religion as a Cultural System," in *The Interpretation of Cultures* (New York: Basic Books, 1973), 89.

8. Isabel V. Hull, *Absolute Destruction: Military Culture and the Practices of War in Imperial Germany* (Ithaca, NY: Cornell University Press, 2004), 97.

9. This theory of cultural change is from Pierre Bourdieu, *Outline of a Theory of Practice*, tr. Richard Nice (Cambridge: Cambridge University Press, 1977).

10. There is substantial disagreement on whether humans are "naturally" reluctant to kill, or whether such reluctance is only socialized, and therefore relatively easily unsocialized. See Robert L. O'Connell, *Ride of the Second Horseman: The Birth and Death of War* (New York: Oxford University Press, 1995), 31; John Keegan, *A History of Warfare* (New York: Knopf, 1993), 227; David Livingstone Smith, *The Most Dangerous Animal: Human Nature and the Origins of War* (New York: St. Martin's, 2007); Dave Grossman, *On Killing: The Psychological Cost of Learning to Kill in War and Society* (Boston: Little, Brown, 1995); Joanna Bourke, *An Intimate History of Killing: Face to Face Killing in 20th Century Warfare* (New York: Basic Books, 1999); Paul Roscoe, "Intelligence, Coalitional Killing, and the Antecedents of War," *American Anthropologist* 109 (2007): 485–95; Randall Collins, *Violence: A Micro-Sociological Theory* (Princeton, NJ: Princeton University Press, 2008).

Chapter 1

Epigraph source: SPH 3:176–77 (language modernized)

1. CCM 3:312.

2. The details of Sidney's 1569 summer campaign and the incident at Clonmel come from Ciaran Brady, ed., *A Viceroy's Vindication? Sir Henry Sidney's Memoir of Service in Ireland, 1556–1578* (Cork, Ireland: Cork University Press, 2002), 62–69; SP 63/29/70; TCD ms. 660, ff. 78–91.

3. A standard account of English cultural prejudice is David B. Quinn, *The Elizabethans and the Irish* (Ithaca, NY: Cornell University Press, 1966); Nicholas P. Canny develops this attitude in its relation to colonization in *The Elizabethan Conquest of Ireland: A Pattern Established, 1565–76* (Hassocks, UK: Harvester, 1976), 122–36.

4. CSPI(NS) 225–26 (quoted in the epigram to chapter 2).

5. The following survey of the English in Ireland leans heavily on Steven G. Ellis, *Ireland in the Age of the Tudors, 1447–1603: English Expansion and the End of Gaelic Rule* (London: Longman, 1998); Colm Lennon, *Sixteenth-Century Ireland: The Incomplete Conquest*, rev. ed. (Dublin: Gill & Macmillan, 2005); T. W. Moody,

F. X. Martin, and F. J. Byrne, eds. *A New History of Ireland*, vol. 3, *Early Modern Ireland, 1534–1691* (Oxford: Clarendon, 1976); S. J. Connolly, *Contested Island: Ireland 1460–1630* (Oxford: Oxford University Press, 2007); Ciaran Brady, *The Chief Governors: The Rise and Fall of Reform Government in Tudor Ireland, 1536–1588* (New York: Cambridge University Press, 1994).

6. Art Cosgrove, ed., *A New History of Ireland*, vol. 2, *Medieval Ireland, 1169–1534* (Oxford: Clarendon, 1987); R. R. Davies, *Domination and Conquest: The Experience of Ireland, Scotland and Wales, 1100–1300* (Cambridge: Cambridge University Press, 1990), 11, 39–40, 88; Davies, *The First English Empire: Power and Identities in the British Isles, 1093–1343* (Oxford: Oxford University Press, 2000), 146–47, 150–53, 176–77, 179–80, 186–87; Brendan Smith, *Colonisation and Conquest in Medieval Ireland: The English in Louth, 1170–1330* (Cambridge: Cambridge University Press, 1999), 51, 56.

7. SPH 2:14, 30–31, 34; Ellis, *Ireland*, 122–23.

8. Steven G. Ellis, "The Tudors and the Origins of the Modern Irish State: A Standing Army," in *A Military History of Ireland*, ed. Thomas Bartlett and Keith Jeffery (Cambridge: Cambridge University Press, 1996), 130; Ellis, *Ireland*, 138–40.

9. David Edwards, "The Escalation of Violence in Sixteenth-Century Ireland," in *Age of Atrocity: Violence and Political Conflict in Early Modern Ireland*, ed. David Edwards, Pádraig Lenihan, and Clodagh Tait (Dublin: Four Courts, 2007), 55–58; Ellis, "Tudors and the Origins," 130–31.

10. Ellis, *Ireland*, 141–43. Canny suggests that Henry was not eager to entangle himself in Ireland, but his hand was forced by the 1534 revolt. *Elizabethan Conquest*, 31. Traditionally, the conversion of Ireland from a lordship to a kingdom in 1541 is seen as a major transformation in the status of the Irish within the English legal system. Henry VIII's instructions about making obedient subjects in the 1520s suggest some continuity, although the surrender and regrant program (discussed below) was a clear indicator of change. See also Brendan Bradshaw, *The Irish Constitutional Revolution of the Sixteenth Century* (Cambridge: Cambridge University Press, 1979), 235. The real significance of this shift is a major point of debate among historians of Ireland. I agree with Brendan Kane, who argues that Henry VIII saw his honor as bound up with the treatment of his new subjects. Brendan Kane, "From Irish *Eineach* to British Honor? Noble Honor and High Politics in Early Modern Ireland, 1500–1650," *History Compass* 7 (2009): 417–20.

11. T. W. Moody and F. X. Martin, eds., *The Course of Irish History*, rev. ed. (Dublin: Mercier, 1994), quotation from 179; SPH 3:347–48; "Book of Howth," in CCM 6:195.

12. SPH 3:176–77.

13. Lennon, *Sixteenth-Century Ireland*, 146, 152–64; Canny, *Elizabethan Conquest*, 33; Ciaran Brady, "England's Defence and Ireland's Reform: The Dilemma of the Irish Viceroys, 1541–641," in *The British Problem, c. 1534–1707: State Formation in the Atlantic Archipelago*, ed. Brendan Bradshaw and John Morrill (New York: St. Martin's, 1996), 95.

14. For a case study of this kind of co-option see David Edwards, "Collaboration without Anglicisation: The MacGiollapadraig Lordship and Tudor Reform," in *Gaelic Ireland, c. 1250–c. 1650: Land, Lordship, and Settlement*, ed. Patrick J. Duffy, David Edwards, and Elizabeth FitzPatrick (Dublin: Four Courts, 2001), 77–97.

15. Guy Halsall, *Warfare and Society in the Barbarian West, 450–900* (London: Routledge, 2003), quote on 18; D. Alan Orr, *Treason and the State: Law, Politics and Ideology in the English Civil War* (Cambridge: Cambridge University Press, 2002), 12–13, 16–24; Penry Williams, *The Tudor Regime* (Oxford: Clarendon, 1979), 352–59, 375–78; John Bellamy, *The Tudor Law of Treason: An Introduction* (London: Routledge & Kegan Paul, 1979). Early modern writers on the law of nations explicitly allowed severe treatment of rebels. Geoffrey Parker, "Early Modern Europe," in *The Laws of War: Constraints on Warfare in the Western World*, ed. Michael Howard, George J. Andreopoulos, and Mark R. Shulman (New Haven, CT: Yale University Press, 1994), 44.

16. The following comes from J. V. Capua, "The Early History of Martial Law in England from the Fourteenth Century to the Petition of Right," *Cambridge Law Journal* 36 (1977): 152–73, and Bellamy, *Tudor Law of Treason*, 228–35.

17. Capua, "Early History," 160–62, 167; Steven G. Ellis, "Promoting 'English Civility' in Tudor Times," in *Tolerance and Intolerance in Historical Perspective*, ed. Csaba Lévai and Vasile Vese (Pisa: Università di Pisa, 2003), 155–69. David Edwards has written extensively on martial-law commissions in Ireland: "Beyond Reform: Martial Law and the Tudor Reconquest of Ireland," *History Ireland* 5 (1997): 16–21; Edwards, "Ideology and Experience: Spenser's *View* and Martial Law in Ireland," in *Political Ideology in Ireland, 1541–1641*, ed. Hiram Morgan (Dublin: Four Courts, 1999), 127–57.

18. An overview of the role of pardons in Tudor government is K. J. Kesselring, *Mercy and Authority in the Tudor State* (Cambridge: Cambridge University Press, 2003). Julius R. Ruff, *Violence in Early Modern Europe* (Cambridge: Cambridge University Press, 2001), 98–108; Williams, *Tudor Regime*, 387. A good study of the late medieval view of rebellion is Theodor Meron, *Henry's Wars and Shakespeare's Laws: Perspectives on the Law of War in the Later Middle Ages* (Oxford: Clarendon, 1993), 191–207. The number of rebels and estimates of numbers executed for English revolts between 1487 and 1569 can be found in Capua, "Early History," 158–59, 161; *Rotuli Parliamentorum*, ed. J. Strachey, STC N013055 (London: 1767–77), 6:397–400, 502–3; Michael Bennett, *Lambert Simnel and the Battle of Stoke* (New York: St. Martin's, 1987), 92, 99–103, 149n8; Anthony Fletcher and Diarmaid MacCulloch, *Tudor Rebellions*, 4th ed. (Harlow, UK: Longmans, 1997), 16, 56, 65–71, 84–90, 102–3; A. L. Rowse, *Tudor Cornwall: Portrait of a Society*, rev. ed. (New York: Scribner, 1969), 124–29, 281–87; Michael Bush, *The Pilgrimage of Grace: A Study of the Rebel Armies of October 1536* (Manchester, UK: Manchester University Press, 1996); Michael Bush and David Bownes, *The Defeat of the Pilgrimage of Grace: A Study of the Postpardon Revolts of December 1536 to March 1537 and Their Effect* (Hull, UK: University of Hull Press, 1999), 383–95, 411–12; S. T. Bindoff, *Ket's*

Rebellion 1549 (London: Philip, 1949); Barrett L. Beer, *Rebellion and Riot: Popular Disorder in England During the Reign of Edward VI* (Kent, OH: Kent State University Press, 1982), 79–80. For discussions of the Irish rebellions of 1534, 1569–73, and 1580–82 clearly showing an escalation in the use of martial-law executions, see Steven G. Ellis, "Henry VIII, Rebellion and the Rule of Law," *Historical Journal* 24 (1981): 513–31; Ellis, "Promoting Civility," 164; Ellis, *Ireland*, 317; Moody, *New History of Ireland*, 3:105–6.

19. Williams, *Tudor Regime*, 388–89; Ellis, "Rebellion and the Rule of Law," 529; Capua, "Early History of Martial Law," 157; Edwards, "Escalation," 61, 63–78; Edwards, "Spenser's *View* and Martial Law," 129.

20. There is a brief overview in Kesselring, *Mercy and Authority*, 193–98; see also Rory Rapple, *Martial Power and Elizabethan Political Culture: Military Men in England and Ireland, 1558–1594* (Cambridge: Cambridge University Press, 2009), 210–11, 217–19, 224–29. See below for further specific evidence of pardons granted in Ireland and the critics thereof.

21. Ellis, *Ireland*, 266, 279, 317, 325–26, 333; Lennon, *Sixteenth-Century Ireland*, 171, 232–33; Nicholas Canny, *Making Ireland British, 1580–1650* (Oxford: Oxford University Press, 2001), 121–64; Canny, *Elizabethan Conquest*, 34–36; Rapple, *Martial Power*, 234–43; personal communication with David Edwards. The interaction of martial law with actual land confiscation and dispossession remains obscure.

22. David Kaiser, *Politics and War: European Conflict from Philip II to Hitler* (Cambridge, MA: Harvard University Press, 1990), 23.

23. Bruce P. Lenman, *England's Colonial Wars, 1550–1688: Conflicts, Empire and National Identity* (Harlow, UK: Longman, 2001), 53–54; K. W. Nicholls, *Gaelic and Gaelicized Ireland in the Middle Ages*, 2nd ed. (Dublin: Lilliput, 2003); Seán Ó Domhnaill, "Warfare in Sixteenth-Century Ireland," *Irish Historical Studies* 5 (1946): 29–54; Cyril Falls, *Elizabeth's Irish Wars* (London: Methuen, 1950); Katharine Simms, "Warfare in the Medieval Gaelic Lordships," *Irish Sword* 12 (1975): 98–109; G. A. Hayes-McCoy, "Strategy and Tactics in Irish Warfare, 1593–1601," *Irish Historical Studies* 2 (1941): 255–79; Hayes-McCoy, "The Army of Ulster, 1593–1601," *Irish Sword* 1 (1949–1953): 105–17; Hiram Morgan, ed., "A Booke of Questions and Answers Concerning the Warrs or Rebellions of the Kingdome of Irelande," *Analecta Hibernia* 36 (1995): 95.

24. No doubt worsening the frustration was that Tudor England saw a decline in serious rebellion over the century, while Ireland experienced a trend in the opposite direction. Ellis, "Promoting Civility."

25. The literature on this war is extensive. Key recent studies are Hiram Morgan, *Tyrone's Rebellion: The Outbreak of the Nine Years War in Tudor Ireland* (Woodbridge, UK: Boydell, 1993); John McGurk, *Elizabethan Conquest of Ireland: The 1590s Crisis* (Manchester, UK: Manchester University Press, 1997); and several essays in Hiram Morgan, ed., *The Battle of Kinsale* (Bray, Ireland: Wordwell, 2004); and Edwards et al., *Age of Atrocity*. Lenman, *England's Colonial Wars*, provides a fine short summary.

26. I am here sidestepping the substantial debate within Irish historiography over "reform" versus "conquest." I am framing the question of atrocity and violence in a way that seems different to me from the terms current in that debate, although many of the issues raised within it inevitably appear in these pages. See Edwards, "Escalation," 35–37, and the sources cited there. See also Vincent Carey, "The Irish Face of Machiavelli: Richard Beacon's *Solon his Follie* (1594) and Republican Ideology in the Conquest of Ireland," in Morgan, *Political Ideology in Ireland*, 83–109; Nicholas Canny, "Edmund Spenser and the Development of an Anglo-Irish Identity," *Yearbook of English Studies* 13 (1983): 1–19; Brady, "England's Defence"; Nicholas Canny, "Spenser's Irish Crisis: Humanism and Experience in the 1590s," *Past and Present* 120 (1988): 201–9; Ciaran Brady, "Spenser's Irish Crisis: Humanism and Experience in the 1590s: Reply," *Past and Present* 120 (1988): 210–15; Brendan Bradshaw, "Robe and Sword in the Conquest of Ireland," in *Law and Government Under the Tudors: Essays Presented to Sir Geoffrey Elton, Regius Professor of Modern History in the University of Cambridge, on the Occasion of his Retirement*, ed. Claire Cross, David Loades, and J. J. Scarisbrick (Cambridge: Cambridge University Press, 1988), 139–62.

27. For a depiction of this vacillation, see Edwards, "Spenser's *View* and Martial Law"; Canny, *Making Ireland British*, 118–20.

28. DNB, s.v. Sidney, Sir Henry (1529–1586).

29. These are treated by historians as the Butler Revolt and the Fitzmaurice Rebellion. They were indeed separately inspired, and the families were themselves normally inveterate enemies. Attempts to cooperate mostly failed.

30. Lennon, *Sixteenth-Century Ireland*, 214.

31. CSPI 1:330.

32. David Edwards and Nicholas Canny have both suggested that the fighting that began in the late 1560s represented a new qualitative escalation in violence in Irish wars. Although I have found other examples of the large-scale killing of women and children from before this period, and the records for earlier years are frankly sketchy, it nevertheless seems that these killings of women and children accompanied by mass rape is indeed new. Edwards, "Escalation," 71–72; Canny, *Elizabethan Conquest*, 141, 154–55.

33. SP 63/26/4 (ix).

34. SP 63/25/86 (ii).

35. Quotation from SP 63/29/70; David Edwards, "The Butler Revolt of 1569," *Irish Historical Studies* 28 (1993): 250; HMC *Salisbury* 1:417; Brady, *Vindication*, 62–63. Atrocity reports against the Butlers and Desmonds would continue to trickle in, including that they burned four young children, and later that they were sending the countrymen into Waterford stripped naked, including the "honest huswives." CSPI 1:416; SP 63/29/5.

36. SP 63/29/40.

37. SP 63/29/1.

38. CCM 1:397–99. For Fitzmaurice's contacts with Spain, see Edwards, "Butler Revolt," 247–48.

39. Brady, *Vindication*, 51, 74; SP 63/27/56; SP 63/11/112 (the table does not include the Lord Deputy's company and so is not complete); SP 63/29/70. For a wider discussion of the English use of Irish troops, see Wayne E. Lee, "Subjects, Clients, Allies or Mercenaries? The British Use of Irish and Indian Military Power, 1500–1815" in *Britain's Oceanic Empire: Projecting Imperium in the Atlantic and Indian Ocean Worlds, ca. 1550–1800*, ed. H. V. Bowen, Elizabeth Mancke, and John G. Reid (Cambridge: Cambridge University Press, forthcoming); Rapple, *Martial Culture*, 219–24.

40. *A Breviate of the Getting of Ireland and the Decaye of the Same*, Harleian MS 35, 210r–216r, British Library. Undated, but from near this period. A summary of local *standing* forces in Munster from 1570 or 1571 shows a similar order of magnitude of men potentially available if the rebels were to combine and sway the countryside in their favor (542 horse, 5874 kerne, 360 galloglass): CCM 1:393–94.

41. SP 63/25/86 (ii). Other estimates of the Butler brothers' numbers are more conservative. "A Breviate of the Getting of Ireland and the Decaye of the Same," Harleian MS 35, 210r–216r, British Library; Edwards, "Butler Revolt," 234.

42. CSPI 1:414; Canny, *Elizabethan Conquest*, 142.

43. SP 63/27/56. 340 men had recently been left to garrison Carrickfergus.

44. SP 63/29/22; SP 63/29/70.

45. Edmunde Campion, *Two Bokes of the Histories of Ireland*, edited by A. F. Vossen (Assen, The Netherlands: Van Gorcum, 1963), 148.

46. SP 63/187/32. For more on the recruitment of English expeditionary forces during the Tudor era, see Williams, *Tudor Regime*, 109–35; C. G. Cruickshank, *Elizabeth's Army*, 2nd ed. (Oxford: Clarendon, 1966), 17–40; Mark Charles Fissel, *English Warfare, 1511–1642* (London: Routledge, 2001), 82–113; McGurk, *Elizabethan Conquest*, 29–134.

47. First quotation quoted in Gervase Phillips, *The Anglo-Scots Wars, 1513–1550: A Military History* (Woodbridge, UK: Boydell, 1999), 49; second quotation from HMC, *Salisbury*, 12:169, quoted in McGurk, *Elizabethan Conquest of Ireland*, 33.

48. CSPI 5:386

49. CSPI 1:307–8, 310, 314; CCM 2:168

50. H. C. [Hugh Collier], *Dialogue of Silvynne and Peregrynne (PRO SP 63/203/119)* Corpus of Electronic Texts, University College Cork, http://www.ucc.ie/celt/published/E590001–001/index.html, 18.

51. SP63/7/61.

52. Hayes-McCoy, *Scots-Mercenary Forces*, 15–76; Nicholls, *Gaelic and Gaelicized Ireland*, 99–104.

53. Dymmok, *A Treatice of Ireland* (Dublin: Irish Archaeological Society, 1842), 7–8. Dymmok's *Treatice* is copied from older versions of a standard text describing Ireland that exists in multiple copies. He added certain details appropriate to 1600. See "A Treatise of Ireland," MS 669, National Library of Ireland, Dublin; HMC *Report 3 for 1872*, appendix, 115; David B. Quinn, "Edward Walshe's 'Conjectures' Concerning the State of Ireland [1552]," *Irish Historical Studies* 5 (1947): 303.

54. Falls, *Elizabeth's Irish Wars*, 67; Nicholls, *Gaelic and Gaelicized Ireland*, 95–98; Ellis, *Ireland*, 42, 44.

55. M. Ó Báille, "The Buannadha: Irish Professional Soldiery of the Sixteenth Century," *Journal of the Galway Archaeological and Historical Society* 22 (1946): 49–94.

56. See note 2.

57. CSPI 1:308.

58. "Lord Chancellor Gerrard's Notes of His Report on Ireland," ed. C. McNeill, *Analecta Hibernia* 2 (1931): 158–60; Falls, *Elizabeth's Irish Wars*, 211–12; CCM 2:122; McGurk, *Elizabethan Conquest*, 206.

59. John Dawtrey, *The Falstaff Saga: Being the Life and Opinions of Captain Nicholas Dawtrey Sometime Seneschal of Claneboye and Warden of the Palace of Carrickfergus, Immortalized by Shakespeare as Sir John Falstaff* (London: Routledge, 1927), 82.

60. John Harington, *Nugae Antiquae: Being a Collection of Original Papers, in Prose and Verse; Written during the Reigns of Henry VIII, Edward VI, Queen Mary, Elizabeth, and King James* (London: J. Wright, 1804), 1:176–77.

61. RV 1:223. Gervase Phillips provides a sound study of the tradition of Tudor military "indiscipline," connecting it closely to a broader English tradition of legitimate complaint. Gervase Phillips, "To Cry 'Home! Home!': Mutiny, Morale, and Indiscipline in Tudor Armies," *Journal of Military History* 65 (2001): 313–32.

62. William Shakespeare, *Henry V*, act 3, scene 3 (emphasis added).

63. Late medieval and early Tudor disciplinary codes prescribed beheading for soldiers who cried havoc too soon and thus endangered the success of the assault. Charles Cruickshank, *Henry VIII and the Invasion of France* (New York: St. Martin's, 1991), 86.

64. Meron, *Henry's Wars*, 101–4; J. R. Hale, *War and Society in Renaissance Europe, 1450–1620* (Leicester, UK: Leicester University Press, 1985), 191–96; Christopher Duffy, *Siege Warfare: The Fortress in the Early Modern World, 1494–1660* (London: Routledge & Kegan Paul, 1979), 50, 66; Parker, "Early Modern Europe," 48–49; John France, "Siege Conventions in Western Europe and the Latin East," in *War and Peace in Ancient and Medieval History*, ed. Philip de Souza and John France (Cambridge: Cambridge University Press, 2008), 158–72.

65. A full discussion of the uncertainty of Irish garrisons' sense of their status is in Rapple, *Martial Power*, 214–15.

66. Neither of the accounts written by Sidney mention this small number. His secretary's journal, however, is clear. The only question is whether the total ward had been eight or if that was the number of men remaining after the English completed their final assault.

67. Andrew Ayton and J. L. Price, "Introduction," in *The Medieval Military Revolution: State, Society and Military Change in Medieval and Early Modern Europe*, ed. Andrew Ayton and J. L. Price (New York: Barnes & Noble, 1998), 7–8, and the citations there; Clifford J. Rogers, "By Fire and Sword: *Bellum Hostile* and 'Civilians' in the Hundred Years' War," in *Civilians in the Path of War*, ed. Mark Grimsley and Clifford J. Rogers (Lincoln: University of Nebraska Press, 2002), 33–78.

68. CCM 1:401.

69. Tomás Ó Laidhin, ed., *Sidney State Papers, 1565–70* (Dublin: Irish Man-sucripts Commission, 1962), 116. Elizabeth took a similar attitude later in 1579, rebuking her officers for proclaiming the earl of Desmond a traitor, an act which may have forced him deeper into rebellion. Rapple, *Martial Power*, 204, 207–8.

70. Thomas Churchyard, *A Generall Rehearsall of Warres, Called Churchyardes Choise*, STC 5235.2 (London: John Kingston for Edward White, [1579]), Q.iii–R.i; David Beers Quinn, ed., *The Voyages and Colonising Enterprises of Sir Humphrey Gilbert* (London: Hakluyt Society, 1940), 1:16–19.

71. Ben Kiernan, "English Conquest of Ireland, 1566–1603," in *Blood and Soil: A World History of Genocide and Extermination From Sparta to Darfur* (New Haven, CT: Yale University Press, 2007), 195–97; Canny, *Elizabethan Conquest*, 121 and passim; David Edwards, "Introduction," in Edwards et al., *Age of Atrocity*, 22–23; Ellis, "Promoting English Civility," 163–64; Roger B. Manning, *An Apprentice-ship in Arms: The Origins of the British Army 1585–1702* (Oxford: Oxford University Press, 2006), 8, 15; Falls, *Elizabeth's Irish Wars*, 345–46.

Chapter 2

Epigraph combines two different readings of the original: CSPI(NS) 225–56; SP 63/38/20, quoted in Rory Rapple, *Martial Power and Elizabethan Political Culture: Military Men in England and Ireland, 1558–1594* (Cambridge: Cambridge University Press, 2009), 248–49.

1. Nicholas P. Canny *The Elizabethan Conquest of Ireland: A Pattern Estab-lished, 1565–76* (Hassocks, UK: Harvester, 1976), 65–69, 88.

2. G. A. Hayes-McCoy, *Scots Mercenary Forces in Ireland (1565–1603): An Account of Their Service during That Period, of the Reaction, of their Activities on Scot-tish Affairs, and of the Effect of their Presence in Ireland, Together with an Examina-tion of the Gallóglaigh or Galloglas* (Dublin: Burns, Oates & Washbourne, 1937), esp. 16–18, 106–7, 117–21.

3. Queen to William Fitzwilliam, July 17, 1573, Bodleian Library, Oxford, Carte MS 56 no. 260, 565v, 566r, 565r, quoted in Ben Kiernan, "English Conquest of Ireland, 1566–1603," in *Blood and Soil: A World History of Genocide and Extermination from Sparta to Darfur* (New Haven, CT: Yale University Press, 2007), 197; Walter Bour-chier Devereux, ed., *Lives and Letters of the Devereux, Earls of Essex, in the Reigns of Elizabeth, James I., and Charles I., 1540–1646* (London: John Murray, 1853), 1:32.

4. CSPI(NS) 804, 832.

5. CSPI(NS) 880–82, 883; Rapple, *Martial Power*, 216–17.

6. Devereux, *Lives*, 1:39. Interestingly, an almost identical quote from a vet-eran of the Irish wars in relation to Indians appears in chapter 5.

7. CSPI(NS) 745; Rapple, *Martial Power*, 229–32; Kiernan, "English Con-quest," 198; Vincent Carey, "John Derricke's *Image of Irelande*, Sir Henry Sidney, and the Massacre at Mullaghmast, 1578," *Irish Historical Studies* 31 (1999): 305–27, especially 323–24, 324n80; Richard Bagwell, *Ireland Under the Tudors: With a Suc-cinct Account of the Earlier History* (London: Holland, 1963) 2:288–89.

8. Except insofar as its threat helped motivate the Irish to rebel in the first place.

9. CSPI(NS) 880–82.

10. Roger B. Manning, *Swordsmen: The Martial Ethos in the Three Kingdoms* (New York: Oxford University Press, 2003); Arthur B. Ferguson, *The Chivalric Tradition in Renaissance England* (Washington, DC: Folger Shakespeare Library, 1986), passim, esp. 72; R. C. McCoy, "'A Dangerous Image': The Earl of Essex and Elizabethan Chivalry," *Journal of Medieval and Renaissance Studies* 13 (1983): 313–29.

11. Philip Benedict, *Rouen during the Wars of Religion* (Cambridge: Cambridge University Press, 1981), 218.

12. DNB, s.v. Devereux, Robert; McCoy, "Dangerous Image," 315. Paul E. J. Hammer provides a modern political biography of Essex through 1597, emphasizing his rational and calculating side, while fully incorporating Essex's sense of himself as a soldier-knight and a noble. *The Polarisation of Elizabethan Politics: The Political Career of Robert Devereux, 2nd Earl of Essex, 1585–1597* (Cambridge: Cambridge University Press, 1999), 3–10, 199–268.

13. Anthony Wingfield, *A True Coppie of a Discourse Written by a Gentleman, Employed in the Late Voyage of Spaine and Portingale*, STC 6790 (London: Thomas Woodcock, 1589), 17, 19, 40. The expedition is narrated by R. B. Wernham, *After the Armada: Elizabethan England and the Struggle for Western Europe, 1588–1595* (Oxford: Clarendon, 1984), 107–30.

14. Retinue strength derived from the following sources: Steven G. Ellis, *Ireland in the Age of the Tudors, 1447–1603: English Expansion and the End of Gaelic Rule* (London: Longman, 1998), 146; SPH 3:350, 433; CCM 1:200; CSPI 1:78; SP 61/2/57; Bruce P. Lenman, *England's Colonial Wars, 1550–1688: Conflicts, Empire and National Identity* (Harlow, UK: Longman, 2001), 55; SP 61/4/75; Ciaran Brady, "The Captains' Games: Army and Society in Elizabethan Ireland," in *A Military History of Ireland*, ed. Thomas Bartlett and Keith Jeffery (Cambridge: Cambridge University Press, 1996), 136–59; SP 63/1/5; SP 63/7/61; SP 63/14/26(i); SP63/13/60(i); SP 63/27/56; CSPI(NS) 1–3; "Lord Chancellor Gerrard's Notes of His Report on Ireland," ed. C. McNeill, *Analecta Hibernia* 2 (1931): 93–291; James Hogan and N. McNeill O'Farrell, eds., *The Walsingham Letter-Book or Register of Ireland, May 1578 to December 1579* (Dublin: Irish Manuscripts Commission, 1959), 51–61; Charles McNeill, "The Perrot Papers," *Analecta Hibernia* 12 (1943): 4; CCM 2:462–64; CCM 3:127. Sources for expenditures from 1557 to 1603: Frederick C. Dietz, "The Exchequer in Elizabeth's Reign," *Smith College Studies in History* 8.2 (1923): 65–118. Data from 1603 on is from Frederick C. Dietz, "The Receipts and Issues of the Exchequer during the Reigns of James I and Charles I," *Smith College Studies in History* 13.4 (1928), 117–71.

15. Cyril Falls, *Elizabeth's Irish Wars* (London: Methuen, 1950), 232; Fynes Moryson, *An Itinerary Containing His Ten Yeeres Travell* (Glasgow: J. MacLehose, 1907), 2:221–23.

16. John McGurk, *Elizabethan Conquest of Ireland: The 1590s Crisis* (Manchester, UK: Manchester University Press, 1997), 122.

17. CSPI 8:16.

18. T. W. Moody, F. X. Martin, and F. J. Byrne, eds. *A New History of Ireland*, vol. 3, *Early Modern Ireland, 1534–1691* (Oxford: Clarendon, 1976), quote at 3:127. Condemnations of Essex's decision to march south abound. A more positive assessment of the campaign is L. W. Henry, "The Earl of Essex and Ireland, 1599," *Bulletin of the Institute of Historical Research* 32 (1959): 1–23.

19. CCM 3:295.

20. Dawtrey produced several analyses: CSPI 7:162–65; Dawtrey, *The Falstaff Saga: Being the Life and Opinions of Captain Nicholas Dawtrey Sometime Seneschal of Claneboye and Warden of the Palace of Carrickfergus, Immortalized by Shakespeare as Sir John Falstaff* (London: Routledge, 1927), 190–200; Hiram Morgan, "A Booke of Questions and Answars Concerning the Warrs or Rebellions of the Kingdome of Irelande [Nicholas Dawtrey, ca. 1597]," *Analecta Hibernia* 36 (1995): 79–134. For Lee there is Thomas Lee, "A Brief Declaration of the Government of Ireland," in *Desiderata Curiosa Hibernica*, ed. John Lodge (Dublin: David Hay, 1772), 1:87–150; Lee, "Discoverye & Recoverye," Sloane MS 1818, BL. The latter text (and other extant copies) is discussed in Hiram Morgan, "Tom Lee: The Posing Peacemaker," in *Representing Ireland: Literature and the Origins of Conflict, 1534–1660*, ed. Brendan Bradshaw, Andrew Hadfield, and Willy Maley (Cambridge: Cambridge University Press, 1993), 132–65. Reade, Carlile, Baynard, and Willis are in CSPI 7:449–52; CSPI 8:327–29, 329–31, 347–52.

21. The first quotation is from CSPI 8:330; the second from CSPI 7:164–65. The "bridle" phrase was common, here it comes from Lee's advice. Although both bridling and devastation were "traditional," the idea of garrisons bridling the countryside had less of a pedigree than devastation as a means of pressuring the enemy to submit. To give just one early example, see SPH 2:323–30.

22. Lee, "Discoverye & Recoverye" f. 75v. For clarity this quote has been modernized and closely paraphrased in a few places.

23. Morgan, "Book of Questions," 100, 123, 111.

24. James Spedding, ed., *The Letters and the Life of Francis Bacon* (London: Longmans, Green, Reader, and Dyer, 1868) 3:50. See also James Perrott, *The Chronicle of Ireland, 1584–1608* (Dublin: Irish Manuscripts Commission, 1933), 10.

25. CSPI 10:191.

26. Humfrey Dyson, *A Booke Containing All Such Proclamations as Were Published during the Raigne of the Late Queene Elizabeth*, STC 7758.3 (London: Bonham Norton and John Bill, 1618), 361–62. Elizabeth rebutted charges that she did not intend to differentiate between rebels and good Irish subjects in CSPI 8:363–65.

27. David Edwards, "Beyond Reform: Martial Law and the Tudor Reconquest of Ireland," *History Ireland* 5 (1997); Edwards, "Ideology and Experience: Spenser's *View* and Martial Law in Ireland," in *Political Ideology in Ireland, 1541–1641*, ed. Hiram Morgan (Dublin: Four Courts, 1999), 152–53.

28. *Laws and Orders Issued by Robert Devereux, 2nd Earl of Essex, for the Good Conduct of the Service in Ireland,* [April 1599], Lambeth Palace, London, MS 247. This order was published as *Lawes and Orders of Warre,* STC 14131 (London, 1599?). Essex cultivated a reputation for controlling his troops, and even provided support for Alberico Gentili, a theorist on laws of war (discussed in chapter 7). Hammer, *Polarisation,* 230, 239. In an additional control measure, Elizabeth, irritated at abuses by the army, decreed in 1598 that soldiers accused of crimes would in future be judged by a civil official rather than military officers. Edwards, "Spenser's *View* and Martial Law," 153.

29. J. H. Leslie, "The Printed Articles of War of 1544," *Journal of the Society of Army Historical Research* 7 (1928): 222–40; *Laws and Ordinances Issued by William Herbert . . . for the Government and Discipline of the Army* [July 1557], MS 247, Lambeth Palace Library, ff. 41r–48v; *Ordinances and Instructions for the Army in Brittany,* c. 1590, MS 247, Lambeth Palace Library, ff. 49r–54r; Robert Hare, "Treatise on Military Discipline," 1556, Cotton MS, Julius F.V., British Library; Leicester's code for his army in the low countries in 1585 is reprinted in C. G. Cruickshank, *Elizabeth's Army,* 2nd ed. (Oxford: Clarendon, 1966), 296–303. A brief introduction to some of these codes is in Margaret Griffin, *Regulating Religion and Morality in the King's Armies, 1639–1646* (Leiden, The Netherlands: Brill, 2004), 3–13.

30. *Certain Ordynaunces to be Observed Dureinge the Warrs of Ireland,* Add. MS 19,831, BL.

31. In contrast to the other codes, Essex's also explicitly excluded all women from the camp (and boy servants for footmen). His attempt to exclude them failed (see CCM 3:309). Leicester's code allowed only wives and some nurses.

32. The sources for Essex's march are primarily his "journals" of the march. May 9 to May 18 is CSPI 8:37–40; May 21 to June 22 and June 22 to July 1 are in CCM 3:301–2; another version covering May 9 to July 14 is MS V.b.214, Folger Shaksespeare Library, f202–5; Essex was back in Dublin by July 11, but headed back out into Leix and Offaly on July 25. He was apparently back in Dublin again by August 5. He marched north into Ulster later in August, covered in the entries from August 28 to September 8 in CCM 3:321–25. The gap in July and August is partly filled by Moryson, *Itinerary,* 2:244; Norman Egbert McClure, ed., *The Letters and Epigrams of Sir John Harington* (Philadelphia: University of Pennsylvania Press, 1930), 72; Perrot, *Chronicle,* 162–72. Other sources for the campaign are John Harington, *Nugae Antiquae: Being a Collection of Original Papers, in Prose and Verse; Written during the Reigns of Henry VIII, Edward VI, Queen Mary, Elizabeth, and King James* (London: J. Wright, 1804), 1:268–293; Dymmok, *Treatise,* 30–51.

33. Lord Deputy Sidney, Earl of Kildare, Sir Nicholas Bagenall, and Frances Agarde to the Queen, November 12, 1566, SP 63/19/43. Sidney seemed to think that this was an unusual amount of destruction.

34. CSPI 9:116.

35. Winter campaigns had long been suggested but had only rarely been implemented. SPH 3:356; R. B. Wernham, *The Return of the Armadas: The Last Years of the Elizabethan War Against Spain, 1595–1603* (Oxford: Oxford University Press,

1994), 402–3; Dawtrey, *Falstaff Saga*, 95–96; Lee, "Discoverye & Recoverye," 80v–81v; *Certain Principall Matters Concerning the State of Ireland*, National Library of Ireland, Dublin, MS 3319; CSPI 10:416–17. David Edwards, "The Escalation of Violence in Sixteenth-Century Ireland," in *Age of Atrocity: Violence and Political Conflict in Early Modern Ireland*, ed. David Edwards, Pádraig Lenihan, and Clodagh Tait (Dublin: Four Courts, 2007), 61–62, argues that English forces even as early as 1534 devastated more systematically and intensively than was traditional in Gaelic warfare.

36. Information for 1599: *The Irish Fiants of the Tudor Sovereigns: During the Reigns of Henry VIII, Edward VI, Philip & Mary, and Elizabeth I*, vol. 3 (Dublin: Edmond Burke, 1994), 6281, 6282, 6285, 6288, 6290, 6291, 6307, 6319, 6337, 6342.

37. Ibid., 6280, 6284, 6294, 6302, 6303, 6305, 6309, 6312, 6314, 6315, 6323, 6329, 6338.

38. Commissions of martial law were indeed frequent, but, as David Edwards has pointed out, there were also periods when Elizabeth recoiled at its cost and consequences. Edwards, "Spenser's *View* and Martial Law," 137–42.

39. Quoted in Vincent Carey, "The Irish Face of Machiavelli: Richard Beacon's *Solon his Follie* (1594) and Republican Ideology in the Conquest of Ireland," in Morgan, *Political Ideology in Ireland* 87–88.

40. *Journal of Things Done in Ireland*, May 8–July 14, 1599, MS V.b.214, f. 203r, Folger Shakespeare Library, Washington, DC (the other accounts do not mention the fire discussed in this version).

41. Standish O'Grady, ed., *Pacata Hibernia: Or, A History of the Wars in Ireland during the Reign of Queen Elizabeth, Especially within the Province of Munster under the Government of Sir George Carew* (London: Downey, 1896), 2:5–7, 25–26. See ibid., 2:71–81 for a classic back-and-forth negotiation between Mountjoy and the Spanish in Kinsale over the terms of their surrender.

42. Ibid., 2:199, 202–4, quotation on 210. For another Gaelic garrison being allowed to surrender, see Perrot, *Chronicle*, 31.

43. Philippe Contamine, *War in the Middle Ages*, tr. Michael Jones (Oxford: Blackwell, 1984), 270–78; Theodor Meron, *Henry's Wars and Shakespeare's Laws: Perspectives on the Law of War in the Later Middle Ages* (Oxford: Clarendon, 1993), 91–95.

44. Quoted in David B. Quinn, *The Elizabethans and the Irish* (Ithaca, NY: Cornell University Press, 1966), 133.

45. Edwards, "Escalation," 71.

46. Falls, *Elizabeth's Irish Wars*, 299.

47. This story is usually told more for the salacious detail that the children had been cooking the body of a dead woman. William Farmer and C. Litton Falkiner, "William Farmer's Chronicles of Ireland from 1594 to 1613," *English Historical Review* 22 (1907): 129–30.

48. Wingfield, *True Coppie*, 40.

49. SPH 3:199; 256–59; 351; Perrot, *Chronicle*, 84–86; SP 63/106/62.

50. Perrot, *Chronicle*, 91–93; Lee, "Discoverye & Recoverye," f. 76v.

51. SP 63/106/62.

52. DNB s.v. Devereux, Robert; Dymmok, *Treatise*, 49–51; CCM 3:323–27.

53. Manning, *Swordsmen*; Lenman, *England's Colonial Wars*, 13–15; Christopher Storrs and H. M. Scott, "Military Revolution and the European Nobility, c. 1600–1800," *War in History* 3 (1996): 1–41; Richard C. McCoy, *The Rites of Knighthood: The Literature and Politics of Elizabethan Chivalry* (Berkeley: University of California Press, 1989), passim, esp. 9.

54. Manning, *Swordsmen*, 93. Manning also shows the extent to which Essex was hardly unique in his attitude to knighthood, although he was perhaps a bit freer with the honor than most. Charles I, over the course of five years of fighting in the English Civil War, knighted a mere sixty-six men. Charles Carlton, *Going to the Wars: The Experience of the British Civil Wars, 1638–1651* (London: Routledge, 1992), 81. Halsall examines the early medieval history of the challenge to duel as part of "military" thinking. Guy Halsall, *Warfare and Society in the Barbarian West, 450–900* (London: Routledge, 2003), 193.

55. Farmer and Falkiner, "Farmer's Chronicles," 106.

56. CSPI 1:417.

57. CSPI(NS) 93, 131.

58. Quotations from *Journal of Things Done in Ireland*, May 8–July 14, 1599, f. 202v; Harington, *Nugae Antiquae*, 1:271–72.

59. CCM 3:306.

60. Ciaran Brady, *The Chief Governors: The Rise and Fall of Reform Government in Tudor Ireland, 1536–1588* (New York: Cambridge University Press, 1994), 136–38.

61. SP 63/31/28 (i).

62. SP 63/27/56. There were 340 men at Carrickfergus, but this was unusual.

63. CSPI(NS) 1–2.

64. "The plat of the forte of Maribroughe," Unsigned, Undated, *ca*. 1560, TCD MS 1209/10.

65. Harington, *Nugae Antiquae*, 1:270.

66. Dymmok, *Treatise*, 32.

67. Coyne and livery was one of the institutions most condemned by early modern English writers. The actual details of its operation are obscure and highly variable. A "bonaght" was a similar imposition originally used to support the Scots mercenaries. Cosgrove, *Medieval Ireland*, 426, 541–42; Dymmok, *Treatise*, 8–9; Nicholls, *Gaelic and Gaelicized Ireland*, 34–40, 104; David Edwards, "The Butler Revolt of 1569," *Irish Historical Studies* 28 (1993), 235–37.

68. Colm Lennon, *Sixteenth-Century Ireland: The Incomplete Conquest*, rev. ed. (Dublin: Gill & Macmillan, 2005), 174, 182–83, Brady, *Chief Governors*, 216–44. A "composition" was an attempt to reform cess into a more regular and centralized system of payments that would support regional armies (especially in Connacht and Munster) and in theory reduce both abuses of the cess and the keeping of private armies via coyne. Ibid., 141–54. To greatly simplify the process, think of coyne replaced by cess replaced by composition. Each step was an attempt at greater centralization, implemented only haltingly, regionally, and over much resistance.

69. Brady, *Chief Governors*, 222–24.

70. McGurk, *Elizabethan Conquest*, 195–219.

71. CSPI 1:441.

72. Nicholas P. Canny, *Making Ireland British, 1580–1650* (Oxford: Oxford University Press, 2001), 101–2.

73. Brady, "Captain's Games." In an important revision, David Edwards argues that many captains preferred a state of unrest, since it allowed them to use martial law to execute offenders and claim one-third of their movable property. "Escalation," 67–68. Rory Rapple's basic argument that the captains used the authority granted to them as the basis to extend their own power fits in this interpretation. Rapple's book did not appear in time for full consideration here. *Martial Power*, 234–37, 305–8.

74. SP 63/96/10 (ii).

75. McGurk, *Elizabethan Conquest*, is the best source for the conditions within the army during the Nine Years War, esp. chaps. 8 and 10.

76. CCM 3:350.

77. McGurk, *Elizabethan Conquest*, 206.

78. English peasants would have had less meat; Irish peasants less bread. This generalization is highly subjective and variable, but not wildly off the mark. Christopher Dyer, *Standards of Living in the Later Middle Ages: Social Change in England c. 1200–1520* (Cambridge: Cambridge University Press, 1989), 152–60.

79. McClure, *Harington's Letters*, 73–74.

80. CSPI 6:195.

81. CSPI 7:59.

82. McGurk, *Elizabethan Conquest,* 241.

83. Hayes-McCoy, "Strategy and Tactics in Irish Warfare."

84. For the 1566 expedition see Falls, *Elizabeth's Irish Wars*, 96–97. Docwra describes his own expedition in Henry Docwra, "A Narration of the Services Done by the Army Imployed to Lough-Foyle," in *Miscellany of the Celtic Society*, ed. John O'Donovan (Dublin: The Celtic Society, 1849), 189–325 (hereafter "Docwra's Narration"). See also Hiram Morgan, "Missions Comparable? The Lough Foyle and Kinsale Landings of 1600 and 1601," in *The Battle of Kinsale*, ed. Hiram Morgan (Bray, Ireland: Wordwell, 2004), 73–89; John McGurk, "The Pacification of Ulster, 1600–3," in *Age of Atrocity: Violence and Political Conflict in Early Modern Ireland*, ed. David Edwards, Pádraig Lenihan, and Clodagh Tait (Dublin: Four Courts, 2007), 119–29.

85. See the troop distribution to those garrisons (excluding Culmore, which may have been abandoned by this time), CSPI 10:265.

86. CSPI 10:259

87. CSPI 10:236

88. McGurk, "Pacification of Ulster," 127.

89. Moryson, *Itinerary*, 2:218–19.

90. Harington, *Nugae Antiquae*, 1:267–68.

91. H. C. [Hugh Collier], *Dialogue of Silvynne and Peregrynne (PRO SP 63/203/119)*, Corpus of Electronic Texts, University College Cork, http://www.ucc.ie/celt/published/E590001–001/index.html, 23, 34, 48, 70.

92. CSPI 1:180. Surviving sixteenth-century maps of Ireland continued to represent the interior of Ulster as something of a forested unknown even as late as 1587 and 1599. See especially TCD MS 1209/2, "Irelande," *ca.* 1587; and Baptista Boazio's 1599, figure 4 in Bernhard Klein, "Partial Views: Shakespeare and the Map of Ireland," *Early Modern Literary Studies* 4.2 (1998): 15.

93. CCM 3:312.

94. CSPI 8:42–43.

95. CSPI 7:449–52.

96. Charles Hughes, ed., *Shakespeare's Europe: Being a Survey of the Condition of Europe at the End of the 16th Century*, 2nd ed. (New York: Benjamin Blom, 1967), 238–39. For more on the Gaelic use of guerrilla war, see note 23 to chapter 1.

97. In addition to the earlier discussion of Irishmen in Sidney's army in Chapter 1, see Wayne E. Lee, "Subjects, Clients, Allies or Mercenaries? The British Use of Irish and Indian Military Power, 1500–1815" in *Britain's Oceanic Empire: Projecting Imperium in the Atlantic and Indian Ocean Worlds, ca. 1550–1800*, ed. H. V. Bowen, Elizabeth Mancke, and John G. Reid (Cambridge: Cambridge University Press, forthcoming).

98. Harington, *Nugae Antiquae*, 1:261–63.

99. James Hogan and N. McNeill O'Farrell, eds., *The Walsingham Letter-Book or Register of Ireland, May 1578 to December 1579* (Dublin: Irish Manuscripts Commission, 1959), 119.

100. *Certain Principall Matters.*

101. Liam Miller and Eileen Power, eds., *Holinshed's Irish Chronicle: The Historie of Irelande from the First Inhabitation thereof, unto the Yeare 1509* (Dublin: Dolmen, 1979), 303.

102. Thomas Churchyard, *A Generall Rehearsall of the Warres*, STC 5235.2 (London: [John Kingston for] Edward White, 1579), f3; CSPI(NS) 411.

103. CSPI(NS) 692; Thomas Lee to Richard Lee, December 22, 1597, Folger Shakespeare Library, Washington, DC, MS V.b.214, f. 246v–47v. Even if these accounts were exaggerated, and admitting that the English were perfectly capable of similar atrocities, that does not change the role of English soldiers' fears in escalating violence.

104. Falls, *Elizabeth's Irish Wars*, 215–17, 225, 244, 326; CSPI 6:231–32, 249, 429; Perrot, *Chronicle*, 140, 143.

105. English ethnic prejudice against the Irish has been examined in great detail. For its medieval roots, see R. R. Davies, *Domination and Conquest: The Experience of Ireland, Scotland and Wales, 1100–1300* (Cambridge: Cambridge University Press, 1990), 21. For the sixteenth century, see Quinn, *Elizabethans and Irish*; Canny, *Elizabethan Conquest*, chap. 7; Patricia Palmer, "'An Headlesse Ladie' and 'A Horses Loade of Heades': Writing the Beheading," *Renaissance Quarterly* 60 (2007): 25–57; Canny, *Making Ireland British*, 42–55. In addition there has been extensive analysis of Edmund Spenser's writings, especially his *A View of the State of Ireland*, ed. Andrew Hadfield and Willy Maley (Oxford: Blackwell, 1997).

106. CSPI 8:354; Edward M. Hinton, "Rych's *Anothomy of Ireland*, with an Account of the Author," *Proceedings of the Modern Language Association* 55 (1940): 83.

107. Dawtrey, *Falstaff Saga*, 191.

108. Quinn, *Elizabethans and Irish*, 162–69.

109. Quotations from Hiram Morgan, ed., "A Booke of Questions and Answars Concerning the Warrs or Rebellions of the Kingdome of Irelande," *Analecta Hibernia* 36 (1995): 93; Harington, *Nugae Antiquae*, 1:176–77. For concerns about the authenticity of submissions, see *Certain Principall Matters*; Lee, "Brief Declaration," 1:111–12.

110. *Certain Principall Matters.* When the captains wrote about the problems with submission, they often advocated forcing submitted rebels to do active service on behalf of the English to prove their loyalty and to separate them from the other rebels by blood service. Lee, "Discoverye & Recoverye," f. 51r; Morgan, "Booke of Questions," 112.

111. Barnabe Rich, *Allarme to England*, STC 20979 (London: Henrie Middleton, 1578), d.ii.

112. CSPI 8:45–51.

113. In this sense, I would argue that martial law in Ireland operated in a way not unlike the *requerimiento* in the Spanish New World, providing a necessary moral and legal foundation for nearly unlimited violence. Patricia Seed, "Taking Possession and Reading Texts: Establishing the Authority of Overseas Empires," *William and Mary Quarterly*, 3rd ser., 49 (1992): 204. Edwards's work on martial law is crucial; see chapter 1, note 17.

114. Robert Hare, *Treatise on Military Discipline* (1556), Cotton MS, Julius F.V., BL.

115. Lee, "Discoverye & Recoverye," f. 76v.

116. CCM 3:311–12.

117. CSPI 8:440–47.

118. CSPI 9:116–17.

119. CSPI 7:383–86.

120. CSPI 8:45–51, 353–56.

121. Brady, "Captains' Games," 158; Vincent Carey, "'What pen can paint or tears atone?': Mountjoy's Scorched Earth Campaign," in *The Battle of Kinsale*, ed. Hiram Morgan (Bray, Ireland: Wordwell, 2004), 205–16; McGurk, "Pacification of Ulster," 121, 128.

122. SP 63/16/71. Elizabeth's instructions to Mountjoy in January 1600 (and afterwards) reflect an abiding desire to use mercy but also a hardening of attitude toward accepting submissions too casually. This was partly a specific response to O'Neill's repeated temporary submissions and truces. CSPI 8:440–47.

123. CSPI 6:490–91; Vincent P. Carey, "Elizabeth I and State Terror in Ireland," in *Elizabeth I and the Sovereign Arts*, ed. Donald Stump and Carol Levin (forthcoming; I am grateful to Vincent Carey for sharing this manuscript with me); Rapple, *Martial Power*, 208–12.

124. Carey documents the strategy of devastation in 1601–3 and Elizabeth's complicity in "Elizabeth I and State Terror" and "What Pen Can Paint." Rapple discusses her ability to demand bloody suppression of rebellion in earlier years in *Martial Power*, 205–7.

125. Geoffrey Parker, "Early Modern Europe," in *The Laws of War: Constraints on Warfare in the Western World*, ed. Michael Howard, George J. Andreopoulos, and Mark R. Shulman (New Haven, CT: Yale University Press, 1994), 43–44; Robert A. Kann, "The Law of Nations and the Conduct of War in the Early Times of the Standing Army," *Journal of Politics* 6 (1944): 79–80.

126. Quoted and discussed in Adam N. McKeown, *English Mercuries: Soldier Poets in the Age of Shakespeare* (Nashville, TN: Vanderbilt University Press, 2009), 113.

127. John Nalson, ed., *An Impartial Collection of the Great Affairs of State*, Wing N106 (London: Printed for S. Mearne, T. Dring, B. Tooke, T. Sawbridge, and C. Mearne, 1682–83), 1:11–12.

Chapter 3

Epigraph from William Waller, *Vindication of the Character and Conduct of Sir William Waller, Knight; Commander in Chief of the Parliament Forces in the West: Explanatory of His Conduct in Taking Up Arms Against King Charles the First* (London: J. Debrett, 1793).

1. *A True Relation of the Manner of the Execution of Thomas Earle of Strafford with the Severall Passages and Circumstances, together with his Speech to the People on the Scaffold, the 12. of May, 1641*, Wing T3002A (London: J. A., 1641), 1. Other accounts are in Earl of Strafford, *The Last Speeches of Thomas Wentworth, Late Earle of Strafford, and Deputy of Ireland: The One in the Tower, the Other on the Scaffold on Tower-Hill, May the Twelfth. 1641*, Wing S5785B (London, 1641); John H. Timmis III, *Thine Is the Kingdom: The Trial for Treason of Thomas Wentworth, Earl of Strafford, First Minister to King Charles I and Last Hope of the English Crown* (University: University of Alabama Press, 1974), 171–72; C. V. Wedgwood, *Thomas Wentworth, First Earl of Strafford, 1593–1641* (London: Cape, 1961), 310–89. The discussion of the role of the London public in his death is from Keith Lindley, *Popular Politics and Religion in Civil War London* (Aldershot, UK: Scolar, 1997), 19–26; Brian Manning, *The English People and the English Revolution, 1640–1649* (London: Heinemann, 1976), 8–18.

2. Anon., *The Petition of the Citizens of London to Both Houses of Parliament wherein is a Demonstration of their Grievances*, Wing P1784 (London: John Aston, 1641).

3. The exact legalities of Strafford's trial are well studied. The most persuasive account is D. Alan Orr, *Treason and the State: Law, Politics, and Ideology in the English Civil War* (Cambridge: Cambridge University Press, 2002), 61–100.

4. Quoted in Lindley, *Popular Politics and Religion*, 21.

5. Barbara Donagan, *War in England, 1642–1649* (Oxford: Oxford University Press, 2008), 1–2.

6. D. Alan Orr, *Treason and the State: Law, Politics and Ideology in the English Civil War* (Cambridge: Cambridge University Press, 2002), 2. Much of the following is based on Orr's analysis of the treason law and of Strafford's trial. See also the discussion in chapter 1.

7. Ibid., 2.

8. Ibid., 3–5, 39; John Morrill, *Revolt in the Provinces: The People of England and the Tragedies of War, 1630–1648*, 2nd ed. (London: Longman, 1999), 20; David Cressy, "Revolutionary England 1640–1642," *Past and Present*, 181 (2003): 35–72. In this article and in *England on Edge: Crisis and Revolution, 1640–1642* (Oxford: Oxford University Press, 2006), David Cressy emphasizes the role of events over some brewing ideological shift.

9. Cressy, "Revolutionary England," 69–71.

10. Donagan, *War in England*, 8, 11–12, chaps. 7–10, and 227. The study of violence in the English Civil War has been well served by historians. Of particular importance is the work of Barbara Donagan (cited here and elsewhere), as well as Charles Carlton, *Going to the Wars: The Experience of the British Civil Wars, 1638–1651* (London: Routledge, 1992); Ian Gentles, *The English Revolution and the Wars in the Three Kingdoms, 1638–1652* (Harlow, UK: Pearson Longman, 2007); Ian Roy, "England Turned Germany? The Aftermath of the Civil War in Its European Context," *Transactions of the Royal Historical Society*, 5th ser., 28 (1978): 127–44; Will Coster, "Massacre and Codes of Conduct in the English Civil War," in *The Massacre in History*, ed. Mark Levene and Penny Roberts (New York: Berghahn Books, 1999), 89–105; Stephen Porter, *Destruction in the English Civil Wars* (Stroud, UK: Sutton, 1994).

11. The fullest treatment of the two "Bishops' Wars" is Mark Charles Fissel, *The Bishops' Wars: Charles I's Campaigns Against Scotland, 1638–1640* (Cambridge: Cambridge University Press, 1994).

12. Quotations from HMC v.2, 3rd report, 1, 3; Charles Carlton, *Charles I: The Personal Monarch* (London: Routledge & Kegan Paul, 1983), 224.

13. Micheál Ó Siochrú, "Atrocity, Codes of Conduct and the Irish in the British Civil Wars 1641–1653," *Past and Present* 195 (2007): 55–86; Nicholas Canny, "What Really Happened in Ireland in 1641?," in *Ireland From Independence to Occupation, 1641–1660*, ed. Jane H. Ohlmeyer (Cambridge: Cambridge University Press, 1995), 24–42; Nicholas P. Canny, *Making Ireland British, 1580–1650* (Oxford: Oxford University Press, 2001), 461–550; essays by Brian Mac Cuarta and Kenneth Nicholls in *Age of Atrocity: Violence and Political Conflict in Early Modern Ireland*, ed. David Edwards, Pádraig Lenihan, and Clodagh Tait (Dublin: Four Courts, 2007), 154–75 and 176–91. For response in England, see Carlton, *Going to the Wars*, 33–35; Keith J. Lindley, "The Impact of the 1641 Rebellion upon England and Wales, 1641–5," *Irish Historical Studies* 18 (1972): 143–76; David A. O'Hara, *English Newsbooks and Irish Rebellion, 1641–1649* (Dublin: Four Courts, 2006), 27–54. The original depositions describing the violence are now available online at http://www.tcd.ie/history/1641/.

14. Tanner MS v.64 f.97, Bodleian Library, Oxford (hereafter Tanner MS); Nehemiah Wallington, *Historical Notices of Events Occurring Chiefly in the Reign of Charles I* (London: Richard Bentley, 1869), 2:41–42.

15. Richard Baxter, *Reliquiae Baxterianae*, Wing B1370 (London: T. Parkhurst, J. Robinson, F. Larence, and F. Dunton, 1696), 33.

16. David Scott, *Politics and War in the Three Stuart Kingdoms, 1637–49* (Houndmills: Palgrave Macmillan, 2004), 33; Ann Hughes, *The Causes of the English Civil War* (New York: St. Martin's, 1991), 173–77; Malcolm Wanklyn and Frank Jones, *A Military History of the English Civil War, 1642–1646* (New York: Pearson Longman, 2005), 8, 39–40.

17. Wanklyn and Jones, *Military History*, 4.

18. Mark Stoyle argues that even within England there was a substantial presence of non-English, and that their presence affected the violence of the war. It remains true, however, that the competing armies in this period were primarily English. *Soldiers and Strangers: An Ethnic History of the English Civil War* (New Haven, CT: Yale University Press, 2005).

19. The history of the "law of going to war," or the *jus ad bellum*, is extensive but for the most part affects the argument here only in the sense that belief in it provided some license to the participants; more below on "license." Frederick H. Russell, *The Just War in the Middle Ages* (Cambridge: Cambridge University Press, 1975); Robert C. Stacey, "The Age of Chivalry," in *The Laws of War: Constraints on Warfare in the Western World*, ed. Michael Howard, George J. Andreopoulos, and Mark R. Shulman (New Haven, CT: Yale University Press, 1994), 27–39; William Ballis, *The Legal Position of War: Changes in its Practice and Theory From Plato to Vattel* (New York: Garland, 1973); Richard Tuck, *The Rights of War and Peace: Political Thought and the International Order From Grotius to Kant* (Oxford: Oxford University Press, 1999); James Turner Johnson, *Ideology, Reason, and the Limitation of War: Religious and Secular Concepts, 1200–1740* (Princeton, NJ: Princeton University Press, 1975); Roland Bainton, *Christian Attitudes Toward War and Peace: A Historical Survey and Critical Re-evaluation* (New York: Abingdon, 1960); Philippe Contamine, *War in the Middle Ages*, tr. Michael Jones (Oxford: Blackwell, 1984), 282–92.

20. Sloane MS 5247, f.20, BL, cited in Donagan, *War in England*, 17.

21. J. V. Capua, "The Early History of Martial Law in England from the Fourteenth Century to the Petition of Right," *Cambridge Law Journal* 36 (1977): 154; Ann Hughes, "The King, the Parliament and the Localities during the English Civil War," *Journal of British Studies* 24 (1985): 243.

22. Anon., *A True and Exact Relation of the Manner of His Maiesties Setting Up of His Standard at Nottingham*, Wing T2452 (London: F. Coles, 1642), quotation at A2r.

23. Wanklyn and Jones, *Military History*, 12–16.

24. Wanklyn and Jones, *Military History*, 217–18. Going into winter quarters was not an inviolable rule, and sieges in particular often continued through the winter. But even when forces were not entirely stood down, activity greatly diminished.

25. Wanklyn and Jones, *Military History*, 39–43.

26. Carlton, *Going to the Wars*, 150–52.

27. Geoffrey Parker, *The Military Revolution: Military Innovation and the Rise of the West, 1500–1800*, 2nd ed. (Cambridge: Cambridge University Press, 1996), 41.

28. John Landers, *The Field and the Forge: Population, Production, and Power in the Pre-Industrial West* (Oxford: Oxford University Press, 2003), 227–75, explains the problems of controlling and moving through the inevitably diffuse "demographic space" of a preindustrial economy. By the seventeenth century the resources necessary for military operations (men, money, food) were just beginning to become concentrated in urban environments (hence the key role of London) but for the most part still emerged from the diffuse and poorly connected countryside. The garrisons were intended to control and tap that space.

29. Donagan, *War in England*, 66.

30. Baxter, *Reliquiae Baxterianae*, 44

31. George Satterfield argues that this kind of warfare could be decisive in shaping not only strategy but ultimate success or failure. *Princes, Posts and Partisans: The Army of Louis XIV and Partisan Warfare in the Netherlands, 1673–1678* (Leiden, The Netherlands: Brill, 2003), passim, esp. 9; Parker, *Military Revolution*, 40–42.

32. Carlton, *Going to the Wars*, 206. Carlton nevertheless believes the larger battles were decisive to the outcome (113).

33. Unfortunately, Waller's papers do not survive. Many of his official letters are in the CSPD, and two autobiographical texts survive: Waller, *Vindication*; Waller, "Recollections," in *The Poetry of Anna Matilda*, ed. Hannah Cowley (London: John Bell, 1788), 97–135 (this is often called "The Experiences"). There is one modern military biography, John Adair, *Roundhead General: The Campaigns of Sir William Waller* (Thrupp, UK: Sutton, 1997), as well as an extensive account of him in the DNB by Barbara Donagan. See also Donagan, "Understanding Providence: The Difficulties of Sir William and Lady Waller," *Journal of Ecclesiastical History* 39 (1988): 433–44. Additional sources are discussed below. Quotation from *The Souldiers Report concerning William Wallers Fight Against Basing-House on Sunday Last November the 12 1643*, Wing S4431 (London: Printed by John Hammond, 1643).

34. Lawrence Stone, *The Crisis of the Aristocracy*, abridged ed. (Oxford: Oxford University Press, 1967), 130–31.

35. Roger B. Manning, *Swordsmen: The Martial Ethos in the Three Kingdoms* (New York: Oxford University Press, 2003), 17–19; Donagan, *War in England*, 40–54, 232–34; Martyn Bennett, "The Officer Corps and Army Command in the British Isles, 1620–1660," in *The Chivalric Ethos and the Development of Military Professionalism*, ed. D. J. B. Trim (Leiden, The Netherlands: Brill, 2003), 291–317; Ian Roy, "The Profession of Arms," in *The Professions in Early Modern England*, ed. Wilfrid Prest (London: Croom Helm, 1987), 187–94.

36. A study of 1,629 senior royalist officers found about half with pre-war experience. P. R. Newman, *Royalist Officers in England and Wales, 1642–1660: A Biographical Dictionary* (New York: Garland, 1981), vii–viii.

37. Donagan, *War in England*, 49; Carlton, *Going to the Wars*, 21. Englishmen also acquired some experience of military activity through the militia, and

London's artillery companies were particularly well trained. Henrik Langelüddecke, "'The chiefest strength and glory of this kingdom': Arming and Training the 'Perfect Militia' in the 1630s," *English Historical Review* 118 (2003): 1264–1303.

38. Donagan, *War in England*, 33–40; Donagan, "Halcyon Days and the Literature of War: England's Military Education before 1642," *Past and Present* 147 (1995): 65–100; John S. Nolan, "The Militarization of the Elizabethan State," *Journal of Military History* 58 (1994), 391–420; David R. Lawrence, *The Complete Soldier: Military Books and Military Culture in Early Stuart England, 1603–1645* (Leiden, The Netherlands: Brill, 2009). See also note 94.

39. Debate continues about how bad the Thirty Years War actually was. Henry Kamen, "The Economic and Social Consequences of the Thiry Year's War," *Past and Present* 39 (1968): 44–61; C. R. Friedrichs, "The War and German Society," in *The Thirty Years' War*, ed. Geoffrey Parker (New York: Barnes and Noble, 1987), 208–15; Geoff Mortimer, *Eyewitness Accounts of the Thirty Years War, 1618–48* (Houndmills, UK: Palgrave, 2002), passim, esp. 150; John Theibault, "The Rhetoric of Death and Destruction in the Thirty Years' War," *Journal of Social History* 27 (1993): 271–90; Bernd Roeck, "The Atrocities of War in Early Modern Art," in *Power, Violence and Mass Death in Pre-Modern and Modern Times*, ed. Joseph Canning, Hartmut Lehmann, and Jay Winter (Aldershot, UK: Ashgate, 2004), 138–39; John Theibault, "The Demography of the Thirty Years War Re-revisited: Günther Franz and his Critics," *German History* 15 (1997): 1–21; Ronald G. Asch, "'Wo der soldat hinkömbt, da is alles sein': Military Violence and Atrocities in the Thirty Years War Re-examined," *German History* 18 (2000): 291–309; Asch, *The Thirty Years War: The Holy Roman Empire and Europe, 1618–48* (Basingstoke, UK: Macmillan, 1997), 177–84. Barbara Donagan has addressed the role of this war in generating a genre of atrocity narrative in England. "Atrocity, War Crime, and Treason in the English Civil War," *American Historical Review* 99 (1994), 1137–66. For the intellectual reaction to the horrors of the war, see Geoffrey Parker, "Early Modern Europe," in *The Laws of War: Constraints on Warfare in the Western World*, ed. Michael Howard, George J. Andreopoulos, and Mark R. Shulman (New Haven, CT: Yale University Press, 1994), 54; Otto Ulbricht, "The Experience of Violence During the Thirty Years War: A Look at the Civilian Victims," in *Power, Violence and Mass Death in Pre-Modern and Modern Times*, ed. Joseph Canning, Hartmut Lehmann, and Jay Winter (Aldershot, UK: Ashgate, 2004), 127.

40. Olivier Chaline, *La bataille de la Montagne Blanche (8 novembre 1620: Un mystique chez les guerriers* (Paris: Noesis, 1999), 374–75, 395. Peter H. Wilson's argument deemphasizing the religious aspects of the war appeared too late to be considered fully here. *The Thirty Years War: Europe's Tragedy* (Cambridge, MA: Harvard University Press, 2009).

41. All three layers overlapped; considering them separately is a convenient device, but it is not one that people then would necessarily always have recognized. Barbara Donagan, "Codes and Conduct in the English Civil War," *Past and Present* 118 (1988); 65–95; Donagan, *War in England*, 128–29, 134–44. A valuable short survey is Parker, "Early Modern Europe," 40–58.

42. John A. Lynn, "The Evolution of Army Style in the Modern West, 800–2000," *International History Review* 18 (1996): 516–18; Lynn, *Women, Armies, and Warfare in Early Modern Europe* (Cambridge: Cambridge University Press, 2008), 21–23, 27–32, 145–50.

43. Donagan, *War in England*, 9–10, discusses providentialism for both puritans and royalists.

44. Robert Monro, *Monro, His Expedition With the Worthy Scots Regiment Called Mac-Keys*, ed. William S. Brockington Jr. (Westport, CT: Praeger, 1999), 52, 196.

45. Waller, "Recollections," 130. Colonel William Brereton feared God's "judgement and curse" as a result of his soldiers' actions. Tanner MS v.62/2A f.420v–21r, cited in Donagan, *War in England*, 232n27.

46. Monro, *Monro, His Expedition*, 25.

47. Ibid., 22, 51–52.

48. Adair, *Roundhead General*, 165. A similar incident occurred at the battle of Newbury. Gentles, *English Revolution*, 190; Manning, *Swordsmen*, 208–10. An overdeveloped sense of honor could also be virtual suicide; the earl of Northampton, unhorsed and surrounded, was offered quarter, but "he scorned to take quarter off such base rogues and Rebels as they were." So they killed him. Quoted in Carlton, *Going to the Wars*, 136–37. Essex, frustrated at not bringing the war to a quick close, and perhaps to jolt his parliamentary masters, suggested offering battle as a trial by ordeal in the summer of 1643. Wanklyn and Jones, *Military History*, 91–92. For "chivalry" in this period see Manning, *Swordsmen*; J. S. A. Adamson, "Chivalry and Political Culture in Caroline England," in *Culture and Politics in Early Stuart England*, ed. Kevin Sharpe and Peter Lake (Stanford, CA: Stanford University Press, 1993), 193–97.

49. Adair doubts a real connection between the accepted challenge and the following battle. *Roundhead General*, 89.

50. This definition is based on Richard Cust, "Honour and Politics in Early Stuart England: The Case of Beaumont *v.* Hastings," *Past and Present* 149 (1995): 57–94 (quotation on 59); Mervyn James, *English Politics and the Concept of Honour 1485–1642* (Oxford: Past and Present Society, 1978); Adamson, "Chivalry and Political Culture"; Richard C. McCoy, *The Rites of Knighthood: The Literature and Politics of Elizabethan Chivalry* (Berkeley: University of California Press, 1989); Donagan, *War in England*, 166–67, 227, 360–61.

51. Waller, *Vindication*, 4.

52. Donagan, "Understanding Providence," 441–42; Wanklyn and Jones, *Military History*, 104.

53. Waller, *Vindication*, 14.

54. Henry Townshend, *Diary of Henry Townshend of Elmley Lovett, 1640–1663*, ed. J. W. Willis Bund (London: Hughes and Clarke, 1915–20), 2:296.

55. John Birch, *Military Memoir of Colonel John Birch, Sometime Governor of Hereford in the Civil War between Charles I. and the Parliament*, ed. T. W. Webb (Westminster, UK: Camden Society, 1873), 217.

56. Adair, *Roundhead General*, 48.

57. Peter Young and Richard Holmes, *The English Civil War: A Military History of the Three Civil Wars 1642–1651* (London: Eyre Methuen, 1974), 181; Wanklyn and Jones, *Military History*, 142–43, 157–58.

58. CCM 3:311–12 (see pp. 58–59, chapter 2).

59. Reprinted in Waller, *Vindication* (quotation on xiii–xiv); Manning, *Swordsmen*, 36; Townshend, *Diary*, 1:149–50; Robert Bell, ed., *Memorials of the Civil War: Comprising the Correspondence of the Fairfax Family: Comprising the Correspondence of the Fairfax Family with the Most Distinguished Personages Engaged in That Memorable Contest* (London: Richard Bentley, 1849), 1:185–86; Eliot Warburton, ed., *Memoirs of Prince Rupert and the Cavaliers* (London: Richard Bentley, 1849), 3:39–40.

60. Edward Walker, "His Majesty's Happy Progress and Success from the Thirtieth of March to the Twenty third of November, 1644," in *Historical Discourses, Upon Several Occasions* (London: Printed by W. B. for Sam. Keble, 1705), 7. Compare Wanklyn and Jones, *Military History*, 157–58, 161–62.

61. Walker, "His Majesty's Happy Progress," 8, quotation on 13.

62. Earl of Clarendon, *The History of the Rebellion and Civil Wars in England* (Oxford: Clarendon Press, 1992), 3:347–48, 351–52.

63. CSPD 1644, 214–15; Clarendon, *History*, 3:348. Strategies shifted year to year. With the outbreak of war in 1642, Parliament hoped in theory to force the king to give up and return to London, but their primary goals in practice were to secure key locations. In 1643 they focused on achieving victory in battle, which would force the king to give up his cause. Wanklyn and Jones, *Military History*, 43, 82.

64. Waller, *Vindication*, 8.

65. Tanner MS v.61 f.32.

66. CSPD 1644, 220; 233.

67. The following is based on Carlton, *Going to the Wars*, 46, 66–69, 340; Gentles, *English Revolution*, 67; J. R. Hale, *War and Society in Renaissance Europe, 1450–1620* (Leicester, UK: Leicester University Press, 1985), 78–79; Donagan, *War in England*, 218–24; Godfrey Davies, "The Parliamentary Army under the Earl of Essex, 1642–5," *English Historical Review* 49 (1934): 32–54.

68. *A True Relation of the Taking of the City, Minister, and Castle of Lincolne*, Wing T3056 (London: R. Cotes, 1644), quotation on A3; Baxter, *Baxterianae*, 53; Carlton, *Going to the Wars*, 174, 242; Barbara Donagan, "Prisoners in the English Civil War," *History Today* 41 (1991): 30.

69. Carlton, *Going to the Wars*, 182; Ronald Hutton, *The Royalist War Effort, 1642–1646*, 2nd ed. (London: Routledge, 1999), 22–23; Bennett, "The Officer Corps," 301–12.

70. County-trained bands were also tapped more or less successfully by both sides. Often they were merely drained of their store of arms, but sometimes the men themselves agreed in a body to serve one side or the other. Young and Holmes, *English Civil War*, 89.

71. Waller, *Vindication*, 17; CSPD 1644, 300–301.

72. Young and Holmes, *English Civil War*, 148.

73. For Waller's army at Cropredy, see Margaret Toynbee and Peter Young, *Cropredy Bridge, 1644: The Campaign and the Battle* (Kineton, UK: Roundwood, 1970), 80–83. For desertion from Walker's army, see Walker, *Happy Progress*, 34; quotations from *Mercurius Aulicus*, June 8, 1644, Thomason E.52(7); and from CSPD 1644, 300–301.

74. John Adair, "The Court Martial Papers of Sir William Waller's Army, 1644," *Journal of the Society for Army Historical Research* 44 (1966): 205–26. As discussed above, the conception of military law as one of three pillars of the "legal" restraints on violence in war is from Donagan, "Codes and Conduct." 76. Donagan greatly expands on these ideas in *War in England*, 136, 141–56 (although preferring the phrase "articles of war"). I prefer "military law," but it should be recognized that this is not a reference to martial law or rule by the military. Also crucial to the following discussion is Margaret Griffin, *Regulating Religion and Morality in the King's Armies, 1639–1646* (Leiden, The Netherlands: Brill, 2004), esp. chap. 1 and 83–97.

75. The following is merely a sample; most exist in multiple editions, and also were substantially copied by other issuing generals: "Articles of War—1642," *Journal of the Society of Army Historical Research* 9 (1930): 117–23; Charles I, *Military Orders And Articles Established by His Majesty, for the Better Ordering and Government of His Majesties Army: Also Two Proclamations, One against Plundring and Robbing. The Other against Selling or Buying of Armes and Horse*, Wing C2497A (Oxford: Leonarch [*sic*] Lichfield, 1644; the first edition is from 1642); *Lawes and Ordinances of Warre, Established for the Better Conduct of the Army by His Excellency the Earle of Essex*, Wing E3314 (London: John Partridge and John Rothwell, 1642); *Orders Established the 14th of this Present January, by His Excellency Sir Thomas Fairfax, for Regulating the Army, and for the Soldiers Paying of Quarters*, Wing E740 (London: Edward Husband, 1647).

76. Charles I, *Military Orders*, clauses 70, 71, 83. Nearly identical clauses can be found in the other codes cited above.

77. Charles I, *Military Orders*, 25–27; Adair, "Court Martial Papers," 219, 223.

78. See chapter 4, note 78.

79. Adair, "Court Martial Papers." A ninth robbery charge was for robbing a fellow soldier.

80. The idea of a contrast between an imagined battlefield and a real one is developed by John A. Lynn, although my interpretation of the medieval vision of battlefield success takes a different path than his. *Battle: A History of Combat and Culture* (Boulder, CO: Westview, 2003).

81. Richard W. Kaeuper, *Chivalry and Violence in Medieval Europe* (Oxford: Oxford University Press, 1999), 172–73; David Crouch, *William Marshal: Knighthood, War and Chivalry, 1147–1219*, 2nd ed. (London: Longman, 2002), 179–206; John France, *Western Warfare in the Age of the Crusades, 1000–1300* (Ithaca, NY: Cornell University Press, 1999), 138; Albert Lynn Winkler, "The Swiss and War: The Impact of Society on the Swiss Military in the Fourteenth and Fifteenth Centuries" (PhD diss., Brigham Young University, 1982), 41. For the "radical

individualism" of the chivalric tradition, see Arthur B. Ferguson, *The Chivalric Tradition in Renaissance England* (Washington, DC: Folger Shakespeare Library, 1986), 17. James Scott Wheeler, *The Making of a World Power: War and the Military Revolution in Seventeenth-Century England* (Stroud, England: Sutton Publishing, 1999), 5–7 presents a balanced discussion of a fourteenth-century shift to more disciplined infantry armies. Crucial also is Clifford Rogers, "Tactics and the Face of Battle," in *European Warfare, 1350–1750*, ed. Frank Tallett and D. J. B. Trim (Cambridge: Cambridge University Press, 2010), 203–35. I am grateful to Clifford Rogers for sharing this essay with me in manuscript, for pointing me to Winkler's useful dissertation, and for his many comments—although our disagreements remain. Stephen Morillo also provided some important insights for this section. In focusing on the imagined battlefield, I do not mean to downplay the importance of sieges in medieval warfare or to overemphasize "chivalry." For a recent statement on these subjects, see Kelly DeVries, "Medieval Warfare and the Value of a Human Life," in *Noble Ideals and Bloody Realities: Warfare in the Middle Ages*, ed. Niall Christie and Maya Yazigi (Leiden, The Netherlands: Brill, 2006), 27–56. For "chivalry" versus "professionalism" see D. J. B. Trim, "Introduction," in *The Chivalric Ethos and the Development of Military Professionalism*, ed. D. J. B. Trim (Leiden, The Netherlands: Brill, 2003), 1–40.

82. Cited in Kaeuper, *Chivalry and Violence*, 130 (see also 129–60 and 172–73).

83. For communalism as the root of the success of late medieval infantry and the aggressiveness of the Swiss pikemen, see Winkler, "The Swiss and War," 24, 54, 70, 137–40, 144, 169–70, 177; Azar Gat, *War in Human Civilization* (Oxford: Oxford University Press, 2006), 292, 458; Thomas F. Arnold, *The Renaissance at War* (London: Cassell, 2001), 84; Dennis E. Showalter, "Caste, Skill, and Training: The Evolution of Cohesion in European Armies from the Middle Ages to the Sixteenth Century," *Journal of Military History* 57 (1993): 425–26.

84. The idea of the individualism of the *Landsknecht*-style pikemen is my own, but see J. R. Hale, *Artists and Warfare in the Renaissance* (New Haven, CT: Yale University Press, 1990), 64–69. Hale documents how Swiss and German pikemen in the late fifteenth and early sixteenth century aped the costumes of the elite, in an allowed exception to the normal sumptuary laws, but he does not explore this as an attempt to participate in the military honor accrued by their betters. For more on the individualism and "libertinism" of the period's mercenary armies and the need for a new kind of discipline, see Lynn, *Women, Armies, and Warfare*, 40–44; Showalter, "Caste, Skill, and Training," 426–29.

85. For examinations of the emergence of a belief in collective discipline, see Arnold, *Renaissance at War*, 85–91; Maury D. Feld, "Middle-Class Society and the Rise of Military Professionalism: The Dutch Army, 1589–1609," in *The Structure of Violence: Armed Forces As Social Systems*, ed. Maury D. Feld (Beverly Hills, CA: SAGE, 1977), 179–85; David Eltis, *The Military Revolution in Sixteenth-Century Europe* (New York: St. Martin's, 1995), 52, 59, 103. Robert L. O'Connell, sees it as a shift in preferring men who could "stand fast" versus those with "ferocious aggressiveness." *Of Arms and Men* (New York: Oxford University Press, 1989), 118–19. For

the complexity of the fully developed pike and shot synchronization, see Wanklyn and Jones, *Military History*, 33; Rogers, "Tactics and the Face of Battle"; Arnold, *Renaissance at War*, 97–100; Wilson, *Thirty Years War*, 139–41.

86. The literature on the military revolution is extensive. As introductions these four works are indispensable: Jeremy Black, *A Military Revolution? Military Change and European Society 1550–1800* (Basingstoke, UK: Macmillan, 1991); Geoffrey Parker, *The Military Revolution: Military Innovation and the Rise of the West, 1500–1800*, 2nd ed. (New York: Cambridge University Press, 1996); Bert S. Hall, *Weapons and Warfare in Renaissance Europe: Gunpowder, Technology, and Tactics* (Baltimore: Johns Hopkins University Press, 1997); Clifford J. Rogers, ed., *The Military Revolution Debate: Readings on the Transformation of Early Modern Europe* (Boulder, CO: Westview, 1995).

87. Honoré Bonet, *The Tree of Battles*, trans. G. W. Coopland (Liverpool: University of Liverpool Press, 1949). These references are scattered throughout his text, but see esp. 119–21, 130–31. Pizan's list of virtues is similarly individualistic. Christine de Pizan, *The Book of Deeds of Arms and of Chivalry*, trans. Sumner Willard, ed. Charity Cannon Willard (University Park: Pennsylvania State University Press, 1999), 25. Later authors would not discard these virtues, but would add the need for officers to have technical skills and to become masters of drill and training. See Fernando González de León, "'Doctors of the Military Discipline': Technical Expertise and the Paradigm of the Spanish Soldier in the Early Modern Period," *Sixteenth Century Journal* 27 (1996): 70, 73, 82.

88. Geoffroi de Charny, *The Book of Chivalry of Geoffroi De Charny: Text, Context, and Translation*, ed. and trans. Richard W. Kaeuper and Elspeth Kennedy (Philadelphia: University of Pennsylvania Press, 1996), 89, 99–104, 151.

89. Kaeuper, *Chivalry and Violence*, 135; Catherine Hanley, *War and Combat, 1150–1270: The Evidence from Old French Literature* (Woodbridge, UK: Brewer, 2003), 112–16, 130, 226–28.

90. Vegetius I.1, I.26, preface to III, *Epitome of Military Science*, trans. N. P. Milner (Liverpool: Liverpool University Press, 1993); Latin text from Flavius Vegetius Renatus, *Epitoma rei militaris*, ed. Carl Lang (Stuttgart, West Germany: Teubner, 1967). I am grateful to Everett Wheeler for detailed discussions of the Latin here, but I have stayed with Milner's translation for the sake of other readers.

91. Pizan, *Book of Deeds*, 26–28, 66–68. A good discussion of the problems of Pizan's sources is in Everett Wheeler, "Christine de Pizan's *Livre des fais d'armes et de Chevalerie*: Gender and the Prefaces," *Nottingham Medieval Studies* 46 (2002): 119–22.

92. Matthew Sutcliffe, *The Practice, Proceedings, and Lawes of Armes* (London: Christopher Barker, 1593), B4r, 85. There are many contemporary manuals that exemplify this new emphasis on discipline, and the fact that Sutcliffe was doing so *before* volley fire is evidence of the evolutionary nature of this process from the fifteenth century into the seventeenth. Roger Ascham, writing in 1545, was still using the word "obedience," but his discussion is about the obedience of the archer, and the emphasis is clearly shifting toward collective obedience or discipline. *Toxophilus: The Schole of Shootinge*, STC 837 (London: Edouard Whytchurch, 1545), 25.

93. The production of volley fire is the subject of much of the military-revolution literature (see note 86). Most recently, see Geoffrey Parker, "The Limits to Revolutions in Military Affairs: Maurice of Nassau, the Battle of Nieuwpoort (1600), and the Legacy," *Journal of Military History* 71 (2007): 331–72.

94. J. R. Hale, "The Military Education of the Officer Class in Early Modern Europe," in *Renaissance War Studies* (London: Hambledon, 1983); Hale, *War and Society*, 56–57; Hale, "Printing and the Military Culture of Renaissance Venice," *Medievalia et Humanistica* 8 (1977): 21–62; González de León, "Technical Expertise," 64–65; Maurice J. D. Cockle, *A Bibliography of English Military Books Up to 1642 and of Contemporary Foreign Books* (London: Simpkin, Marshall, Hamilton, Kent, 1900); Carlton, *Going to the Wars*, 69–78; Donagan, *War in England*, 38–40; S. Porter, "The Fire-Raid in the English Civil War," *War and Society* 2 (1984): 331–32.

95. Donagan, *War in England*, 260–61; Donagan, "Did Ministers Matter? War and Religion in England, 1642–1649," *Journal of British Studies* 33 (1994): 130–37; *The Souldiers Pocket Bible*, Wing S4428 (London: G. B. and R. W. for G. C., 1643); John Eacherd, *Good Newes for All Christian Souldiers*, Wing E48 (London: Matthew Simmons, 1645); Samel Kem, *Orders Given Out; the Word, Stand Fast*, Wing K254 (London: I. M. for Michael Spark, 1646); Robert Ram, *The Soldiers Catechisme, Composed for the Parliaments Army*, Wing R196A (London: J. Wright, 1645). Royalists also used religion as a spur to governing the conduct of their soldiers. Griffin, *Regulating Religion and Morality*.

96. Hale cites an ideal of 150 men, two sergeants, and six corporals, but few of the muster lists I examined for Ireland showed so many noncommissioned officers. *War and Society*, 60. However, see the caution in James Raymond, *Henry VIII's Military Revolution: The Armies of Sixteenth-Century Britain and Europe* (London: Tauris Academic Studies, 2007), 59–61.

97. RV 1:132 and n2. As this source indicates, the movement to smaller companies was contested, but the end results were clear. Essex, during his 1599 campaign, created regiments on an ad hoc basis after the start of the campaign. John Harington, *Nugae Antiquae: Being a Collection of Original Papers, in Prose and Verse; Written during the Reigns of Henry VIII, Edward VI, Queen Mary, Elizabeth, and King James* (London: J. Wright, 1804), 1:269.

98. Carlton, *Going to the Wars*, 97–98. Units were rarely at full strength.

99. This account is derived from Richard Coe, *An Exact Diarie or a Breife Relation of the Progress of Sir William Waller's Army* (London: Humphrey Tuckey, 1644); Walker, *Happy Progress*, 22; CSPD 1644, 213, 219–20; *Mercurius Aulicus*, June 15, 1644, Thomason 9:E.53[5].

100. Other examples are in Porter, *Destruction*, 17–28; Gentles, *English Revolution*, 188.

101. Waller defined these terms precisely when he captured the castle of Arundel the previous fall. William Waller, *Certain Propositions Made by Sir William Waller, at the Surrender of Arundell-Castle*, Wing W540 (London: John

Field, 1644). Ignoring the role of the codes of war, Anthony Fletcher imposes a class-based interpretation of "gentry solidarity" on Waller's kind behavior to the Bishop family at Arundel. *A County Community in Peace and War: Sussex, 1600–1660* (London: Longman, 1975), 289.

102. Wanklyn and Jones, *Military History*, 87; Donagan, *War in England*, 148, 151–52, 187–88.

103. [Robert Devereaux], *Lawes and Ordinances of Warre, Established for the Better Conduct of the Army*, Wing E3314, (London:John Partridge and John Rothwell, 1642) facing B.

104. [Robert Devereaux], *Lawes and Ordinances of Warre, Established for the Better Conduct of the Army*, Wing E3316 (London: Luke Fawne, 1643), B1r-B1v.

105. Carlton summarizes several examples in *Going to the Wars*, 172–79.

106. See the discussions in Mortimer, *Eyewitness Accounts*, 40–43; Christopher Duffy, *Siege Warfare: The Fortress in the Early Modern World, 1494–1660* (London: Routledge & Kegan Paul, 1979), 66; Parker, *Military Revolution*, 59, each of which, however, are also at pains to argue that the frequency of such bloody storms has been exaggerated to some extent.

107. Carlton, *Going to the Wars*, 167; Coster, "Massacre," 102.

108. Examples of violated terms in Carlton, *Going to the Wars*, 170. See Donagan's discussions of how the system remained intact: *War in England*, 163–64; "Atrocity," 1152–55.

109. Donagan, *War in England*, 131–32, 157–58.

110. See CSPD 1644, 351; *The Two State Martyrs, or, The Murther of Master Robert Yeomans, and Master George Bowcher*, Wing T3535 (Oxford: H. Hall, 1643).

111. Charles's single clause is a vestigial version of the complex rules in the medieval-through-sixteenth-century regulations (clause 136 in Charles I, *Military Orders*). Donagan discusses this shift more generally. *War in England*, 143–44, 153. For the enslavement, killing, or ransoming of prisoners during the Middle Ages, see Clifford J. Rogers, *Soldiers' Lives Through History: The Middle Ages* (Westport, CT: Greenwood Press, 2007), 222–24; Contamine, *War in the Middle Ages*, 256–57. Ransom continued into the seventeenth century but yielded more and more to exchange. Hale, *War and Society*, 116–17; Geoffrey Parker, *The Army of Flanders and the Spanish Road, 1567–1659: The Logistics of Spanish Victory and Defeat in the Low Countries' Wars*, 2nd ed. (Cambridge: Cambridge University Press, 2004), 143–44; André Corvisier, *Armies and Societies in Europe, 1494–1789*, trans Abigail T. Siddall (Bloomington: Indiana University Press, 1979), 71–72; M. S. Anderson, *War and Society in Europe of the Old Regime, 1618–1789* (Leicester, UK: Leicester University Press, 1988), 55–56.

112. Devereaux, *Lawes and Ordinances of Warre*, Wing E3314, 20.

113. See note 68.

114. Carlton, *Going to the Wars*, 242–43.

115. *A True Relation of the Great and Glorious Victory Through Gods Providence, Obtained by Sir William Waller*, Wing T2958 (London: Edward Husbands, 1643).

116. Richard Atkyns, *The Vindication of Richard Atkyns Esquire*, Wing V489 (London: 1669), 42–43; *Weekly Account* (May 14–21, 1645), Thomason E.284 (25); Donagan, *War in England*, 229.

117. Carlton, *Going to the Wars*, 206; Gentles, *War of Three Kingdoms*, 166, 178; John Gell, "A True Relation of What Service hath beene Done by Colonell Sir John Gell," in *The History, Gazetteer, and Directory of the County of Derby*, ed. Stephen Glover (Derby, UK: Henry Mozley, 1829), 67.

118. Monro, *His Expedition*, 202.

119. Wanklyn and Jones, *Military History*, 164–66.

120. Clarendon, *History*, 3:358–59.

121. Clarendon, *History*, 3:363.

122. CSPD 1644, 247.

123. The battle is analyzed, and contemporaneous descriptions reprinted, in Toynbee and Young, *Cropredy Bridge*. For a reasonable discussion of battle tactics and their evolution during the war, see Wanklyn and Jones, *Military History*, 27–35.

124. Wanklyn and Jones, *Military History*, 270–72; Donagan, *War in England*, 76.

125. CSPD 1644, 298. At least a partial exchange did occur. CSPD 1644, 325.

126. CSPD 1644, 300–301. Waller's court-martial records indicate a distinct spike in soldiers accused of mutiny or disobeying orders in the early part of July after the battle at Cropredy. Adair, "Court Martial Papers," 216–17, 220.

127. CPSD 1644, 324–25.

128. Porter provides an extensive study of this problem. *Destruction*, 17–28.

129. Gat, *War in Human Civilization*, 471.

130. For more on the intolerance of neutralism, see Donagan, *War in England*, 7.

Chapter 4

Epigraphs from Brereton to the House of Commons, April 26, 1645, in *Memorials of the Civil War in Cheshire*, ed. Thomas Malbon (London: Record Society, 1889), 253–54 (modernized); *Kingdomes Weekly Intelligencer* 117, September 9–16, 1645, Thomason E.301(13).

1. The following description comes from Ian Gentles, *The English Revolution and the Wars in the Three Kingdoms, 1638–1652* (Harlow, UK: Pearson Longman, 2007), 88–90; Brian Manning, *The English People and the English Revolution, 1640–1649* (London: Heinemann, 1976), 95–98; *The Order of the House of Commons Declaring the High Breach of Priviledge of Parliament, by His Majesties Coming in Person, Attended with Great Numbers of Persons, Armed with Halberds, Swords and Pistolls, to the Commons House of Parliament*, Wing E2654 (London: Joseph Hunscott, 1642); CJ 2:367–69; CSPD 1641–43, 237, 240–43.

2. CSPD 1641–43, 243–44

3. David Kaiser, *Politics and War: European Conflict from Philip II to Hitler* (Cambridge, MA: Harvard University Press, 1990), chap. 1, esp. 2, 7, 15, 110–11.

4. David Cressy, *England on Edge: Crisis and Revolution, 1640–1642* (Oxford: Oxford University Press, 2006), esp. 281, 290–302, 379–95; Gentles, *English Revolution*, 82–83, 87, 91; Manning, *English People*, 233–43.

5. Charles Carlton, *Going to the Wars: The Experience of the British Civil Wars, 1638–1651* (London: Routledge, 1992), 268–69; Wanklyn and Jones, *Military History*, 56–61; Peter Young and Richard Holmes, *The English Civil War: A Military History of the Three Civil Wars 1642–1651* (London: Eyre Methuen, 1974), 82.

6. For example, at the outset of the struggle, royalists in Chichester used their preexisting social authority to trick the populace into supporting their capture of the city for the king. The trickery did not, however, generate sustained support. Anthony Fletcher, *A County Community in Peace and War: Sussex, 1600–1660* (London: Longman, 1975), 260–62.

7. See Barbara Donagan, *War in England, 1642–1649* (Oxford: Oxford University Press, 2008), 68–70, for the extent of mobility of the war's armies.

8. The following march description is creatively constructed from the sources for Waller's 1644 march cited in the previous chapter, but also from the march diaries of Nehemiah Wharton (August 16 to October 7, 1642) and Henry Foster (August 23 to September 28, 1643). Both men were private foot soldiers from London who provided detailed accounts of their peregrinations through much of the same country as Waller. This description is true to the tone and frequency of incidents in those sources, and avoids compressing an unrealistic number of incidents into too short a period of time. Wharton's diary, for example, covers approximately fifty-two days, during which he indicated nine nights of "good" or pleasant quarter and eight nights of poor or no quarter. Most nights he made no comment. He also recorded eight incidents of plundering for religious purposes, three incidents of plundering being prevented by army discipline, and six incidents of more or less hostile plundering, mostly of "malignants." Foster's diary covered thirty-six days, and he had good quarter for seven, poor or none for ten, and eight nights for which he mentions having quarter but made no comment on its quality. He recorded only one incident of plundering, and that was of a "malignant." Wharton's letters are in CSPD 1641–43, 371–73, 379–88, 391–400. For Foster, see *A True and Exact Relation of the Marchings of the Two Regiments of the Trained Bands of the City of London*, Wing F1625 (London: Benjamin Allen, 1643).

9. Carlton, *Going to the Wars*, 109.

10. For more on the provisioning system on the march and local frictions, see Donagan, *War in England*, 70–71. Soldiers could give tickets indicating that they had taken free quarter. At the end of the war, at least some householders were reimbursed, although some claims were not resolved for years. Wheeler, *Making of a World Power*, 83; David Underdown, *Revel, Riot and Rebellion: Popular Politics and Culture in England, 1603–1660* (Oxford: Oxford University Press, 1985), 150; Henry Townshend, *Diary of Henry Townshend of Elmley Lovett, 1640–1663*, ed. J. W. Willis Bund (London: Hughes and Clarke, 1915–20), 2:203. The right of a monarch to demand free quarter for his army had legal roots as old as the late Roman Theodosian Code, although that code also imposed some limits on the right. Guy Halsall, *Warfare and Society in the Barbarian West, 450–900* (London: Routledge, 2003), 42.

11. Julie Spraggon, *Puritan Iconoclasm during the English Civil War* (Woodbridge, UK: Boydell, 2003), 200–216; Robert Ram, *The Soldiers Catechisme, Composed for the Parliaments Army*, Wing R196A (London: J. Wright, 1645).

12. Complaints about specific garrisons or garrison commanders were legion (more below in the section on the Clubmen). Occasionally one would be moved or removed as a result. David Underdown, "The Chalk and the Cheese: Contrasts among the English Clubmen," *Past and Present* 85 (1979): 39n44; CSPD 1645–47, 30.

13. Philip Tennant, *Edgehill and Beyond: The People's War in the South Midlands, 1642–1645* (Phoenix Mill, UK: Sutton, 1992), 161–70. All of the following is taken from Tennant's study except where otherwise noted.

14. The classic account is Fritz Redlich, *The German Military Enterpriser and His Work Force: A Study in European Economic and Social History* (Wiesbaden, West Germany: Steiner, 1964–65), see esp. 1:239–69, 306–71. See also Geoffrey Parker, ed., *Thirty Years' War* (New York: Barnes and Noble, 1987), 195–98; Azar Gat, *War in Human Civilization* (Oxford: Oxford University Press, 2006), 473; Kaiser, *Politics and War*, 92; Herbert Langer, *The Thirty Years' War*, trans. C. S. V. Salt (New York: Hippocrene Books, 1978, 1990), 136, 153; Geoff Mortimer, "War by Contract, Credit and Contribution: The Thirty Years War," in *Early Modern Military History, 1450–1815*, ed. Geoff Mortimer (New York: Palgrave Macmillan, 2004), 101–17.

15. Kaiser, *Politics and War*, 22, 35, quotation on 74.

16. Malbon, *Civil War in Cheshire*, 253.

17. Langer, *Thirty Years' War*, 61.

18. This is based on the 1642 population of England, calculated at 5,112,000, and Firth's annual estimate for the size of the armies, maxing out at sixty thousand each. In general armies were not at that peak strength for long. Population is from E. A. Wrigley and R. S. Schofield, *The Population History of England, 1541–1871: A Reconstruction* (London: Arnold, 1981), 532; maximum army size from C. H. Firth, *Cromwell's Army* (London: Methuen, 1962), 22. Gat discusses the 1 percent problem in *War in Human Civilization*, 364, 474–76. Gat compares Louis XIV's peak army strength of 350,000 to 400,000 at 2 percent of the population; Spain at 1 percent in 1555 and 2 percent in 1630, largely financed by New World silver, but still (famously) unsustainable. Gat does not address this period in British history, but notes that the combined British army and navy at the beginning of the eighteenth century was around 1 percent, and only hit 2 percent again during the peak years of the Seven Years War and the American Revolution. In this context the figure of 2.45 percent in 1644 suggests considerable strain. The problem of mobilization as a percentage of total population in the early modern period deserves further consideration. During the American Civil War, the prewar Confederate states had a population of 9,101,090 (3,521,110 of whom were slaves). Peak Confederate-army strength was in June, 1863, at 307,464 "aggregate present," representing 3.4 percent of the total population. U.S. Bureau of the Census, *Population of the United States in 1860, Compiled from the Original Returns of the Eighth Census* (Washington, DC: GPO, 1864); *The War of the Rebellion: A Compilation of*

the Official Records of the Union and Confederate Armies (Washington, DC: GPO, 1880–1901), ser. IV, 2:615.

19. Contractors were used for logistics, but not for providing whole armies. Some regiments raised by gentry or aristocrats were at least initially paid for by those officers, but that is a different phenomenon from contracting. Gentles, *English Revolution*, 115–16.

20. Wheeler, *Making of a World Power*, v–vi, 207–12. Wheeler argues that Parliament's wartime innovations became the basis for England's postwar rise to imperial prominence. For the later period, see John Brewer, *The Sinews of Power: War, Money and the English State, 1688–1783* (New York: Alfred A. Knopf, 1989). Crucial to Brewer's argument was the continued acknowledgment of local values in what was otherwise a highly centralized system of revenue collection. For a summary covering both Wheeler's era and the beginnings of Brewer's, see Michael J. Braddick, *The Nerves of State: Taxation and the Financing of the English State, 1558–1714* (Manchester, UK: Manchester University Press, 1996).

21. Gentles, *English Revolution*, 115–16.

22. Wheeler, *Making of a World Power*, 80.

23. Quotation from CSPD 1644, 351; John Gell, "A True Relation of What Service hath beene Done by Colonell Sir John Gell," in *The History, Gazetteer, and Directory of the County of Derby*, ed. Stephen Glover (Derby, UK: Henry Mozley, 1829); Robert Bell, ed., *Memorials of the Civil War: Comprising the Correspondence of the Fairfax Family* (London: Richard Bentley, 1849), 1:79–80; Donagan, *War in England*, 239, 264–65.

24. Carlton, *Going to the Wars*, 279; Eliot Warburton, ed., *Memoirs of Prince Rupert and the Cavaliers* (London: Richard Bentley, 1849), 193–94; Henry Slingsby, *The Diary of Sir Henry Slingsby*, ed. Daniel Parsons (London: Longman, 1836), 94–95. Other examples in S. Porter, "The Fire-Raid in the English Civil War," *War and Society* 2 (1984): 335.

25. Geoff Mortimer, *Eyewitness Accounts of the Thirty Years War, 1618–48* (Houndmills, UK: Palgrave, 2002), 47–49.

26. Martyn Bennett, *The Civil Wars in Britain and Ireland, 1638–1651* (Cambridge, MA: Blackwell, 1997), 170–79, provides a summary of the many local studies of taxation systems, including his own unpublished dissertation on the subject. Porter, "Fire-Raid."

27. Underdown, *Revel, Riot and Rebellion*, 149–53; Donald Pennington, "The War and the People," in *Reactions to the English Civil War, 1642–1649*, ed. John Morrill (New York: St. Martin's, 1983), 128–32.

28. John A. Lynn II, *Women, Armies, and Warfare in Early Modern Europe* (Cambridge: Cambridge University Press, 2008), 36–44; Robert Monro, *Monro, His Expedition With the Worthy Scots Regiment Called Mac-Keys*, ed. William S. Brockington Jr. (Westport, CT: Praeger, 1999), 44; Slingsby, *Diary*, 84–85 (modernized).

29. The argument here is not about soldier cohesion, in combat or otherwise; it is about culture formation.

30. Donagan examines at length the stability of the English belief in war as a necessary institution. Once justified, war was legitimate, although the English also held an abiding fear of its consequences. *War in England*, 15–23. For more on the *jus ad bellum*, see chapter 3.

31. William Waller, *Vindication of the Character and Conduct of Sir William Waller, Knight; Commander in Chief of the Parliament Forces in the West: Explanatory of His Conduct in Taking Up Arms Against King Charles the First* (London: J. Debrett, 1793), 7.

32. Donagan, *War in England*, 22–23.

33. Matthias Milward, *The Souldiers Triumph and the Preachers Glory*, Wing M2186 (London: W. E. and I. G. for John Clark, 1641), 11.

34. Ram, *Soldiers Catechisme*, 1–2. See also John Bond, *A Sermon Preached in Exon*, Wing B3575 (London: T. B. for F. Eglesfeild, 1643).

35. Edward Symmons, *A Militarie Sermon*, Wing S6347 (Oxford: Henry Hall, 1644), 17; Donagan, *War in England*, 21–22.

36. This is an argument I develop at greater length for a different era in Wayne E. Lee, *Crowds and Soldiers in Revolutionary North Carolina: The Culture of Violence in Riot and War* (Gainesville: University Press of Florida, 2001), 189–90.

37. Quoted in Geoffrey Parker, "Early Modern Europe," in *The Laws of War: Constraints on Warfare in the Western World*, ed. Michael Howard, George J. Andreopoulos, and Mark R. Shulman (New Haven, CT: Yale University Press, 1994), 48.

38. Gell, "True Relation," 62, 71. The second account (on p. 71) specifies that the soldiers believed some of their own had been mistreated when captured during the first assault. It is also telling that the earl felt comfortable leaving his wife behind in the house.

39. The soldiers apparently had communicated their discontent with the surrender terms, leading to Essex's cash offer. Philip Stapleton, *An Exact Relation of the Delivering up of Reading to His Excellencie the Earl of Essex*, Wing S5256 (London: Edw. Husbands, 1643), 6; *Victory Proclaymed in an Exact Relation of the Valiant Proceedings of the Parliament Forces in their Seige before Reading from April 15 to 27*, Wing V351 (London: Benjamin Allen, 1643); *Mercurius Aulicus*, April 29, 1643, Thomason E.101(10); Carlton, *Going to the Wars*, 278–79.

40. George Monck, *Observations upon Military & Political Affairs*, Wing A864 (London: A. C. for Henry Mortlocke and James Collins, 1671), 2.

41. Matthew Sutcliffe, *The Practice, Proceedings, and Lawes of Armes* (London: Christopher Barker, 1593), B4r, 93.

42. Lois G. Schwoerer, *No standing armies! The Antiarmy Ideology in Seventeenth-Century England* (Baltimore: Johns Hopkins University Press, 1974), 8–32; Donagan, *War in England*, 24–32, 226, 261–63.

43. Malbon, *Civil War in Cheshire*, 253–54 (modernized). Ian Roy, "The English Civil War and English Society," *War and Society* 1 (1975): 32.

44. Key treatments of the Clubmen are: Oliver Warner, "The Clubmen and the English Civil War," *Army Quarterly* 38 (1936): 287–99; Morrill, *Revolt*, 132–51;

Underdown, "Chalk and Cheese." Other studies exist in county-specific publications, especially David Underdown, *Somerset in the Civil War and Interregnum* (Newton Abbot, UK: David & Charles, 1973), 86–120. Specific references to the major primary accounts follow below.

45. LJ 7:485.

46. Malbon, *Civil War in Cheshire*, 254 (modernized).

47. John Morrill, *Revolt in the Provinces: The People of England and the Tragedies of War, 1630–1648*, 2nd ed. (London: Longman, 1999), 141; LJ 7:485.

48. Morrill, *Revolt*, 142n42.

49. The literature on traditional protest behavior is extensive. A good introduction is Julius R. Ruff, *Violence in Early Modern Europe* (Cambridge: Cambridge University Press, 2001), 184–215. For more specific work on protest in England, see John Bohstedt, "The Moral Economy and the Discipline of Historical Context," *Journal of Social History* 26 (1992): 265–84; E. P. Thompson, *Customs in Common: Studies in Traditional Popular Culture* (New York: New Press, 1992); Roger B. Manning, *Village Revolts: Social Protest and Popular Disturbances in England, 1509–1640* (Oxford: Clarendon, 1988); John Walter, "'A Rising of the People'? The Oxfordshire Rising of 1596," *Past and Present* 107 (1985), 90–143; Roger B. Manning, "Patterns of Violence in Early Tudor Enclosure Riots," *Albion* 6 (1974): 120–33; Underdown, *Revel, Riot and Rebellion*, 157; John Walter, *Understanding Popular Violence in the English Revolution: The Colchester Plunderers* (Cambridge: Cambridge University Press, 1999).

50. The Wiltshire men seem to have issued their statement first in late May, and a virtually identical version appeared as a joint association in June. *True Informer*, June 14, 1645, Thomason E.288(18); *The Desires and Resolutions of the Club-Men of the Counties of Dorset and Wilts*, Thomason E.292(24) (London: Thos. Forcet, 1645); Tanner MS v.60 f.163–64; LJ 7:485–86. See also *Weekly Account*, May 14–21, 1645, Thomason E.284(25); *Kingdomes Weekly Intelligencer*, (June 17, 1645) Thomason E.288(31); also collected in A. R. Bayley, *The Great Civil War in Dorset, 1642–1660* (Taunton, UK: Wessex, 1910), 472–79.

51. The following resolutions are available: Henry Townshend, *Diary of Henry Townshend of Elmley Lovett, 1640–1663*, ed. J. W. Willis Bund (London: Printed for the Worcestershire Historical Society by Mitchell Hughes and Clarke, 1915–20), 2:221–23; Tanner MS v.60 f.52–55; *Perfect Occurrences*, July 4–11, 1645, Thomason E.262(20); Edward Bowles, *The Proceedings of the Army under the command of Sir Thomas Fairfax*, Wing P3573A (London: Samuel Gellibrand, 1645). The differences are discussed in Morrill, *Revolt*, 142–47; Underdown, "Chalk and Cheese."

52. This reliance on the constable for a paramilitary function dovetailed with their increased role in supervising and administering the parish militia quota during the 1630s. Henrik Langelüddecke, "'The chiefest strength and glory of this kingdom': Arming and Training the 'Perfect Militia' in the 1630s," *English Historical Review* 118 (2003): 1269.

53. These actions of the Dorset and Wiltshire Clubmen are taken from their statements above, and also from LJ 7:484; Joshua Sprigge, *Anglia Rediviva: Englands Recovery Being the History of the Motions, Actions, and Successes of the Army Under the Immediate Conduct of His Excellency Sr. Thomas Fairfax, Kt., Captain General of All the Parliaments Forces in England*, Wing S5070 (London: Printed by R.W. for John Partridge, 1647), 56–60, 75–81; *True Informer*, June 14, 1645, Thomason E.288 (18); *Perfect Occurrences*, June 20–27, 1645, Thomason E.262(13). For another example of villagers accepting soldiers' quarter if "soberly taken," see Roy, "English Civil War," 32.

54. For the strategic situation in early summer 1645, see Young and Holmes, *English Civil War*, 251–52.

55. *Kingdomes Weekly Intelligencer*, June 17, 1645, Thomason E.288(31).

56. Sprigg, *Anglia*, 55; *Perfect Occurrences*, June 27–July 4, 1645, Thomason E.262(16).

57. Sprigg, *Anglia*, 81; *A Letter Sent to the Right Honourable William Lenthall, Esquire, Speaker to the Honourable Speaker to the House of Commons*, Thomason E.292(22).

58. The royalists were also worried by the potential of the clubmen movement. For examples of both sides' fears, see Morrill, *Revolt*, 138; CSPD 1645–47, 13, 30, 146, 152; Bell, *Correspondence of the Fairfax Family*, 1:244–45.

59. Sprigg, *Anglia*, 75; CSPD 1645–47, 30.

60. John Rushworth, *Historical Collections of Private Passages of State, Weighty Matters in Law, Remarkable Proceedings in Five Parliaments: The Fourth and Last Part, 1645–1648*, ESTC N033540 (London: Richard Chiswell and Thomas Cockerill, 1701), 1:59–60.

61. Sprigg, *Anglia*, 78–81; Bowles, *Proceedings of the Army*; Rushworth, *Historical Collections* 1:61–62; Oliver Cromwell, *The Writings and Speeches of Oliver Cromwell*, ed. Wilbur C. Abbott (Cambridge, MA: Harvard University Press, 1937), 1:368–69.

62. For other counties: *Mercurius Civicus* 116 (August 7–14, 1645), Thomason E.296(23); *A Diary or an Exact Journall* 3 (October 2–9, 1645), Thomason E.304(13); Warner, "The Clubmen"; Fletcher, *County Community*, 272–73; Andrew J. Hopper, "The Clubmen of the West Riding of Yorkshire during the First Civil War: 'Bradford Club-Law,'" *Northern History* 36 (2000): 59–72.

63. Carlton, *Going to the Wars*, 285. On improved pay, see Wheeler, *Making of a World Power*, 81–85.

64. *Perfect Passages*, August 20–26, 1645, Thomason E.262(51); *Kingdomes Weekly Intelligence*, September 30–October 7, 1645, Thomason E.304(7); CSPD 1645–47, 200–201, 128; *Perfect Occurrences*, October 24–31, 1645, Thomason E.266(10); Morrill, *Revolt*, 149; *A Diary or an Exact Journall* 3 (October 2–9, 1645), Thomason E.304(13).

65. *Kingdomes Weekly Intelligencer*, September 9–16, 1645, Thomason E.301(13).

66. Carlton, *Going to the Wars*, 295, 204; Young and Holmes, *English Civil War*, 249. I am not suggesting that parliament counted the Clubmen as their

own losses, but merely that these numbers would have been considered militarily significant.

67. Charles Tilly, *Coercion, Capital, and European States, AD 990–1990* (Oxford: Blackwell, 1990), 100–103; William Beik, *Urban Protest in Seventeenth-Century France: The Culture of Retribution* (Cambridge: Cambridge University Press, 1997); Beik, "The Absolutism of Louis XIV as Social Collaboration," *Past and Present* 188 (2005): 195–224; Roy L. McCullough, *Coercion, Conversion and Counterinsurgency in Louis XIV's France* (Leiden, The Netherlands: Brill, 2007); Wayne te Brake, *Shaping History: Ordinary People in European Politics, 1500–1700* (Berkeley: University of California Press, 1998); Gat, *War in Human Civilization*, 473n41. For the general lack of coercive power within the English state, see Cynthia Herrup, "The Counties and the Country: Some Thoughts on Seventeenth-Century Historiography," *Social History* 8 (1983): 169–81. This kind of localist resistance is much of the argument of Morrill, *Revolt*, esp. p. 20. A similar argument about localist resistance to either royalist or parliamentarian control is in Alan Everitt, *The Community of Kent and the Great Rebellion, 1640–60* (Leicester, UK: Leicester University Press, 1966), 190–93, 230–41.

68. J. R. Hale, *War and Society in Renaissance Europe, 1450–1620* (Leicester, UK: Leicester University Press, 1985), 97, 100.

69. Monro, *His Expedition*, 19, 17, 60, 252, 74–75. See also note 81.

70. Gentles, *English Revolution*, 64, summarizes a vast body of work on the growth of an English political public. See also Steve Pincus, "'Coffee Politicians Does Create': Coffeehouses and Restoration Political Culture," *Journal of Modern History* 67 (1995): 807–34; Andy Wood, "Subordination, Solidarity and the Limits of Popular Agency in a Yorkshire Valley, c. 1596–1615," *Past and Present* 193 (2006): 41–72.

71. Mark Charles Fissel, *English Warfare, 1511–1642* (London: Routledge, 2001), 91–95, 98; Stephen J. Stearns, "Conscription and English Society in the 1620s," *Journal of British Studies* 11 (1972): 1–24; Lindsay Boynton, *The Elizabethan Militia, 1558–1638* (London: Routledge & Kegan Paul, 1967).

72. Gentles, *English Revolution*, 160, 141; Keith Lindley, *Popular Politics and Religion in Civil War London* (Aldershot, UK: Scolar, 1997), 3 and passim.

73. Tanner MS v.60 f.547; LJ 7:484.

74. Morrill, *Revolt*, passim, esp. 54–55, 133, 136; Gentles, *English Revolution*, 129–35; Fletcher, *County Community*, 271–72, 284–86; Lawrence Stone, *The Causes of the English Revolution, 1529–1642* (New York: Harper & Row, 1972), quotation on 54; Underdown, *Revel, Riot and Rebellion*, 152–54, 174–76; Underdown, *Somerset*, 117–20. In the latter work Underdown also argues that localism could produce partisan loyalty if the local community determined that one side would be more effective than the other in advancing their interests.

75. For continued worries about the problem even after the conclusion of the first civil war, see J. S. Morrill, "Mutiny and Discontent in English Provincial Armies 1645–1647," *Past and Present* 56 (1972), 51.

76. Glanmor Williams, ed., *Glamorgan County History* (Cardiff, UK: Glamorgan County History Trust, 1974), 267; CSPD 1644, 309; CSPD 1645–47, 15, 93; Wanklyn and Jones, *Military History*, 18–19.

77. Tanner MS v.60 f.32.

78. Baxter, *Baxterianae*, 44 (quote). Carlton, *Going to the Wars*, 284–88. Ann Hughes argues more generally that Parliament was more effective at meshing national and local interests. "The King, the Parliament and the Localities during the English Civil War," *Journal of British Studies* 24 (1985), 241, 278–63. In contrast, Ronald Hutton argues that the greater ruthlessness of the royalist army in 1645 almost brought them victory. *The Royalist War Effort, 1642–1646*, 2nd ed. (London: Routledge, 1999), 202–3. For the power of public opinion, see Underdown, *Revel, Riot, and Rebellion*, 152–53; Roy, "English Civil War," 32.

79. For just a few examples of spontaneous "peasant revenge" (unlike the organized and associated Clubmen), see Carlton, *Going to the Wars*, 30, 286–87, 326; *Special and Remarkable Passages*, January 2–9, 1646, Thomason E.315(7); Warburton, *Memoirs of Prince Rupert*, 2:191–94; Ronan Bennett, "War and Disorder: Policing the Soldiery in Civil War Yorkshire," in *War and Government in Britain, 1598–1650*, ed. Mark Charles Fissel (Manchester, UK: Manchester University Press, 1991), 262; Fletcher, *County Community*, 274–75.

80. Slingsby, *Diary*, 108.

81. See also the development of this sense among the mercenaries on the continent, described by Redlich, *German Military Enterpriser*, 515–32; Langer, *Thirty Years' War*, 61–64, 89, 103–6; Hale, *War and Society*, 127–52, 189–91; Myron P. Gutmann, *War and Rural Life in the Early Modern Low Countries* (Princeton, NJ: Princeton University Press, 1980). Gunn et al. see a growth of "generalized hostility" to soldiers, and argue that "soldierly attitudes inflamed such breaches, especially amongst the noble men-at-arms of the *ordonnaces*, who would kill passers-by for mocking or murmuring at them." Steven Gunn, David Grummitt, and Hans Cools, *War, State, and Society in England and the Netherlands, 1477–1559* (Oxford: Oxford University Press, 2007), 275. Yves-Marie Bercé examines riots against soldiers in seventeenth-century France and finds a marked and violent peasant hostility, often in cooperation with local authorities, even against the soldiers raised in their province. He further sees that new soldiers consciously rejected their home values, "scorn[ing] and abuse[ing] their old environment" by "displaying a constant aggressiveness towards the villagers." *History of Peasant Revolts: The Social Origins of Rebellion in Early Modern France*, trans. Amanda Whitmore (Ithaca, NY: Cornell University Press, 1990), 179–96 (quotation on 194). Thanks to Julia Osman for suggesting this reference. See also Otto Ulbricht, "The Experience of Violence During the Thirty Years War: A Look at the Civilian Victims," in *Power, Violence and Mass Death in Pre-Modern and Modern Times*, ed. Joseph Canning, Hartmut Lehmann, and Jay Winter (Aldershot, UK: Ashgate, 2004), 107–8. There are many examples of soldiers' revenge for peasant hostility from the Thirty Years War, but Goring's royalists apparently took revenge

for Clubmen attacks, and several of the spontaneous peasant attacks mentioned in this note also inspired army retaliation. *Perfect Passages*, August 20–26, 1645, Thomason E.262(51); Bennett, "War and Disorder," 252, 261, quotation on 251.

82. Sprigg, *Anglia*, 79.

83. Donagan, *War in England*, 134.

84. Carlton, *Going to the Wars*, 314–16, 320, 328. Carlton and others also note that much of the fighting in the second and third civil wars was against Scots or Irish, where the standards of conduct were already markedly more violent.

85. Compare Gat, *War in Human Civilization*, 319–20.

86. One apparent late-war order to burn corn fields in Wales was singled out by Parliament's propaganda machine as exceptional and egregious. *Perfect Passages*, August 6–13, 1645, Thomason E.262(42). Bennett, "War and Disorder," 254, suggests that some deliberate devastation occurred in Yorkshire.

87. Schwoerer, *No Standing Armies*, 51–71; Carlton, *Going to the Wars*, 348; Donagan, *War in England*, 260.

88. George Percy, "'A Trewe Relacyon': Virginia from 1609 to 1612," *Tyler's Quarterly Historical and Genealogical Magazine* 3 (1922): 271–73.

Chapter 5

Epigraphs from Roger Williams, *A Key into the Language of America* (London: 1643; Bedford, MA: Applewood Books, 1997), 186–87 (page numbers refer to reprint); *The Winthrop Papers* (Boston: Massachusetts Historical Society, 1863–1892), 3:413–14.

1. The Roanoke voyages and related documents have been the subject of intense scrutiny, primarily by David Beers Quinn. The following account is based on the documents in RV; Quinn, *Set Fair for Roanoke: Voyages and Colonies, 1584–1606* (ChapelHill: University of North Carolina Press, 1985); Karen Ordahl Kupperman, *Roanoke: The Abandoned Colony* (Savage, MD: Rowman & Littlefield, 1984).

2. It is not known whether Manteo and Wanchese departed voluntarily. The English were known to kidnap, but they also claimed on occasion to have taken volunteers.

3. Bruce P. Lenman, *England's Colonial Wars 1550–1688: Conflicts, Empire, and National Identity* (Harlow, UK: Longman, 2001), 77–80.

4. Gilbert's son, Raleigh Gilbert, would also be a major sponsor of the Popham colony in Maine.

5. American National Biography Online, s.v. Lane, Sir Ralph; DNB, s.v. Lane, Sir Ralph; On the charges against Lane, see CSPI 6:190, 196, 236, 292, 305, 314, 318, 391; 7:7:39–40, 62–63; Lane's replies: CSPI 6:214–15, 252, 263, 337, 392, 464; quotation on 7:71–73; Michael Leroy Oberg, *Dominion and Civility: English Imperialism and Native America, 1585–1685* (Ithaca, NY: Cornell University Press, 1999), 29.

6. Richard Hakluyt the Elder, "Inducements to the Liking of the Voyage," in *New American World: A Documentary History of North America to 1612*, ed. David B. Quinn (New York: Arno, 1979) 3:64, 65, 67. For a more expansive set of similar

sentiments see Richard Hakluyt the Younger, "Discourse of Western Planting (1584)," in ibid., 3:71–123. For the disciplinary codes, see RV 1:138–39. Quinn suggests that this code was well within the normal codes of the day, but the specific punishments for relatively mild intrusions on "civilians" have no real parallel in the sixteenth-century disciplinary codes discussed in chapter 1 (or even those of the English civil war in chapter 2).

7. Karen Ordahl Kupperman, *The Jamestown Project* (Cambridge, MA: Belknap Press of Harvard University Press, 2007), 172.

8. RV 1:384.

9. RV 1:255–94, 381.

10. Cortes and Pizarro, in Mexico and Peru respectively, gained crucial time to understand their predicament and create a plan by seizing the emperor.

11. Kupperman, *Roanoke*, 76.

12. RV 1:204 (quotation), 229, 185.

13. Karen Ordahl Kupperman, "Presentment of Civility: English Response to American Self-Presentation, 1580–1640," *William and Mary Quarterly*, 3rd ser., 54 (1997): 193–228; Kupperman, *Jamestown Project*, 315. In addition to Lane's comment about Menatonon quoted earlier, see RV 1:209.

14. Oberg, *Dominion*, 6–7, 50–54; Melanie Perreault, *Early English Encounters in Russia, West Africa, and the Americas, 1530–1614* (Lewiston, NY: Mellen, 2004), 11–14; Alison Games, *The Web of Empire: English Cosmopolitans in an Age of Expansion, 1560–1660* (Oxford: Oxford University Press, 2008). For the focus on dispossession over incorporation, see Anthony Pagden, *Lords of All the World: Ideologies of Empire in Spain, Britain and France c. 1500–c. 1800* (New Haven, CT: Yale University Press, 1995), chaps. 1–3, esp. 91–92, 94; Patricia Seed, "Taking Possession and Reading Texts: Establishing the Authority of Overseas Empires," *William and Mary Quarterly*, 3rd ser., 49 (1992): 183–209; Seed, *Ceremonies of Possession in Europe's Conquest of the New World, 1492–1640* (Cambridge: Cambridge University Press, 1995).

15. RV 1:371–72.

16. Louis B. Wright, ed., *A Voyage to Virginia in 1609: Two Narratives; Strachey's "True Reportory" and Jourdain's Discovery of the Bermudas* (Charlottesville: University Press of Virginia, 1964), 88–89.

17. After an entire book on the parallels or similarities between European and Native American culture, Nancy Shoemaker in the end confesses that the competition for land inevitably generated conflict. *A Strange Likeness: Becoming Red and White in Eighteenth-Century North America* (New York: Oxford University Press, 2004), 141. Ben Kiernan focuses on land as the core of the issue in *Blood and Soil: A World History of Genocide and Extermination from Sparta to Darfur* (New Haven, CT: Yale University Press, 2007), 216–19. We will return to the problem of land and its use in chapter 8.

18. For the epidemic in New England, see Neal Salisbury, *Manitou and Providence: Indians, Europeans, and the Making of New England, 1500–1643* (New York: Oxford University Press, 1982), 101–6.

19. The role of disease in the depopulation of these two regions remains debated. For a discussion of the issues see HNAI 15:357–61; Paul Kelton, *Epidemics and Enslavement: Biological Catastrophe in the Native Southeast, 1492–1715* (Lincoln: University of Nebraska Press, 2007); Daniel K. Richter, *Facing East from Indian Country: A Native History of Early America* (Cambridge, MA: Harvard University Press, 2001), 35–36.

20. For two contrasting opinions on whether disease escalated or restrained Native American war, see Daniel K. Richter, "War and Culture: The Iroquois Experience," *William and Mary Quarterly*, 3rd ser., 40 (1983): 528–59; Helen C. Rountree, "Summary and Implications," in *Powhatan Foreign Relations, 1500–1722*, ed. Helen C. Rountree (Charlottesville: University Press of Virginia, 1993), 221.

21. The reader should assume from this point on that all references to Native Americans refer to peoples living in the eastern woodlands between 1500 and 1800 unless otherwise specified. Specifically excluded, therefore, are the peoples of the Great Plains, the arctic and subarctic, the Pacific Northwest, and the desert Southwest. Warfare patterns in these places were very different.

22. There were striking exceptions to this town-cluster based description of a tribe. The Shawnees, for example, sustained an ethnic identity across extremely widely scattered settlements.

23. Most tribal names found in history texts are either derivations of the word "people" from that group's own language or an epithet applied to them by their enemies and taught to the Europeans. For a brief discussion of the problem of vocabulary, see April Lee Hatfield, "Colonial Southeastern Indian History," *Journal of Southern History* 73 (2007): 575–76.

24. A complex subject, perhaps best summarized in Daniel K. Richter, "Stratification and Class in Eastern Native America," in *Class Matters: Early North America and the Atlantic World*, ed. Simon Middleton and Billy G. Smith (Philadelphia: University of Pennsylvania Press, 2008), 35–48.

25. This discussion of Mississippian warfare and society derives from Charles Hudson, *Knights of Spain, Warriors of the Sun: Hernando De Soto and the South's Ancient Chiefdoms* (Athens, GA: University of Georgia Press, 1997); Timothy R. Pauketat, *Chiefdoms and Other Archaeological Delusions* (Lanham, MD: AltaMira, 2007); David G. Anderson, "Fluctuations between Simple and Complex Chiefdoms: Cycling in the Late Prehistoric Southeast," in *Political Structure and Change in the Prehistoric Southeastern United States*, ed. John F. Scarry (Gainesville: University Press of Florida, 1996), 245–46. The question of Mississippian cultural continuity into the historic period is contentious and complex. See Adam King, "The Historic Period Transformation of Mississippian Societies," in *Light on the Path: The Anthropology and History of the Southeastern Indians*, ed. Thomas J. Pluckhahn and Robbie Ethridge (Tuscaloosa: University of Alabama Press, 2006), 179–95. For warfare the crucial works are David H. Dye, "The Transformation of Mississippian Warfare: Four Case Studies from the Mid-South," in *The Archaeology of Warfare: Prehistories of Raiding and Conquest*, ed. Elizabeth N. Arkush and Mark W. Allen (Gainesville: University Press of Florida, 2006), 101, 47; Dye, "The

Art of War in the Sixteenth-Century Central Mississippi Valley," in *Perspectives on the Southeast: Linguistics, Archaeology and Ethnohistory*, ed. Patricia B. Kwachka (Athens, GA: University of Georgia Press, 1994), 54–56; Dye, "Warfare in the Sixteenth-Century Southeast: The de Soto Expedition in the Interior," in *Columbian Consequences*, vol. 2, *Archaeological and Historical Perspectives on the Spanish Borderlands East*, ed. David Hurst Thomas (Washington, DC: Smithsonian Institution Press, 1990), 211–22; Dye, "Warfare in the Protohistoric Southeast, 1500–1700," in *Between Contacts and Colonies: Archaeological Perspectives on the Protohistoric Southeast*, ed. Cameron Wesson and Mark A. Rees (Tuscaloosa: University of Alabama Press, 2002), 131–32; Karl T. Steinen, "Ambushes, Raids, and Palisades: Mississippian Warfare in the Interior Southeast," *Southeastern Archaeology* 11 (1992): 132–39.

26. Jacques Cartier, *The Voyages of Jacques Cartier* (Toronto: University of Toronto Press, 1993), 67–68. John Smith described another early surprise and massacre. John Smith, *Captain John Smith: A Select Edition of His Writings*, ed. Karen Ordahl Kupperman (Chapel Hill: University of North Carolina Press, 1998), 157.

27. RV 1:113–14.

28. For prisoner taking equipment, see José António Brandão, *Your fyre shall burn no more: Iroquois Policy towards New France and Its Native Allies to 1701* (Lincoln: University of Nebraska Press, 1997), 34–35.

29. The lethality and violence of precontact Native American warfare is a subject of some controversy among scholars. One of the most frequently cited works on native warfare, after describing the death of fifty Mohawks in a 1669 battle, calmly asserts the unknowable: "Such heavy losses in a single action were unheard of before the arrival of the white man and his weapons." Patrick M. Malone, *The Skulking Way of War: Technology and Tactics Among the New England Indians* (Lanham, MD: Madison Books, 1991; reprint, Baltimore: Johns Hopkins University Press, 1993), 65. In a similar vein, see Nathaniel Knowles, "The Torture of Captives by the Indians of Eastern North America," *American Philosophical Society Proceedings* 82 (1940): 151–225; Daniel P. Barr, "'This Land Is Ours and Not Yours': The Western Delawares and the Seven Years' War in the Upper Ohio Valley, 1755–1758," in *The Boundaries Between Us: Natives and Newcomers Along the Frontiers of the Old Northwest Territory, 1750–1850*, ed. Daniel P. Barr (Kent, OH: Kent State University Press, 2006), 32. Archaeological and anthropological work is increasingly challenging this view. See George R. Milner, "Warfare in Prehistoric and Early Historic Eastern North America," *Journal of Archaeological Research* 7 (1999): 126–27; Patricia M. Lambert, "The Archaeology of War: A North American Perspective," *Journal of Archaeological Research* 10 (2002): 227–29; Keith F. Otterbein, "A History of Research on Warfare in Anthropology," *American Anthropologist* 101 (2000): 800; William Divale, *Warfare in Primitive Societies: A Bibliography* (Santa Barbara, CA: ABC-Clio, 1973), xxi–xxii; Thomas B. Abler, "European Technology and the Art of War in Iroquoia," in *Cultures in Conflict: Current Archaeological Perspectives; Proceedings of the Twentieth Annual Conference of the Archaeological Association of the University of Calgary*, ed. Diana Tkaczuk and Brian C. Vivian (Calgary: University of Calgary Archaeology Association, 1989), 278–79; David H.

Dye, *War Paths, Peace Paths: An Archaeology of Cooperation and Conflict in Native Eastern North America* (Lanham, MD: AltaMira, 2009), 8–11, 67, 111–13; Richard J. Chacon and Rubén G. Mendoza, eds., *North American Indigenous Warfare and Ritual Violence* (Tucson: University of Arizona Press, 2007); Azar Gat, *War in Human Civilization*, (Oxford: Oxford University Press, 2006), 116–32. Lawrence H. Keeley's emphasis on the per capita lethality of primitive warfare in general is an important recent stimulus to this argument. *War Before Civilization* (New York: Oxford University Press, 1996). For a continued assertion of the paucity of evidence for prestate warfare, see R. Brian Ferguson, "Archaeology, Cultural Anthropology, and the Origins and Intensification of War," in *The Archaeology of Warfare: Prehistories of Raiding and Conquest*, ed. Elizabeth N. Arkush and Mark W. Allen (Gainesville: University Press of Florida, 2006), 469–523.

30. *Winthrop Papers*, 3:413–14.

31. In many cultures, a fourth motive functioned in parallel with the political, blood feud, and status motives, and that was the acquisition of prisoners for adoption. This is discussed more fully in the next chapter.

32. Anthony F. C. Wallace, *The Death and Rebirth of the Seneca* (New York: Vintage Books, 1972), 44.

33. A sophisticated and detailed discussion of the blood feud is John Phillip Reid, *A Law of Blood: The Primitive Law of the Cherokee Nation* (New York: New York University Press, 1970). Among many others, see Charles Hudson, *The Southeastern Indians* (Knoxville: University of Tennessee Press, 1976), 230–32, 239–40; Richard White, *The Middle Ground: Indians, Empires, and Republics in the Great Lakes Region, 1650–1815* (Cambridge: Cambridge University Press, 1991), 80; Gordon M. Sayre, *Les Sauvages Américains: Representations of Native Americans in French and English Colonial Literature* (Chapel Hill: University of North Carolina Press, 1997), 279–80; Dye, *War Paths*, 53–56, 177.

34. Dean R. Snow, "Iroquois-Huron Warfare," in Chacon and Mendoza, *Indigenous Warfare*, 151–52.

35. Hudson, *Southeastern Indians*, 242, based on James Adair, *History of the American Indians* (New York: Argonaut Press, (1930), 1966), 407. See also Cadwallader Colden's description of the possibility of such inter-people resolution of blood feud between the Adirondacks and the Iroquois. *The History of the Five Indian Nations of Canada* (London: 1747), 22. For other examples of intergroup attempts to assuage a blood feud before it got started see Karen Ordahl Kupperman, Indians and English: Facing Off in Early America (Ithaca, NY: Cornell University Press, 2000), 106; Tobias Fitch, "Captain Fitch's Journal to the Creeks, 1725," in *Travels in the American Colonies*, ed. Newton D. Mereness (New York: Macmillan, 1916), 175–214, 203; Reid, *Law of Blood*, 171–72; James H. Merrell, *Into the American Woods: Negotiators on the Pennsylvania Frontier* (New York: Norton, 1999), 116–21; Colin G. Calloway, *The Western Abenakis of Vermont, 1600–1800: War, Migration, and the Survival of an Indian People* (Norman: University of Oklahoma Press, 1990), 165, 189–90; Wendell S. Hadlock, "War among the Northeastern Woodland Indians," *American Anthropologist*, new ser., 49 (1947): 213–14.

36. Talk of Tistoe and The Wolf of Keowee to Governor Lyttelton, March 5, 1759, William Lyttelton Papers, William L. Clements Library, University of Michigan, Ann Arbor.

37. John Phillip Reid, *A Better Kind of Hatchet: Law, Trade, and Diplomacy in the Cherokee Nation During the Early Years of European Contact* (University Park: Pennsylvania State University Press, 1976), 9.

38. Ferguson argues that this lack of coercive structures was the most fundamental limitation on prestate warfare. R. Brian Ferguson, "Violence and War in Prehistory," in *Troubled Times: Violence and Warfare in the Past*, ed. Debra L. Martin and David W. Frayer (Amsterdam: Gordon and Breach, 1997), 336.

39. For one example of this ideology at work see John Winthrop, *The Journal of John Winthrop, 1630–1649*, unabridged ed., ed. Richard S. Dunn, James Savage, and Laetitia Yeandel (Cambridge, MA: Harvard University Press, 1996), 252. See also chapter 8, note 48.

40. Edward Waterhouse, *A Declaration of the State of the Colony in Virginia* (London: 1622; repr., New York: Da Capo, 1970), 22–23.

41. HNAI 15:315; Leroy V. Eid, "'National War' among Indians of Northeastern North America," *Canadian Review of American Studies* 16 (1985): 125–54; Brandão, *Your Fyre*, 32–33.

42. Quotation is from Joseph François Lafitau, *Customs of the American Indians Compared with the Customs of Primitive Times*, trans. and ed. William N. Fenton and Elizabeth L. Moore (Toronto: Champlain Society, 1974), 2:101. A good example of a "grand war" was the Iroquois attack on the Hurons in 1648–49, which nearly destroyed the Hurons and forced their displacement. For the debate on the Iroquois' motives, see: Brandão, *Your Fyre*, 5–18; Thomas S. Abler, "Iroquois Policy and Iroquois Culture: Two Histories and an Anthropological Ethnohistory," *Ethnohistory* 47 (2000): 483–91; William A. Starna and José António Brandão, "From the Mohawk-Mahican War to the Beaver Wars: Questioning the Pattern," *Ethnohistory* 51 (2004): 725–50. The attack is narrated in Keith F. Otterbein, "Huron vs. Iroquois: A Case Study in Inter-Tribal Warfare," *Ethnohistory* 26 (1979): 141–52, which is basically summarizing the French Jesuit account found in JR 34:123–37.

43. Salisbury, *Manitou and Providence*, 229; Frederic W. Gleach, *Powhatan's World and Colonial Virginia: A Conflict of Cultures* (Lincoln: University of Nebraska Press, 1997), 51–54; Thomas C. Parramore, "The Tuscarora Ascendancy," *North Carolina Historical Review* 59 (1982): 322–23.

44. Hence the Keowee Cherokees' threat to draw in other towns, quoted above.

45. *Winthrop Papers* 3:413. For a duel example, the Montagnais and the Iroquois once agreed to "spare the blood of our followers" and submit to the judgment of a wrestling contest. JR 1:269–70. More on "conquest by harassment" follows in chapter 6.

46. Hudson, *Southeastern Indians*, 257; Daniel K. Richter, *The Ordeal of the Longhouse: The Peoples of the Iroquois League in the Era of European Colonization* (Chapel Hill: University of North Carolina Press, 1992), 40. See the section on making peace in chapter 6.

47. Kupperman, *Captain John Smith*, 165. For a modern parallel, consider the United States's bombing of Libya in 1986, or the cruise missile strikes in Sudan and Afghanistan in 1998.

48. For such disavowals, see Peter Wraxall, *An Abridgment of the Indian Affairs Contained in Four Folio Volumes: Transacted in the Colony of New York, From the Year 1678 to the Year 1751* (Cambridge, MA: Harvard University Press, 1915), 88, 100; Claudio Saunt, *A New Order of Things: Property, Power, and the Transformation of the Creek Indians, 1733–1816* (Cambridge: Cambridge University Press, 1999), 23. See chapter 6 for a more developed example. Saunt argues that such contradictions were a functional aspect of native society, allowing Indians to explore multiple policy paths until eventually settling on the one that seemed to have the clearest advantage (23–24).

49. Daniel K Richter, "Native Peoples of North America," in *The Oxford History of the British Empire*, vol. 2, *The Eighteenth Century*, ed. P. J. Marshall (Oxford: Oxford University Press, 1998), 357.

50. Wayne E. Lee, "Fortify, Fight, or Flee: Tuscarora and Cherokee Defensive Warfare and Military Culture Adaptation," *Journal of Military History* 68 (2004): 761–62.

51. This analysis comes from Tom Hatley, *The Dividing Paths: Cherokees and South Carolinians through the Era of Revolution* (New York: Oxford University Press, 1995), 44–45. The original document is available in *Calendar of State Papers, Colonial Series, American and West Indies, 1574–1739*, CD-ROM, ed. Karen Ordahl Kupperman, John C. Appleby, and Mandy Banton (London: Routledge, 2000), Item 429v, 34:280.

Chapter 6

Epigraphs from John Norton, *The Journal of Major John Norton, 1816*, ed. Carl F. Klinck and James J. Talman (Toronto: Champlain Society, 1970), 262; Peter Wraxall, *An Abridgment of the Indian Affairs Contained in Four Folio Volumes: Transacted in the Colony of New York, From the Year 1678 to the Year 1751* (Cambridge, MA: Harvard University Press, 1915), 88.

1. Steven C. Hahn provides the best outline of Old Brims's career. *The Invention of the Creek Nation, 1670–1763* (Lincoln: University of Nebraska Press, 2004), 66–148. In addition to Hahn, the discussion here of Creek political structure and other narrative components of the following is from David H. Corkran, *The Creek Frontier, 1540–1783* (Norman: University of Oklahoma Press, 1967); Claudio Saunt, *A New Order of Things: Property, Power, and the Transformation of the Creek Indians, 1733–1816* (Cambridge: Cambridge University Press, 1999); Vernon J. Knight Jr., "The Formation of the Creeks," in *The Forgotten Centuries: Indians and Europeans in the American South, 1521–1704*, ed. Charles Hudson and Carmen Chaves Tesser (Athens: University of Georgia Press, 1994), 373–92; John E. Worth, "The Lower Creeks: Origins and Early History," in *Indians of the Greater Southeast: Historical Archaeology and Ethnohistory*, ed. Bonnie G. McEwan (Gainesville: University Press of Florida, 2000), 265–98; Gregory A. Waselkov and

Marvin T. Smith, "Upper Creek Archaeology," in *Indians of the Greater Southeast: Historical Archaeology and Ethnohistory*, ed. Bonnie G. McEwan (Gainesville: University Press of Florida, 2000), 242–64; Verner Crane, *The Southern Frontier, 1670–1732* (Durham, NC: Duke University Press, 1929), 259–71; Steven C. Hahn, "The Mother of Necessity: Carolina, the Creek Indians, and the Making of a New Order in the American Southeast, 1670–1763," in *The Transformation of the Southeastern Indians, 1540–1760*, ed. Robbie Ethridge and Charles Hudson (Jackson: University Press of Mississippi, 2002), 79–114; and, especially for the 1725 raid itself, Tobias Fitch, "Captain Fitch's Journal to the Creeks, 1725," in *Travels in the American Colonies*, ed. Newton D. Mereness (New York: Macmillan, 1916), 175–214.

2. The war devastated South Carolina and greatly reduced the colony's power for a decade or more. Alan Gallay, *The Indian Slave Trade the Rise of the English Empire in the American South, 1670–1717* (New Haven, CT: Yale University Press, 2002), 338–41. Two recent treatments are Steven J. Oatis, *A Colonial Complex: South Carolina's Frontiers in the Era of the Yamasee War, 1680–1730* (Lincoln: University of Nebraska Press, 2004); William L. Ramsey, *The Yamasee War: A Study of Culture, Economy, and Conflict in the Colonial South* (Lincoln: University of Nebraska Press, 2008).

3. The story of the murdered delegation is in George Chicken, "Journal of the March of the Carolinians into the Cherokee Mountains, in the Yemassee Indian War, 1715–16," *Yearbook of the City of Charleston* (1894): 315–54.

4. Fitch, "Captain Fitch's Journal," 182.

5. Lee, "Fortify, Fight, or Flee," 753–57; Tom Hatley, *The Dividing Paths: Cherokees and South Carolinians through the Era of Revolution* (New York: Oxford University Press, 1995), 93; David Corkran, *The Cherokee Frontier, 1740–1762* (Norman: University of Oklahoma Press, 1962), 35–37.

6. Hahn contends that Chipacasi was not really interested in damaging the Yamasees but instead merely hoped to get the equivalent of Brims's commission from Fitch. *Invention*, 133–35. If that is so, Chipacasi took a fairly large number of casualties (below), and the war party followed generally standard procedure as detailed in the following pages. For more on the international aspects of this situation see, Crane, *Southern Frontier*, 265–67.

7. See note 1, but especially Knight, "Formation of the Creeks," 386–88.

8. Fitch, "Captain Fitch's Journal," 178. Others also blamed the young men for excesses. Ibid., 191. See also Saunt, *New Order*, 16–17.

9. Joseph François Lafitau, *Customs of the American Indians Compared with the Customs of Primitive Times*, trans. and ed. William N. Fenton and Elizabeth L. Moore (Toronto: Champlain Society, 1974), 2:101–3; Colin G. Calloway, *The Western Abenakis of Vermont, 1600–1800: War, Migration, and the Survival of an Indian People* (Norman: University of Oklahoma Press, 1990), 172; Saunt, *New Order*, 23–25.

10. Hudson, *Southeastern Indians*, 223–24.

11. HNAI 15:192, 314–15. One European observer, for example, exactly reversed the real power relationship among the Cherokees: "Every Town has a

Head Warrior, who is in great Esteem among them, and whose Authority seems to be greater than their Kings, because their King is looked upon as little else than a Civil Magistrate, except it so happens that he is at the same Time a Head Warrior." Sir Alexander Cuming, "Journal of Sir Alexander Cuming (1730)," in *Early Travels in the Tennessee Country, 1540–1800*, ed. Samuel Cole Williams (Johnson City, TN: Watauga, 1928), 122.

12. The English "appointed" at least two such "emperors" for the Creeks, both of whom the Creeks ignored. Saunt, *New Order*, 26; Rennard Strickland, *Fire and the Spirits: Cherokee Law From Clan to Court* (Norman: University of Oklahoma Press, 1975), 47–48; Hudson, *Southeastern Indians*, 222; HNAI 15:192, 315; Anthony F. C. Wallace, *The Death and Rebirth of the Seneca* (New York: Vintage Books, 1972), 40; James Axtell, "Making Do," in *Natives and Newcomers: The Cultural Origins of North America* (New York: Oxford University Press, 2001), 139; Fred Gearing, *Priests and Warriors: Social Structures for Cherokee Politics in the 18th Century*, American Anthropological Association Memoir 93 (Menasha, WI: American Anthropological Association, 1962); Colin G. Calloway, *New Worlds for All: Indians, Europeans, and the Remaking of Early America* (Baltimore: Johns Hopkins University Press, 1997), 111.

13. Fitch, "Captain Fitch's Journal," 190.

14. Hahn, *Invention*, 123.

15. John Phillip Reid, *A Better Kind of Hatchet: Law, Trade, and Diplomacy in the Cherokee Nation During the Early Years of European Contact* (University Park: Pennsylvania State University Press, 1976), 9–10; Alfred A. Cave, *The Pequot War* (Amherst: University of Massachusetts Press, 1996), 157.

16. Ian K. Steele, "Surrendering Rites: Prisoners on Colonial North American Frontiers," in *Hanoverian Britain and Empire: Essays in Memory of Philip Lawson*, ed. Stephen Taylor, Richard Connors, and Clyve Jones (Woodbridge, UK: Boydell, 1998), 138; Hatley, *Dividing Paths*, 94; Daniel K. Richter, *The Ordeal of the Longhouse: The Peoples of the Iroquois League in the Era of European Colonization* (Chapel Hill: University of North Carolina Press, 1992), 111; Reid, *Better Kind of Hatchet*, 111; Bruce G. Trigger, *The Huron: Farmers of the North*, 2nd ed. (Fort Worth: Holt, Rinehart and Winston, 1990), 53; Cave, *Pequot War*, 66–67; Wayne E. Clark and Helen C. Rountree, "The Powhatans and the Maryland Mainland," in *Powhatan Foreign Relations, 1500–1722*, ed. Helen C. Rountree (Charlottesville: University Press of Virginia, 1993), 132. R. Demere to Gov. Lyttelton, July 23, 1757; Demere to Lyttelton, June 26, 1757; White Outerbridge to Gov. Lyttelton, March 8, 1757; all in William Lyttelton Papers, William L. Clements Library, University of Michigan, Ann Arbor. Reid points out, however, that such resident aliens could also be loose cannons, acting on their own behalf and thus creating trouble for the community in which they lived. John Phillip Reid, *A Law of Blood: The Primitive Law of the Cherokee Nation* (New York: New York University Press, 1970), 163–72.

17. The Jesuit Jean de Brébeuf saw this warning function as the main intent of sending and receiving visitors to and from the village. JR 10:229. There are many

examples of warnings passed between peoples to prevent major surprise. Calloway, *Western Abenakis*, 178, 212; Fintan·O'Toole, *White Savage: William Johnson and the Invention of America* (New York: Farrar, Straus and Giroux, 2005), 205–6; Jon Parmenter, "After the Mourning Wars: The Iroquois as Allies in Colonial North American Campaigns, 1676–1760," *William and Mary Quarterly*, 3rd ser., 64.1 (2007): 39–82. The divided members of the Iroquois Confederacy apparently passed warnings back and forth during the American Revolution. Karim M. Tiro, "A 'Civil' War? Rethinking Iroquois Participation in the American Revolution," *Explorations in Early American Culture* 4 (2000): 148–65; Tiro, "The Dilemmas of Alliance: The Oneida Indian Nation in the American Revolution," in *War and Society in the American Revolution: Mobilization and Home Fronts*, ed. John Resch and Walter Sargent (DeKalb: Northern Illinois University Press, 2007), 215–34.

18. Fitch, "Captain Fitch's Journal," 182, 194.

19. Joshua Piker, *Okfuskee: A Creek Indian Town in Colonial America* (Cambridge, MA: Harvard University Press, 2004), 22–23.

20. Frederic W. Gleach, *Powhatan's World and Colonial Virginia: A Conflict of Cultures* (Lincoln: University of Nebraska Press, 1997), 49–53; Kupperman, *Indians and English*, 196.

21. "The Examination and Relation of James Quannapaquait," in *The Sovereignty and Goodness of God*, by Mary Rowlandson (Boston: Bedford Books, 1997), 120.

22. Lee, "Fortify, Fight, or Flee," 757. The Europeans viewed such "go-betweens" much differently. For an extended treatment of the initial successes and ultimate failures of intermediaries between Europeans and Indians, see James H. Merrell, *Into the American Woods: Negotiators on the Pennsylvania Frontier* (New York: Norton, 1999). Note, however, that I am treating long-term "resident aliens" as a much more specific category than Merrell's "go-betweens." See also Kupperman's treatment of the early "go-betweens," Karen Ordahl Kupperman, *The Jamestown Project* (Cambridge, MA: Belknap Press of Harvard University Press, 2007), 233–35, 289–90.

23. Hudson, *Southeastern Indians*, 243–44; Adair, *History*, 167–78; John Gyles, "Memoirs of Odd Adventures, Strange Deliverances, etc.," in *Puritans among the Indians: Accounts of Captivity and Redemption, 1676–1724*, ed. Alden T. Vaughan and Edward W. Clark (Cambridge, MA: Harvard University Press, 1981), 120; Mary Rowlandson, *The Sovereignty and Goodness of God* (Boston: Bedford Books, 1997), 100; HNAI 15:315–16, 685–86; Jason Baird Jackson, "A Yuchi War Dance in 1736," *European Review of Native American Studies* 16 (2002): 27–32; Cave, *Pequot War*, 22; J. Frederick Fausz, "Fighting 'Fire' with Firearms: The Anglo-Powhatan Arms Race in Early Virginia," *American Indian Culture and Research Journal* 3 (1979): 41–42; Corkran, *Cherokee Frontier*, 155; José António Brandão, *Your fyre shall burn no more: Iroquois Policy towards New France and Its Native Allies to 1701* (Lincoln: University of Nebraska Press, 1997), 33.

24. Gregory Evans Dowd, *A Spirited Resistance: The North American Indian Struggle for Unity, 1745–1815* (Baltimore: Johns Hopkins University Press, 1992), 1–22.

25. George Turner to Governor Lyttelton, July 2, 1758, in DRIA 2: 471. In 1778 the Creeks delayed a war party until the completion of the green corn ceremony. Colin G. Calloway, *The American Revolution in Indian Country: Crisis and Diversity in Native American Communities* (Cambridge: Cambridge University Press, 1995), 62.

26. JR 47:227. War parties also carried medicine bundles and other sacred objects whose loss could send the warriors home. Hudson, *Southeastern Indians*, 244, 247; Adair, *History*, 409; Gleach, *Powhatan's World*, 53; HNAI 15:685, 695–96; Barbara Graymont, *The Iroquois in the American Revolution* (Syracuse, NY: Syracuse University Press, 1972), 139.

27. James Axtell, "The White Indians of Colonial America," *William and Mary Quarterly*, 3rd ser., 32 (1975): 67; Rowlandson, *Sovereignty and Goodness*, 107n82; Thomas S. Abler, "Scalping, Torture, Cannibalism and Rape: An Ethnohistorical Analysis of Conflicting Cultural Values in War," *Anthropologica* 34 (1992): 13; Dowd, *Spirited Resistance*, 9–10; Armstrong Starkey, *European and Native American Warfare, 1675–1815* (Norman: University of Oklahoma Press, 1998), 28, 81; Cave, *Pequot War*, 20; Elizabeth Hanson, "God's Mercy Surmounting Man's Cruelty," in *Puritans among the Indians: Accounts of Captivity and Redemption, 1676–1724*, ed. Alden T. Vaughan and Edward W. Clark (Cambridge, MA: Harvard University Press, 1981), 242 (see also the editors' comments in ibid., 14); Adair, *History*, 171–72; James Drake, "Restraining Atrocity: The Conduct of King Philip's War," *New England Quarterly* 70.1 (1997): 50.

28. Hatley, *Dividing Paths*, 107; Adair, *History*, 260–61; Richard White, *The Middle Ground: Indians, Empires, and Republics in the Great Lakes Region, 1650–1815* (Cambridge: Cambridge University Press, 1991), 345; O'Toole, *White Savage*, 264.

29. Hudson, *Southeastern Indians*, 252; Starkey, *European and Native American*, 28; HNAI 15:628. See also the Miamis' post-raid ritual, which, although not as directly confining, deprived the warriors of their personal sacred bundles for several days, presumably preventing them from returning to war immediately. HNAI 15:685.

30. For limited coercive power within Creek communities prior to the late eighteenth century, see Saunt, *New Order*, 1, 21–22.

31. In addition, the group's religious figure or shaman wielded a separate authority through his more extensive contact with the spirit world. Native American societies that had combined, or partially combined, this tripartite power structure into one person were typically more militant and aggressive. Gleach, *Powhatan's World*, 31; Charles Hudson, *Knights of Spain, Warriors of the Sun: Hernando De Soto and the South's Ancient Chiefdoms* (Athens, GA: University of Georgia Press, 1997), 17; David H. Dye, "The Transformation of Mississippian Warfare: Four Case Studies from the Mid-South," in *The Archaeology of Warfare: Prehistories of Raiding and Conquest*, ed. Elizabeth N. Arkush and Mark W. Allen (Gainesville: University Press of Florida, 2006), 101, 47; Dye, "The Art of War in the Sixteenth-Century Central Mississippi Valley," in *Perspectives on the Southeast: Linguistics, Archaeology*

and Ethnohistory, ed. Patricia B. Kwachka (Athens: University of Georgia Press, 1994), 54–56; Dye, "Warfare in the Sixteenth-Century Southeast: The de Soto Expedition in the Interior," in *Columbian Consequences*, vol. 2, *Archaeological and Historical Perspectives on the Spanish Borderlands East*, ed. David Hurst Thomas (Washington, DC: Smithsonian Institution Press, 1990), 213–14.

32. Adair, *History*, 167–68; Cadwallader Colden, *The History of the Five Indian Nations of Canada* (London: 1747), 6–7; Gyles, "Memoirs," 120; JR 47:221–31; Gordon M. Sayre, *Les Sauvages Américains: Representations of Native Americans in French and English Colonial Literature* (Chapel Hill: University of North Carolina Press, 1997), 275–76.

33. See, for example, Richter, *Ordeal*, 34–35; Cave, *Pequot War*, 3; Bruce G. Trigger, *The Children of Aataentsic: a History of the Huron People to 1660* (Montreal: McGill-Queen's University Press, 1976), 69; Saunt, *New Order*, 21; Hadlock, "War," 211.

34. Fitch, "Captain Fitch's Journal," 194.

35. Pierre Francois Xavier de Charlevoix, *Journal of a Voyage to North America* (London: R. and J. Dodsley, 1761), 1:360, quoted in Stephen Brumwell, *Redcoats: The British Soldier and War in the Americas, 1755–1763* (Cambridge: Cambridge University Press, 2002), 204. See also the Iroquois mobilization process described in José António Brandão, ed., *Nation Iroquoise: A Seventeenth-Century Ethnography of the Iroquois* (Lincoln: University of Nebraska Press, 2003), 67, 73, 75.

36. Adair, *History*, 416.

37. HNAI 15:676. See also Stephen Aron, *How the West Was Lost: The Transformation of Kentucky from Daniel Boone to Henry Clay* (Baltimore: Johns Hopkins University Press, 1996), 34; Sayre, *Sauvages Américains*, 270.

38. Cave, *Pequot War*, 157.

39. Fitch, "Captain Fitch's Journal," 202–5.

40. Chicken, "Journal of the March," 342.

41. Lee, "Fortify, Fight, or Flee," 718–23.

42. There is an excellent example in Christina Snyder, "Conquered Enemies, Adopted Kin, and Owned People: The Creek Indians and Their Captives," *Journal of Southern History* 73 (2007): 268–69. See also chapter 5, note 45.

43. Discussion in Wayne E. Lee, "Peace Chiefs and Blood Revenge: Patterns of Restraint in Native American Warfare in the Contact and Colonial Eras," *Journal of Military History* 71 (2007): 701–41, especially 707–9. Four hundred is clearly a rounded number, but the point is that only three men escaped.

44. Chicken, "Journal of the March," quotation on 345.

45. Bruce G. Trigger, *Natives and Newcomers: Canada's "Heroic Age" Reconsidered* (Kingston, ON: McGill-Queen's University Press, 1985), 175–76.

46. Gleach, *Powhatan's World*, 43–44; John Smith, *Captain John Smith: A Select Edition of His Writings*, ed. Karen Ordahl Kupperman (Chapel Hill: University of North Carolina Press, 1998), 84.

47. Jacques Le Moyne De Morgues, *The Work of Jacques Le Moyne De Morques: A Huguenot Artist in France, Florida, and England*, ed. Paul Hulton (London:

British Museum Publications, 1977), 1:144, and plate 105; cf. Roger Williams, *A Key into the Language of America* (London: 1643; Bedford, MA: Applewood Books, 1997), 188–89.

48. Richter, *Ordeal*, 35. This result is in keeping with how ethnohistorians have described battles by other tribal peoples witnessed in the twentieth century. Lawrence H. Keeley, *War Before Civilization* (New York: Oxford University Press, 1996), 59–61.

49. The most detailed account of the battle is unfortunately from an eighteenth-century letter written by Richard Hyde, quoted in full in Daniel Coit Gilman, *A Historical Discourse Delivered in Norwich, Connecticut, September 7, 1859, at the Bi-centennial Celebration of the Settlement of the town* (Boston: Rand and Avery, 1859), 82–84. The battle is also mentioned in Edward Johnson, *Wonder Working Providence of Sions Saviour*, Wing J771 (London: Nath. Brooke, 1654), 182–85; John Winthrop, *The Journal of John Winthrop, 1630–1649*, unabridged ed., ed. Richard S. Dunn, James Savage, and Laetitia Yeandel (Cambridge, MA: Harvard University Press, 1996), 236–37; William Bradford, *Of Plymouth Plantation, 1620–1647* (New York: Random House, 1981), 367; Michael Leroy Oberg, *Uncas: First of the Mohegans* (Ithaca, NY: Cornell University Press, 2003), 102–3; Herbert Milton Sylvester, *Indian Wars of New England* (Boston: Clarke, 1910; repr., New York: Arno, 1979), 1:390–97.

50. Craig S. Keener, "An Ethnohistorical Analysis of Iroquois Assault Tactics Used against Fortified Settlements of the Northeast in the Seventeenth Century," *Ethnohistory* 46 (1999): 777–807; Le Moyne De Morgues, *Work of Jacques Le Moyne*, 1:149 and plate 123. Contra Malone, *Skulking Way of War*, 14; Alden T. Vaughan, *New England Frontier: Puritans and Indians, 1620–1675*, 3rd ed. (Norman: University of Oklahoma Press, 1995), xxv.

51. See chapter 5, note 29.

52. Keener, "Ethnohistorical Analysis," 789–91; Lee, "Peace Chiefs," 707–9.

53. This does not mean that "ritualism" in war disappeared, just the ritual battle. Fausz, "Fighting Fire"; Starkey, *European and Native American*, 24–25; Richter, *Ordeal*, 54; Otterbein, "Why the Iroquois Won," 59–60. Note that these sources argue that the shift in tactics was due to firearms. My argument, significantly influenced by Otterbein and Divale, is for a broader shift from a duality of war styles (ritual battle and deadly ambush) to a stricter reliance on the ambush based on the shifting balance of offense versus defense (see chapter 5, note 29). For a conclusive discussion of why a musket wound was so much worse than one from an arrow, see Clifford Rogers, "Tactics and the Face of Battle across the Divide," in *European Warfare, 1350–1750*, ed. David Trim and Frank Tallett (Cambridge: Cambridge University Press, 2010).

54. Leroy V. Eid, "'A Kind of Running Fight': Indian Battlefield Tactics in the late Eighteenth Century," *Western Pennsylvania Historical Magazine* 71 (1988): 147–71; DRIA 2:467, 468; Gleach, *Powhatan's World*, 43.

55. Williams, *Key into the Language*, 184–86. As an indication of its importance, the half-moon formation is described by a single word, "onúttug."

56. Paul A. Robinson, "Lost Opportunities: Miantonomi and the English in Seventeenth-Century Narragansett Country," in *Northeastern Indian Lives, 1632–1816*, ed. Robert S. Grumet (Amherst: University of Massachusetts Press, 1996), 22–23.

57. The main sources for the attack are in Charles Orr, ed., *History of the Pequot War: The Contemporary Accounts of Mason, Underhill, Vincent and Gardener* (Cleveland, OH: Helman-Taylor, 1897); Mason's account is on 26–31, Underhill's on 78–84, (quote on 84), and Gardener's on 137. For the application of "laws of war" to the Pequots, see Drake, "Restraining Atrocity," 36–37. Accounts vary from three hundred to seven hundred killed, with few to no prisoners, although Gardener's figure of three hundred killed and "many prisoners" is probably the most reliable (some prisoners may have been taken in fighting after the initial attack on Mystic.)

58. Orr, *Pequot War*, 138.

59. Karen Anderson, *Chain Her by One Foot: The Subjugation of Native Women in Seventeenth-Century New France* (New York: Routledge, 1993), 169–78; Richter, *Ordeal*, 35–36; Hudson, *Southeastern Indians*, 254–55. Cf. Trigger, *Children*, 1:73–74. For a survey of the different torture rituals see Nathaniel Knowles, "The Torture of Captives by the Indians of Eastern North America," *American Philosophical Society Proceedings* 82 (1940): 151–255.

60. Brandão, *Your Fyre*, passim, 130–31; Daniel K. Richter, "War and Culture: The Iroquois Experience," *William and Mary Quarterly*, 3rd ser., 40 (1983). Jon Parmenter's work is a crucial addition to their analysis. Parmenter, "After the Mourning Wars."

61. There are few discussions of the role of prisoners in Indian societies as captured laborers, or even as slaves, whose lives might be preserved but who were not necessarily fully adopted members of the captors' society. Snyder, "Conquered Enemies"; William A. Starna and Ralph Watkins, "Northern Iroquoian Slavery," *Ethnohistory* 38 (1991): 34–57; Brett Rushforth, *Savage Bonds: Indigenous and Atlantic Slaveries in New France* (Chapel Hill: University of North Carolina Press, forthcoming), chapter 1. Thanks to Brett Rushforth for allowing me to cite his manuscript.

62. Axtell, "White Indians," 59–61.

63. For the various uses of prisoners and the implications for tactics, see Steele, "Surrendering Rites," 138–42. Richter argues that the whole nature of Iroquois warfare (reliance on surprise, avoidance of assaulting forts, ritualized battles) was determined by this desire to take prisoners. *Ordeal*, 37–38.

64. Increase Mather, "Quentin Stockwell's Relation of his Captivity and Redemption," in *Puritans among the Indians: Accounts of Captivity and Redemption, 1676–1724*, ed. Alden T. Vaughan and Edward W. Clark (Cambridge, MA: Harvard University Press, 1981), 81; Ian K. Steele, *Betrayals: Fort William Henry and the "Massacre"* (New York: Oxford University Press, 1990), 121.

65. Richter, *Ordeal*, 40.

66. James Axtell, "The Scholastic Philosophy of the Wilderness," in *The European and the Indian: Essays in the Ethnohistory of Colonial North America* (New

York: Oxford University Press, 1981), 138–50; Wayne E. Lee, *Crowds and Soldiers in Revolutionary North Carolina: The Culture of Violence in Riot and War* (Gainesville: University Press of Florida, 2001), 119–29.

67. John Lawson, *A New Voyage to Carolina* (London: 1709; repr., Chapel Hill: University of North Carolina Press, 1967), 207.

68. John Demos, *The Unredeemed Captive* (New York: Vintage Books, 1995), 24, 29, 33, 38–39.

69. JR 34:135–37.

70. Adair, *History*, 427.

71. Steele, *Betrayals*, 113, 131, 184.

72. For Indian diplomatic practices and alternative peacekeeping structures, see Robert A. Williams Jr., *Linking Arms Together: American Indian Treaty Visions of Law and Peace, 1600–1800* (New York: Oxford University Press, 1997); Rountree, *Powhatan Foreign Relations*; Daniel K. Richter and James H. Merrell, eds., *Beyond the Covenant Chain: The Iroquois and Their Neighbors in Indian North America, 1600–1800* (University Park: Pennsylvania State University Press, 2003); Dye, *War Paths*, 106, 162–63, Thomas Vennum, *American Indian Lacrosse: Little Brother of War* (Washington, DC: Smithsonian Institution Press, 1994).

73. Chicken, "Journal of the March," 345–46. The Narragansett dictionary is again useful in helping us imagine this process, including such phrases as "let us parley" or "let us cease Armes." Williams, *Key into the Language*, 189. See also JR 70:195.

74. JR 27:246–73. For more details on the peacemaking process, see Richter, *Ordeal*, 41; and Hudson, *Southeastern Indians*, 257. Alternatively, a third party not implicated in the ongoing cycle of revenge could be invoked to serve as an intermediary for this initial and most dangerous establishment of contact. Cave, *Pequot War*, 69–70; John Easton, "A Relacion of the Indyan Warre," in *Narratives of the Indian Wars, 1675–1699*, ed. Charles H. Lincoln (New York: Scribner, 1913), 8–9; Francis Jennings, *The Ambiguous Iroquois Empire: The Covenant Chain Confederation of Indian Tribes with English Colonies from its Beginnings to the Lancaster Treaty of 1744* (New York: Norton, 1984), 197.

75. Quotes from Gardener's narrative in Charles Orr, ed., *History of the Pequot War: The Contemporary Accounts of Mason, Underhill, Vincent and Gardener* (Cleveland, OH: Helman-Taylor, 1897), 131–32. For another example of the protection of embassies, see Rowlandson, *Sovereignty and Goodness*, 102–3, especially where the Wampanoag leadership expressed regret that some "matchit" [bad] Indian had stolen provisions from the English embassy.

76. HNAI 15:314–15; Gleach, *Powhatan's World*, 34–35; Aron, *West Was Lost*, 34.

77. For pushing men to war, see Wallace, *Death and Rebirth*, 101; Richter, *Ordeal*, 60, 224. For pushing an end to war, see Saunt, *New Order*, 25; Hudson, *Southeastern Indians*, 187.

78. The role of Europeans in diminishing women's influence within Native societies forms a substantial literature; see Anderson, *Chain Her by One Foot*; Calloway, *New Worlds for All*, 191.

79. Saunt, *New Order*, 25.

80. Reid, *Better Kind of Hatchet*, 10; Richter, *Ordeal*, 40, 44–45.

81. Nathaniel Sheidley presents a new twist on how older Cherokee chiefs used payment for land cessions to keep their young men at peace in "Hunting and the Politics of Masculinity in Cherokee Treaty-Making, 1763–75," in *Empire and Others: British Encounters with Indigenous Peoples, 1600–1850*, ed. Martin Daunton and Rick Halpern (Philadelphia: University of Pennsylvania Press, 1999), 167–85.

82. Kupperman, *Indians and English*, 198–99. For a discussion of diplomatic marriages in another context see: Hudson, *Southeastern Indians*, 234. For the broader role of interlocking familial ties helping to keep communications open between peoples, see Stephen Brumwell, *White Devil: A True Story of War, Savagery, and Vengeance in Colonial America* (Cambridge, MA: Da Capo Press, 2005), 193–94.

83. Kupperman, *Indians and English*, 197–99.

84. Ralph Hamor, *A True Discourse of the Present State of Virginia* (London: 1615; repr., New York: Da Capo, 1971), 38.

85. Gleach, *Powhatan's World*, 152–54; Helen C. Rountree, "The Powhatans and Other Woodland Indians as Travelers," in Rountree, *Powhatan Foreign Relations*, 50.

86. Based on still controversial evidence that warfare by nonstate peoples could produce an annual casualty rate of 25 percent, Azar Gat has argued that the invention of the state vastly reduced the exposure of most of the population to war, and indeed that the state itself, until the worst wars of the twentieth century, was the greatest structural restraint on the lethality of warfare. Azar Gat, *War in Human Civilization* (Oxford: Oxford University Press, 2006), 408.

87. Francis Jennings and Patrick Malone have even argued that Indians were taken aback by the European destruction of villages and crops, although I believe this pushes the argument too far. Francis Jennings, *The Invasion of America: Indians, Colonialism, and the Cant of Conquest* (New York: Norton, 1976), 152–53; Malone, *Skulking Way of War*, 103–4. The "collision" phrase is from Adam J. Hirsch, "The Collision of Military Cultures in Seventeenth-Century New England," *Journal of American History* 74 (1988): 1187–1212. See also Peter Way, "The Cutting Edge of Culture: British Soldiers Encounter Native Americans in the French and Indian War," in *Empire and Others: British Encounters with Indigenous Peoples, 1600–1850*, ed. Martin Daunton and Rick Halpern (Philadelphia: University of Pennsylvania Press, 1999), 123–48; Gregory T. Knouff, "Soldiers and Violence on the Pennsylvania Frontier," in *Beyond Philadelphia: The Pennsylvania Hinterland in the American Revolution*, ed. John B. Frantz and William Pencak (University Park: Pennsylvania State University Press, 1998), 171–93; Lee, *Crowds and Soldiers*, 117–29; Matthew H. Jennings, "'This Country Is Worth the Trouble of Going to War to Keep It': Cultures of Violence in the American Southeast to 1740" (PhD diss., University of Illinois, 2007). A variant of this argument postulates that Europeans arrived in North America during the horribly violent European wars of religion, and that they brought that form of unrestrained warfare with them. In

the face of an "uncivilized" enemy, the colonists preserved that way of war despite changes in Europe itself. John Ferling, *A Wilderness of Miseries: War and Warriors in Early America* (Westport, CT: Greenwood, 1980), 29–54; John Morgan Dederer, *War in America to 1775: Before Yankee Doodle* (New York: New York University Press, 1990), 127, 129–36; Ronald Dale Karr, "'Why Should You Be So Furious?': The Violence of the Pequot War," *Journal of American History* 85 (1998): 876–909.

88. The idea that European technology increased lethality is a commonplace in histories of Native American warfare; for example, Malone, *Skulking Way of War*, 65; Donald E. Worcester and Thomas E. Schilz, "The Spread of Firearms among the Indians on the Anglo-French Frontiers," *American Indian Quarterly* 8 (1984): 103; Calloway, *Western Abenakis*, 56, 61, 88. The role of desire for European trade goods in increasing the frequency of war for control of trade routes or trade items (notably fur) is more complex. Brandão, *Your Fyre*, 5–18; Thomas S. Abler, "Iroquois Policy and Iroquois Culture: Two Histories and an Anthropological Ethnohistory," *Ethnohistory* 47 (2000); William A. Starna and José António Brandão, "From the Mohawk-Mahican War to the Beaver Wars: Questioning the Pattern," *Ethnohistory* 51 (2004); George T. Hunt, *The Wars of the Iroquois: A Study in Intertribal Trade Relations* (Madison: University of Wisconsin Press, 1940); Richter, *Ordeal*, 55–74; Robbie Ethridge, "Creating the Shatter Zone: Indian Slave Traders and the Collapse of the Southeastern Chiefdoms," in *Light on the Path: The Anthropology and History of the Southeastern Indians*, eds. Thomas J. Pluckhahn and Robbie Ethridge (Tuscaloosa: University of Alabama Press, 2006), 207–18. For a general theory of European-induced escalation, see R. Brian Ferguson and Neil L. Whitehead, eds., *War in the Tribal Zone: Expanding States and Indigenous Societies* (Santa Fe: School of American Research Press, 1992); R. Brian Ferguson, "Violence and War in Prehistory," in *Troubled Times: Violence and Warfare in the Past*, ed. Debra L. Martin and David W. Frayer (Amsterdam: Gordon and Breach, 1997), 339–42.

89. Jill Lepore, *The Name of War: King Philip's War and the Origins of American Identity* (New York: Knopf, 1998), 118–19.

90. James D. Drake argues that there was even a basic misunderstanding of the meaning of "hostage" during King Philip's War. *King Philip's War: Civil War in New England, 1675–1676* (Amherst: University of Massachusetts Press, 1999), 115–16; Cynthia J. Van Zandt, *Brothers Among Nations: The Pursuit of Intercultural Alliances in Early America, 1580–1660* (Oxford: Oxford University Press, 2008), 101.

91. Hamor, *True Discourse*, 44.

92. Fitch assumed she was a "slave." Fitch, "Captain Fitch's Journal," 193. There are many such examples.

93. Important studies of whites living among Indians and English reaction include, Axtell, "White Indians"; Demos, *Unredeemed Captive*. The notoriety and popularity of the Indian captivity narrative speaks to the popular imagination's obsession with the subject, but also to the taint applied to any who failed to rejoin their natal culture. For suspicion of returned captives, see Lepore, *Name of War*, 126–36.

94. Theda Perdue, "'A Sprightly Lover is the Most Prevailing Mission-ary': Intermarriage between Europeans and Indians in the Eighteenth-Century South," in *Light on the Path: The Anthropology and History of the Southeastern Indians*, ed. Thomas J. Pluckhahn and Robbie Ethridge (Tuscaloosa: University of Alabama Press, 2006), 165–66.

95. Philip D. Morgan, "Encounters Between British and 'Indigenous' Peoples, c. 1500–c. 1800," in *Empire and Others: British Encounters with Indigenous Peoples, 1600–1850*, ed. Martin Daunton and Rick Halpern (Philadelphia: University of Pennsylvania Press, 1999), 63; Louise A. Breen, "Praying with the Enemy: Daniel Gookin, King Philip's War, and the Dangers of Intercultural Mediatorship," in ibid., 101, 109, 117; David J. Silverman, "The Curse of God: An Idea and Its Origins Among the Indians of New York's Revolutionary Frontier," *William and Mary Quarterly*, 3rd ser., 66 (2009): 501–5. Van Zandt is more optimistic about English visions of intermarriage. *Brothers Among Nations*, 70–73.

96. Lepore, *Name of War*, 21.

97. Bradford, *Plymouth Plantation*, 108–9. Kathleen J. Bragdon suggests that Hobomock lived in Plymouth for about twenty years. *Native People of Southern New England, 1500–1650* (Norman: University of Oklahoma Press, 1996), 290. Another possible exception is Connecticut's relationship with local Indians during King Philip's War. Jason W. Warren argues that Connecticut's more moderate and inclusive policy built relationships that helped spare that colony the worst of the war. "Connecticut Unscathed: An Examination of Connecticut Colony's Success During King Philip's War, 1675–1676" (MA thesis, Ohio State University, 2009), esp. 133.

98. The progressive legal deterioration of the status of Indian "nations" within the United States lies outside the scope of this chapter. For an introduction see Francis Paul Prucha, *American Indian Treaties: The History of a Political Anomaly* (Berkeley: University of California Press, 1994); Reginald Horsman, *Expansion and American Indian Policy, 1783–1812* (East Lansing: Michigan State University Press, 1967)

99. For one of Washington's statements on protecting Indians and treating them with justice, see Washington's 1791 address to Congress, in George Washington, *The Papers of George Washington: Presidential Series*, vol. 9, *October 1791–February 1792*, ed. Philander D. Chase (Charlottesville: University Press of Virginia, 2000), 111–12. Although he sought here to "attach [Native Americans] firmly to the United States," the meaning is more of allies than citizens. His policy was to consider them separate nations who should be maintained within sovereign "homelands," with an eventual eye to their incorporation as citizens. Washington's hopes failed in the face of a wider American rejection of the possi-bility of Indian citizenship and desire for their land. Joseph J. Ellis, *His Excellency George Washington* (New York: Knopf, 2004), 211–14, 237–38.

100. White, *Middle Ground*. For other studies emphasizing long periods of coexistence and acculturation (without denying conflict), see Daniel H. Usner Jr, *Indians, Settlers, and Slaves in a Frontier Exchange Economy: The Lower Mississippi*

Valley before 1783 (Chapel Hill: University of North Carolina Press, 1992); Jane T. Merritt, *At the Crossroads: Indians and Empires on a Mid-Atlantic Frontier, 1700–1763* (Chapel Hill: University of North Carolina Press, 2003). For the Creeks in specific, see J. Russell Snapp, *John Stuart and the Struggle for Empire on the Southern Frontier* (Baton Rouge: Louisiana State University Press, 1996), chap. 1.

101. Drake, *King Philip's War*, 87; Lepore, *Name of War*, 156–58. For two Praying Indians who struggled to prove their loyalty in the face of persistent doubt, see "The Examination and Relation of James Quannapaquait," in Rowlandson, *Sovereignty and Goodness*, 118–28. For the despair of an eighteenth-century missionary at the apparent failure of civilizing, see Silverman, "Curse of God," 528.

102. Jenny Hale Pulsipher, *Subjects Unto the Same King: Indians, English, and the Contest for Authority in Colonial New England* (Philadelphia: University of Pennsylvania Press, 2005); Wayne E. Lee, "Subjects, Clients, Allies or Mercenaries? The British Use of Irish and Indian Military Power, 1500–1815," in *Britain's Oceanic Empire: Projecting Imperium in the Atlantic and Indian Ocean Worlds, ca. 1550–1800*, ed. H. V. Bowen, Elizabeth Mancke, and John G. Reid (Cambridge: Cambridge University Press, forthcoming). James Drake provides an argument for English restraint toward the Indians during King Philip's War based on a vision of them as subjects, but this restraint certainly broke down in the following years. Drake, "Restraining Atrocity."

103. Two famous examples in Pennsylvania were the Paxton Boys massacre in 1763 and the Gnadenhutten massacre in 1782; see Benjamin Franklin, "Narrative of the Late Massacres," in *The Papers of Benjamin Franklin*, vol. 11, *January 1, 1764 through December 31, 1764*, ed. Leonard W. Labaree (New Haven, CT: Yale University Press, 1967), 42–69; White, *Middle Ground*, 389–91. For other similar incidents see Jenny Hale Pulsipher, "Massacre at Hurtleberry Hill: Christian Indians and English Authority in Metacom's War," *William and Mary Quarterly*, 3rd ser., 53 (1996): 459–86; Dowd, *Spirited Resistance*, 65–87; Aron, *West Was Lost*, 49; Drake, "Restraining Atrocity," 44–46.

104. To be clear, this delay was not a conscious choice for the purpose of allowing warning to spread, but it nevertheless had that result.

105. Lee, *Crowds and Soldiers*, 150.

106. John Grenier, *The First Way of War: American War Making on the Frontier, 1607–1814* (Cambridge: Cambridge University Press, 2005); Lee, "Subjects, Clients, Allies or Mercenaries?"

107. A related problem emerged as Englishmen encouraged Indian allies to procure Indian prisoners for the slave trade. By doing so they changed the dynamic of the prisoner-taking complex and lent it a new materialist motive that proved expansible, with frightful consequences. Gallay, *Indian Slave Trade*; Chicken, "Journal of a March," 344.

108. Steele, "Surrendering Rites," 152–56.

109. Gleach calls the attack a "coup," in an effort to emphasize the limited goals of this style of warfare. Gleach, *Powhatan's World*, 148–58. A similar argument

has been made for the Tuscarora War in 1712. Thomas C. Parramore, "The Tuscarora Ascendancy," *North Carolina Historical Review* 59 (1982): 322–23.

110. Philippe Contamine, *War in the Middle Ages*, tr. Michael Jones (Oxford: Blackwell, 1984), 280.

111. Kathleen DuVal, "Cross-Cultural Crime and Osage Justice in the Western Mississippi Valley, 1700–1826," *Ethnohistory* 54 (2007): 702–3.

112. For two other nuanced arguments about mismatches in styles of violence, see Karr, "So Furious," passim, esp. 888; and Evan Haefeli, "Kieft's War and the Cultures of Violence in Colonial America," in *Lethal Imagination: Violence and Brutality in American History*, ed. Michael A. Bellesiles (New York: New York University Press, 1999), 17–40.

113. Lepore, *Name of War*, 118–19.

114. Compare the detailed effort to move beyond simple racial prejudice in Rob Harper, "Looking the Other Way: The Gnadenhutten Massacre and the Contextual Interpretation of Violence," *William and Mary Quarterly*, 3rd ser., 64 (2007): 621–44. Grenier also critiques the focus on race, in *First Way of War*, 11–12. There is a complex discussion of the role of race and racism (and its limits) in the mid-Atlantic colonies in Peter Silver, *Our Savage Neighbors: How Indian War Transformed Early America* (New York: Norton, 2008).

115. Patricia Seed provides a now-classic discussion of the Spanish and English ideology of conquest, and how visions of land and religion affected the conscience of the individuals involved. "Taking Possession and Reading Texts: Establishing the Authority of Overseas Empires," *William and Mary Quarterly*, 3rd ser., 49 (1992): 183–209. Some legal writers sought to establish rules for the treatment of non-Christians, in terms of both property and war, but they had limited purchase, especially among the English. Andrew Fitzmaurice, "Moral Uncertainty in the Dispossession of Native Americans," in *The Atlantic World and Virginia, 1550–1624*, ed. Peter C. Mancall (Chapel Hill: University of North Carolina Press, 2007), 383–409; Anthony Pagden, *Lords of All the World: Ideologies of Empire in Spain, Britain and France c. 1500–c. 1800* (New Haven, CT: Yale University Press, 1995). For another examination that moves beyond ideology to incorporate questions of power in generating violence, see David J. Weber, *Bárbaros: Spaniards and Their Savages in the Age of Enlightenment* (New Haven, CT: Yale University Press, 2005)

Chapter 7

Epigraph from George Weedon, *Valley Forge Orderly Book of General George Weedon of the Continental Army under command of Genl. George Washington* (New York: Arno, 1971), 78.

1. Captain Johann Ewald, *Diary of the American War: A Hessian Journal*, trans. and ed. Joseph P. Tustin (New Haven, CT: Yale University Press, 1979), 77.

2. Joseph Plumb Martin, *Ordinary Courage: The Revolutionary War Adventures of Joseph Plumb Martin*, ed. Kirby Martin, 2nd ed. (New York: Brandywine, 1999), 83; Ewald recounts what is likely the same incident later in his diary. If so it would seem that one *Jäger* was killed and two captured in the exchange. Ewald, *Diary*, 137.

3. Wayne Bodle, *The Valley Forge Winter: Civilians and Soldiers in War* (University Park: Pennsylvania State University Press, 2002), 109–11, 132.

4. Humphrey Bland, *Treatise of Military Discipline: In which is Laid Down and Eplained the Duty of the Officer and Soldier, thro' the Several Branches of the Service*, 4th ed. (London: Printed for Sam. Buckley, 1740), 186. Washington had read Bland, along with many other practical military manuals. Don Higginbotham, *George Washington and the American Military Tradition* (Athens: University of Georgia Press, 1985), 14–15; Oliver L. Spaulding Jr., "The Military Studies of George Washington," *American Historical Review* 29 (1924): 675–80; Sandra L. Powers, "Studying the Art of War: Military Books Known to American Officers and their French Counterparts during the Second Half of the Eighteenth Century," *Journal of Military History* 70 (2006): 781–814. The British army replaced Bland's manual in 1764, but he remained widely cited. The most popular military manual among American officers at the outset of the war was probably Thomas Simes, *The Military Guide for Young Officers* (London: 1772; repr., Philadelphia: J. Humphreys, R. Bell, and R. Aitken, 1776). It contained similar admonitions against marauding (108). Emer de Vattel pointed in more general terms to a treaty of Louis XIV that limited small parties in the manner suggested by Bland. *The Law of Nations, or, Principles of the Law of Nature, Applied to the Conduct and Affairs of Nations and Sovereigns, with Three Early Essays on the Origin and Nature of Natural Law and on Luxury*, ed. Béla Kapossy and Richard Whatmore; trans. Thomas Nugent (Indianapolis, IN: Liberty Fund, 2008), 569.

5. Martin, *Ordinary Courage*, 64; Weedon, *Valley Forge Orderly Book*, 27. Later during the Valley Forge encampment, the foraging detachments were reduced to twelve men, a subaltern, and a commissary, but this may have been a result of the rapid decline in the army's size. Ibid., 166. Steuben's written regulations also addressed the responsibility of sergeants to prevent excesses in the unusually small groups sent out to get water. United States Army, *Regulations for the Order and Discipline of the Troops of the United States* (Boston: I. Thomas and E. T. Andrews, 1794), 85.

6. Pension Record of John McCasland, in *The Revolution Remembered: Eyewitness Accounts of the War for Independence*, ed. John C. Dann (Chicago: University of Chicago Press, 1980), 156–57.

7. Vattel, *Law of Nations*, 644–45. More on Vattel below.

8. Stephen Conway, "To Subdue America: British Army Officers and the Conduct of the Revolutionary War," *William and Mary Quarterly*, 3d Series, 43 (1986): 381–407; Armstrong Starkey, "War and Culture, A Case Study: The Enlightenment and the Conduct of the British Army in America, 1755–1781," *War and Society* 8 (1990): 1–28.

9. Quoted in Ira D. Gruber, "British Strategy: The Theory and Practice of Eighteenth-Century Warfare," in *Reconsiderations on the Revolutionary War: Selected Essays*, ed. Don Higginbotham (Westport, CT: Greenwood, 1978), 24.

10. David Hackett Fischer, *Washington's Crossing* (New York: Oxford University Press, 2004), 346–62. Fischer ably shows the impact of the militia harassment

on British capabilities in New Jersey in the winter of 1776–77. Thomas J. McGuire, *The Philadelphia Campaign*, vol. 1, *Brandywine and the Fall of Philadelphia* (Mechanicsburg, PA: Stackpole Books, 2006), 5–61. It is probable that this model of harassment was on Washington's mind when he crafted his strategy the following winter.

11. For Washington's vacillation over Howe's intentions (and the resulting marches and countermarches), see Edward G. Lengel, *General George Washington: A Military Life* (New York: Random House, 2005), 215–23. Gruber sees Howe as vacillating between desiring a decisive battle and waging a war of attrition through capturing key posts. I think instead that his vision was always to achieve battle, but that the restrictions on the mobility of eighteenth-century armies forced him into more conservative strategies, much the way described by John A. Lynn, *The Wars of Louis XIV, 1667–1714* (London: Longman, 1999), 375–76; Gruber, "British Strategy," 25–27.

12. McGuire, *Philadelphia Campaign*, 134–35.

13. Joseph Townsend, *Some Account of the British Army, under the Command of General Howe: And of the Battle of Brandywine, on the Memorable September 11th, 1777, and the Adventures of that Day, which Came to the Knowledge and Observation of Joseph Townsend* (Philadelphia: Townsend Ward, 1846), reprinted as *The Battle of Brandywine* (New York: Arno, 1969), 18, 21–26.

14. Alexander Hamilton, *The Papers of Alexander Hamilton*, ed. Harold C. Syrett, vol. 1, 1768–1778 (New York: Columbia University Press, 1961), 275.

15. Lengel, *General George Washington*, 224; Bodle, *Valley Forge Winter*, quotation on 34. For Washington as a Fabian strategist (a reference to the Roman general who had carefully avoided battle with the seemingly invincible Hannibal), see McGuire, *Philadelphia Campaign*, 57; Russell F. Weigley, *The American Way of War: A History of United States Military Strategy and Policy* (Bloomington: Indiana University Press, 1973), 3–17.

16. Dennis Showalter, *The Wars of Frederick the Great* (London: Longman, 1996), 6–7.

17. The Continental army changed who dressed the lines from the rear several times. Weedon, *Orderly Book*, 51–52, 194, 267; United States Army, *Regulations for Order*, plate 1.

18. Brent Nosworthy, *The Anatomy of Victory: Battle Tactics, 1689–1763* (New York: Hippocrene Books, 1990). For attitudes toward discipline in the era, see John A. Lynn, *Battle: A History of Combat and Culture* (Boulder, CO: Westview, 2003), 120–29; Christopher Duffy, *The Military Experience in the Age of Reason* (New York: Atheneum, 1988), 110–15; Scott N. Hendrix, "The Spirit of the Corps: The British Army and the Pre-National Pan-European Military World and the Origins of American Martial Culture, 1754–1783" (PhD diss., University of Pittsburgh, 2005), 107–47; Jeremy Black, *European Warfare, 1660–1815* (New Haven, CT: Yale University Press, 1994), 224–28. The image of Frederick's soldiers as "automatons" has perhaps been overplayed, but there is little doubt of the comparative severity and expectation of linearity and formality relative to the preceding two

centuries or more. For a new look at Frederick's disciplinary system, see Sascha Möbius, *Mehr Angst vor dem Offizier als vor dem Feind? Eine mentalitätsgeschichtliche Studie zur preußischen Taktick im Siebenjährigen Krieg* (Saarbrücken, Germany: Verlag Dr. Müller, 2007).

19. "Articles of War of James II (1688)," in *Military Law and Precedents*, ed. William Winthrop (New York: Arno, 1979), 922–23.

20. Some of these changes were an attempt to end the practice of "dead pays." "British Articles of War of 1765," in *Military Law and Precedents*, ed. William Winthrop (New York: Arno, 1979), 933–35, 939.

21. For desertion in eighteenth-century armies, see Hew Strachan, *European Armies and the Conduct of War* (London: Allen & Unwin, 1983), 9; Michael Sikora, *Disziplin und Desertion: Strukturprobleme militärischer Organisation im 18. Jahrhundert* (Berlin: Duncker & Humblot, 1996); Lorraine White, "The Experience of Spain's Early Modern Soldiers: Combat, Welfare and Violence," *War in History* 9 (2002): 1–38; Jörg Muth, *Flucht aus dem militärischen Alltag: Ursachen und individuelle Ausprägung der Desertion in der Armee Friedrichs des großen; Mit besonderer Bücksichtigung der Infanterie-Regimenter der Potsdamer Garnison* (Freiburg, Germany: Rombach Verlag, 2003); Duffy, *Military Experience*, 172–73; Sylvia R. Frey, *The British Soldier in America: A Social History of Military Life in the Revolutionary Period* (Austin: University of Texas Press, 1981), 72–73; Stephen Brumwell, *Redcoats: The British Soldier and War in the Americas, 1755–1763* (Cambridge: Cambridge University Press, 2002), 103–4.

22. Robert K. Wright Jr., *The Continental Army* (Washington, DC: Center of Military History, U.S. Army, 1983), 126–27.

23. The battle has been described many times. Two recent accounts are in Lengel, *General George Washington*, 229–42 (esp. 237); and McGuire, *Philadelphia Campaign*, 169–261.

24. These numbers are derived from Howard H. Peckham's calculations in Peckham, ed., *The Toll of Independence* (Chicago: University of Chicago Press, 1975). Only draws and losses are considered because the winning side rarely loses prisoners, which skews a perception of how willing the other side is to take prisoners.

25. These figures include British regular and loyalist losses, but not naval engagements. Ira D. Gruber, "British Casualties in the War for American Independence: How Numerous and Important Were They?" Unpublished conference paper delivered at the Society for Military History, April 10, 1994. I am grateful to Ira Gruber for sharing not only his paper but also his painstaking tabulations of casualty data.

26. John S. Pancake, *This Destructive War: The British Campaign in the Carolinas 1780–1782* (University: University of Alabama Press, 1985), 70–71; Peckham, *Toll of Independence*.

27. Armstrong Starkey, "Paoli to Stony Point: Military Ethics and Weaponry during the American Revolution," *Journal of Military History* 58 (1994): 7–27.

28. PGW-RW 8:50; 14:161, 675; JCC 10:203.

29. A study of Tarleton's Legion at Cowpens found that 10–30 percent of his strength had been recruited from Continentals captured at Camden. Lawrence E. Babits and Joshua B. Howard, "Continentals in Tarleton's British Legion, May 1780–October 1781" (forthcoming). I'm grateful to Lawrence Babits for sharing this paper with me. There is substantial anecdotal evidence of prisoners being recruited into British service especially, and deserters even more so, but nothing like the hundreds or even thousands of prisoners changing sides as in the English civil war. John A. Nagy provides examples of soldiers serving on the other side after having been captured. *Rebellion in the Ranks: Mutinies of the American Revolution* (Yardley, PA: Westholme, 2007), 128.

30. See below on exchange; Daniel Krebs, "Approaching the Enemy: German Captives in the American War of Independence, 1776–1783" (PhD diss., Emory University, 2007) is the most recent and important statement on many of these issues, and I am grateful to the author for discussing this section with me.

31. Persifor Frazer, *General Persifor Frazer* (Philadelphia: 1907), iii, 157–60. I first encountered this story in Bodle, *Valley Forge Winter*.

32. Bruce E. Burgoyne, ed., *Enemy Views: The American Revolutionary War As Recorded by the Hessian Participants* (Bowie, MD: Heritage Books, 1996), 131. The broadside is only preserved as *Draft of an Address of the Pennsylvania Council of Safety, Philadelphia, December 31, 1776,* Pennsylvania Division of Archives and Manuscripts, Records of Pennsylvania's Revolutionary Governments, 1775–1790, film 24, roll 11 (graciously provided to me by Daniel Krebs).

33. Frazer and his descendants who collected his letters were careful to make clear that he escaped while under guard. Frazer, *General Persifor Frazer*, 167, 170–72, 176–80, 178–80; "James Morris Memoirs," in *A Salute to Courage: The American Revolution As Seen Through Wartime Writings of Officers of the Continental Army and Navy*, ed. Dennis P. Ryan (New York: Columbia University Press, 1979), 103–6.

34. Krebs, "Approaching the Enemy"; Betsy Knight, "Prisoner Exchange and Parole in the American Revolution," *William and Mary Quarterly*, 3rd ser., 48 (1991): 200–22; Charles H. Metzger, *The Prisoner in the American Revolution* (Chicago: Loyola University Press, 1971); Paul Joseph Springer, "American Prisoner of War Policy and Practice from the Revolutionary War to the War on Terror" (PhD diss., Texas A&M University, 2006), 16–45; Reginald Savoy, "The Convention of Écluse, 1759–1762," *Journal of the Society of Army Historical Research* 42 (1964): 68–77; Edwin G. Burrows, *Forgotten Patriots: The Untold Story of American Prisoners During the Revolutionary War* (New York: Basic Books, 2008). Caroline Cox also analyzes how American perception of social status affected the degree of effort put into securing the exchange of common soldiers versus officers. *A Proper Sense of Honor: Service and Sacrifice in George Washington's Army* (Chapel Hill: University of North Carolina Press, 2004), 199–235. Much has been made of American rebel militia mistreatment of Loyalist prisoners (and vice versa), especially in the southern campaigns, but see Wayne E. Lee, "Restraint and Retaliation: The North Carolina Militias and the Backcountry War of 1780–1782," in *War and Society*

in the American Revolution: Mobilization and Home Fronts, ed. John Resch and Walter Sargent (DeKalb: Northern Illinois University Press, 2007), 168–71. For Frazer's letters and the renewal of the exchange cartel in 1778, see Frazer, *General Persifor Frazer*, 167, 170–72, 176–80; PGW-RW 14:583. See also note 99.

35. William H. McNeill, *The Pursuit of Power: Technology, Armed Force and Society since A.D. 1000* (Chicago: University of Chicago Press, 1982), 161. This general case finds strong support in the detailed case study by Myron P. Gutmann, *War and Rural Life in the Early Modern Low Countries* (Princeton, NJ: Princeton University Press, 1980), 54–71.

36. M. S. Anderson, *War and Society in Europe of the Old Regime, 1618–1789* (Leicester, UK: Leicester University Press, 1988), 137–38; David Kaiser, *Politics and War: European Conflict From Philip II to Hitler* (Cambridge, MA: Harvard University Press, 1990), 146, 153; John A. Lynn, "How War Fed War: The Tax of Violence and Contributions during the *Grand Siècle*," *Journal of Modern History* 65 (1993): 286–310; repeated and expanded in Lynn, *Giant of the Grand Siècle: The French Army, 1610–1715* (Cambridge: Cambridge University Press, 1997), 184–217.

37. Martin Van Creveld, *Supplying War: Logistics from Wallenstein to Patton* (Cambridge: Cambridge University Press, 1977), 17–39. See the responses to Van Creveld in John A. Lynn, ed., *Feeding Mars: Logistics in Western Warfare from the Middle Ages to the Present* (Boulder, CO: Westview, 1993). Of almost equal importance was the fact that most eighteenth-century governments proved much more capable of paying their soldiers.

38. Geoffrey Best, *Humanity in Warfare* (New York: Columbia University Press, 1980), 33–53. See also James Turner Johnson, *Ideology, Reason, and the Limitation of War: Religious and Secular Concepts, 1200–1740* (Princeton, NJ: Princeton University Press, 1975); Johnson, *Just War Tradition and the Restraint of War: A Moral and Historical Inquiry* (Princeton, NJ: Princeton University Press, 1981); Armstrong Starkey, *War in the Age of the Enlightenment, 1700–1789* (Westport, CT: Praeger, 2003).

39. Vattel, *Law of Nations*, 562.

40. Hugo Grotius, *The Rights of War and Peace*, ed. Richard Tuck (Indianapolis, IN: Liberty Fund, 2005). For his influence in England, see Barbara Donagan, "Atrocity, War Crime, and Treason in the English Civil War," *American Historical Review* 99 (1994): 1143. Alberico Gentili, writing in the late sixteenth century, said much the same as Grotius, enunciating perhaps an even more all-encompassing right of the victor to transfer both movable and real property. Alberico Gentili, *De iure belli libri tres* (New York: Oceana, 1964), 304–6. Samuel von Pufendorf, writing after Grotius in 1675, affirmed most of his precepts on the transfer of property in war, although he put more limits on soldiers' rights to plunder. *De jure naturae et gentium libri octo* (New York: Oceana, 1964), 2:584, 3:1085, 1113–14, 1309–12.

41. Grotius, *Rights of War*, 3.V.1, 3.XII.1, 3.XII.8, 3.VI.2(5), 3.VI.12, 3.XIII.4(2); see 3.XIII.1(2) for the definition of "surety" as subject.

42. Vattel, *Law of Nations*, 566–69, 594–98. See also the 1738 commentary on Grotius by Jean Barbeyrac, printed in Grotius, *Rights of War*, 3:1316–17 note II(1).

Later eighteenth-and nineteenth-century writers generally confirmed Vattel's limits on the transfer of private property, at least under the laws of war. Henry Wheaton simply stated that "private rights are unaffected by conquest." Other commentators admitted that a civil legislature could legislate confiscation of real property after the conquest. Henry Wheaton, *Elements of International Law*, ed. George Grafton Wilson (New York: Oceana, 1964), 362–63; William E Birkhimer, *Military Government and Martial Law*, 3rd ed. (Kansas City, MO: Franklin Hudson, 1914), 175–98 (esp. 178, 191–92); Sharon Korman, *The Right of Conquest: The Acquisition of Territory by Force in International Law and Practice* (Oxford: Clarendon, 1996), 29–40; Paul Keal, *European Conquest and the Rights of Indigenous Peoples: The Moral Backwardness of International Society* (Cambridge: Cambridge University Press, 2003), 35, 38. See a practical application of this principle in the French transfer of their dominions and Indian subjects to the British in their September 1760 capitulation, described in Anthony Pagden, *Lords of All the World: Ideologies of Empire in Spain, Britain and France c. 1500–c. 1800* (New Haven, CT: Yale University Press, 1995), 89.

43. Geoffrey Parker, "Early Modern Europe," in *The Laws of War: Constraints on Warfare in the Western World*, ed. Michael Howard, George J. Andreopoulos, and Mark R. Shulman (New Haven, CT: Yale University Press, 1994), 41.

44. *Proposals to Prevent Scalping, &c. Humbly Offered to the Consideration of a Council of War* (New York: Parker and Weyman, 1755), 6; Charles Molloy, *De jure maritimo et navali, or, A Treatise of Affairs Maritime and of Commerce*, 4th ed. (London: Printed for John Bellinger, 1688), 26; Grotius, *Rights of War and Peace*, 3.II.3–4. The author's use of the reprisal concept is a bit garbled but is nevertheless interesting as an appeal to legal authority.

45. Lt. Gov. Bull to Archibald Montgomery, May 23, 1760, James Grant of Ballindalloch Papers, Army Series, Box 32, National Archives of Scotland, Edinburgh, Scotland, Microfilm ed., Library of Congress.

46. John A. Lynn, "The Evolution of Army Style in the Modern West, 800–2000," *International History Review* 18 (1996): 517–19.

47. George Satterfield, *Princes, Posts and Partisans: The Army of Louis XIV and Partisan Warfare in the Netherlands, 1673–1678* (Leiden, The Netherlands: Brill, 2003), 19, 40n61.

48. Lynn, *Battle*, 128–31. Note that battles often did *not* in fact prove to be decisive. Contemporary generals nevertheless continued to believe that they might be, and for this reason they both hoped for and feared them. British generals in particular pursued a more aggressive style than many of their continental contemporaries. Gruber, "British Strategy," 22, 24; Duffy, *Military Experience*, 189–90; Stephen Conway, *The War of American Independence 1775–1783* (London: Arnold, 1995), 23, 34; Russell F. Weigley, *Age of Battles* (Bloomington: Indiana University Press, 1991), xii, 536; Jamel Ostwald, "The 'Decisive' Battle of Ramillies, 1706: Prerequisites for Decisiveness in Early Modern Warfare," *Journal of Military History* 64 (2000): 649–78; Michael Howard, *War in European History* (London: Oxford University Press, 1976), 70–71; Matthew H. Spring, *With Zeal and With Bayonets*

Only: The British Army on Campaign in North America, 1775–1783 (Norman: University of Oklahoma Press, 2008), 10–13.

49. Kaiser, *Politics and War*, 200–201; Lynn, *Wars of Louis XIV*, 367–76. My emphasis on the limited goals of the dynastic wars is not without controversy. See the discussion in Lynn, *Wars of Louis XIV*, 362–67; Black, *European Warfare*, 67–86; Showalter, *Frederick the Great*, 1. It is my sense, however, that mid-eighteenth-century North Americans imagined *conventional* war in this way. They had experienced unconventional war that was far more violent and even existentially threatening with the French, Spanish, and Indians, but they also continuously contrasted that experience with war in Europe.

50. John A. Lynn, "A Brutal Necessity? The Devastation of the Palatinate, 1688–1689," in *Civilians in the Path of War*, ed. Mark Grimsley and Clifford J. Rogers (Lincoln: University of Nebraska Press, 2002), 79–110; Vattel, *Law of Nations*, 571.

51. JCC 5:762–63 and 6:944–45; James Kirby Martin and Mark Edward Lender, *A Respectable Army: The Military Origins of the Republic, 1763–1789* (Arlington Heights, IL: Davidson, 1982), 76

52. Washington rejected the suggestion of using a purely guerrilla strategy. John Shy, "American Strategy: Charles Lee and the Radical Alternative," in *A People Numerous and Armed: Reflections on the Military Struggle for American Independence*, rev. ed. (Ann Arbor: University of Michigan Press, 1990), 133–62. David A. Bell rightly distinguishes between Enlightenment codification of laws of war and the preexisting aristocratic culture of war. He finds the latter more significant in limiting the frightfulness of war prior to the French Revolution. *The First Total War: Napoleon's Europe and the Birth of Warfare As We Know It* (Boston: Houghton Mifflin, 2007). I argue that the two worked in tandem, and, crucially, in North America the revolutionary experience and the lack of an aristocracy encouraged the diffusion of aristocratic notions of honor into a broader population.

53. Wayne E. Lee, *Crowds and Soldiers in Revolutionary North Carolina: The Culture of Violence in Riot and War* (Gainesville: University Press of Florida, 2001), 212–19. Further evidence is examined below.

54. Townsend, *Battle of Brandywine*, 29. For more on British devastation in and around Philadelphia, see Bodle, *Valley Forge Winter*, 63, 77, 182–83, 217; Steven Rosswurm, *Arms, Country, and Class: The Philadelphia Militia and "Lower Sort" during the American Revolution, 1775–1783* (New Brunswick, NJ: Rutgers University Press, 1987), 151–52.

55. For Washington's repeated efforts to find an opportunity to make a "general & vigorous attack," or at least confront the British army in September and early October 1777, see PGW-RW, 11:222, 227, 295, 339, 346, 359.

56. John S. Pancake, *1777: The Year of the Hangman* (University: University of Alabama Press, 1977), 192–99.

57. Quoted in Harry M. Ward, *General William Maxwell and the New Jersey Continentals* (Westport, CT: Greenwood, 1997), 86. See also Bodle, *Valley Forge Winter*, 55–67; Ira D. Gruber, "The Anglo-American Military Tradition and the War for American Independence," in *Against All Enemies: Interpretations of*

American Military History from Colonial Times to the Present, ed. Kenneth J. Hagan and William R. Roberts (New York: Greenwood, 1986), 35–36.

58. Gates to Jefferson, July 19, 1780, Horatio Gates Papers, 1726–1828, Microfilm edition (New York Historical Society and the National Historical Records and Publications Commission) (hereafter Gates's Papers).

59. First quotation from General Orders, September 6, 1776, PGW-RW 6:229; Holly A. Mayer, *Belonging to the Army: Camp Followers and Community During the American Revolution* (Columbia: University of South Carolina Press, 1996), 41; second quotation from Martin and Lender, *Respectable Army*, 173. See also Higginbotham, *George Washington*, 52–53, 94–95; Ronald Hoffman, "The 'Disaffected' in the Revolutionary South," in *The American Revolution*, ed. Alfred F. Young (Dekalb: Northern Illinois University Press, 1976), 310; Mark A. Clodfelter, "Between Virtue and Necessity: Nathanael Greene and the Conduct of Civil-Military Relations in the South, 1780–1782," *Military Affairs* 52 (1988): 173; Harold E. Selesky, "Colonial America," in *The Laws of War: Constraints on Warfare in the Western World*, ed. Michael Howard, George J. Andreopoulos, and Mark R. Shulman (New Haven, CT: Yale University Press, 1994), 75–76.

60. PNG 1:295; PGW-RW 6:200–201, 252, 273; JCC 5:733. Benjamin L. Carp has argued that the rebels did in fact secretly burn the city and denied it later. His argument is hard to prove conclusively, but he does show that Washington would seem to have preferred the city to be burned. "The Night the Yankees Burned Broadway: The New York City Fire of 1776," *Early American Studies* (Fall 2006): 496. Philip Schuyler also deliberately laid waste the countryside in front of Burgoyne's invading army prior to the battle at Saratoga in 1777. Richard M. Ketchum, *Saratoga: Turning Point of America's Revolutionary War* (New York: Holt, 1997), 330.

61. Bodle, *Valley Forge Winter*, 214–15.

62. Pension record of William Hutchinson, in John C. Dann, ed., *The Revolution Remembered: Eyewitness Accounts of the War for Independence* (Chicago: University of Chicago Press, 1980), 151.

63. Bodle, *Valley Forge Winter*, 210.

64. JCC 9:784–85 (the order only authorized its use in a thirty-mile radius around places under British control in Pennsylvania, New Jersey, or Delaware).

65. Bodle, *Valley Forge Winter*, 210–11; PGW-RW 14:476–77.

66. *Valley Forge Orderly Book of General George Weedon* (New York: Arno Press, 1971), 226–27, 247, 261. The more violent side of martial law, especially as enforced by the local militia, can be seen in Nicholas Collin, "The Journal and Biography of Nicholas Collin," in *New Jersey in the American Revolution, 1763–1783: A Documentary History*, ed. Larry R. Gerlach (Trenton: New Jersey Historical Commission, 1975), 302–6.

67. JCC 9:971. A table of the various state confiscation laws is in Claude Halstead Van Tyne, *The Loyalists in the American Revolution* (New York: Peter Smith, 1929), 327–41.

68. The confiscation of land by the legislatures has generated a large body of literature, but Washington's use of martial law against civilians has attracted less

attention. Mayer, *Belonging to the Army*, 243–62, Don Higginbotham, *The War of American Independence: Military Attitudes, Policies, and Practice, 1763–1789* (Boston: Northeastern University Press, 1983), 279–80; G. Norman Lieber, "Martial Law during the Revolution," *Magazine of American History* 1 (1877): 538–41; Asa Bird Gardner, "Martial Law during the Revolution," *Magazine of American History* 1 (1877): 705–19.

69. PGW-RW 12:620; Bodle, *Valley Forge Winter*, 57–71.

70. Brian Fagan, *The Little Ice Age: How Climate Made History, 1300–1850* (New York: Basic Books, 2000).

71. Albigence Waldo, "Valley Forge 1777–1778, Diary of Surgeon Albigence Waldo, of the Connecticut Line," *Pennsylvania Magazine of History and Biography* 21 (1897): 308. The Continentals' suffering at Valley Forge is so often described as to be cliché and needs no repeating here. For a recent treatment, see Thomas Fleming, *Washington's Secret War: The Hidden History of Valley Forge* (New York: HarperCollins, 2005), esp. 129–46.

72. Elijah Fisher, *Elijah Fisher's Journal While in the War for Independence, and Continued Two Years after He Came to Maine* (Augusta, ME: Badger and Manley, 1880), 7.

73. Higginbotham, *War of American Independence*, 304.

74. Dorothy Denneen Volo and James M. Volo, *Daily Life during the American Revolution* (Westport, CT: Greenwood, 2003), 178, 170.

75. E. Wayne Carp, *To Starve the Army at Pleasure: Continental Army Administration and American Political Culture, 1775–1783* (Chapel Hill: University of North Carolina Press, 1984); James A. Huston, *Logistics of Liberty: American Services of Supply in the Revolutionary War and After* (Newark: University of Delaware Press, 1991); John Shy, "Logistical Crisis and the American Revolution: A Hypothesis," in Lynn, *Feeding Mars*, 161–69.

76. PNG 2:283.

77. Higginbotham, *George Washington*, 94–95; Carp, *To Starve the Army*, 77.

78. PGW-RW 11:324, 326, 331.

79. *Valley Forge Orderly Book*, 174, 250; Jefferson to Greene, February 18, 1781, in *The Papers of Thomas Jefferson*, vol. 4, *October 1780–February 1781*, ed. Julian P. Boyd (Princeton, NJ: Princeton University Press, 1950), 648; PNG 6:598.

80. In previous work I studied the Continentals' use of quartering officers during the southern campaign. The Moravian community of North Carolina clearly benefited from knowing that troops were arriving. See Adelaide L. Fries, *Records of the Moravians in North Carolina*, vol. 4, *1780–1783* (Raleigh, NC: North Carolina Historical Commission, 1968), 1554, 1571, 1666–67, 1674. See also Greene's order to a subordinate to make sure he did send such an officer. PNG 6:577–78.

81. The "carving" phrase was common. For local recognition of the gradations of purchase, impressment, and plunder, see Lee, *Crowds and Soldiers*, 180–85.

82. The Massachusetts articles of 1775 (the basis for those adopted by the Continentals) and the Continental articles of 1775 and 1776 are conveniently reprinted in William Winthrop, ed., *Military Law and Precedents* (New York:

Arno, 1979), 947–52, 953–60, 961–71. As one example, the 1776 articles added a clause (section X, article 1) requiring soldiers accused of a capital crime or other violence against the civilian population to be turned over for a civil trial. Robert H. Berlin, "The Administration of Military Justice in the Continental Army during the American Revolution, 1775–1783" (PhD diss., University of California, Santa Barbara, 1976); S. Sidney Bradford, "Discipline in the Morristown Winter Encampments," *New Jersey Historical Society Proceedings* 80 (1962): 1–30.

83. Winthrop, *Military Law*, 967. Cf. the British 1765 articles, in ibid., 940. The requirement to turn soldiers over to a civil magistrate also appeared in the British articles of 1765, but not in the initial Massachusetts or Continental versions of 1775 (see the preceding note).

84. Examples abound. A few include General Orders, July 25, 1777, Jacob Turner Order Book, North Carolina State Archives, Raleigh, NC; entry for August 21, 1776, Gates' General Orders When in Command of the Northern Army, July 10, 1776–June 3, 1777; and entries for August 6, 1780 and September 20, 1780, Copy Book of Orders Issued by Major General Gates While Commanding the Southern Army, July 26–December 4, 1780, the last two in Gates's Papers; Almon W. Lauber, ed., *Orderly Books of the Fourth New York Regiment, 1778–1780, the Second New York Regiment, 1780–1783 by Samuel Tallmadge and Others, with Diaries of Samuel Tallmadge, 1780–1782 and John Barr, 1779–1782* (Albany: State University of New York Press, 1932), 114–16.

85. *Valley Forge Order Book*, 31–32.

86. James C. Neagles's extensive survey of courts-martial recorded in order books found 194 examples of soldiers charged with plunder or theft from civilians, constituting 7.3 percent of the of the court-martial charges he surveyed. There were undoubtedly more, but the order books often did not include enough details to determine if certain offenses like stealing were committed against civilians or other soldiers. *Summer Soldiers: A Survey & Index of Revolutionary War Courts-Martial* (Salt Lake City: Ancestry. 1986), 34. Mayer argues that Washington rarely commuted sentences for men convicted of plundering. *Belonging to the Army*, 41. In addition, the Continental army also used summary punishment for plunderers and occasionally even summarily sentenced them to death. See entry for August 26, 1780 in Israel Angell, *Diary of Colonel Israel Angell, Commanding the Second Rhode Island Continental Regiment during the American Revolution, 1778–1781*, ed. Edward Field (Providence, RI: Preston & Rounds, 1899; repr., New York: Arno, 1971), 108–9; Jeremiah Greenman, *Diary of a Common Soldier in the American Revolution, 1775–1783: An Annotated Edition of the Military Journal of Jeremiah Greenman*, ed. Robert C. Bray and Paul E. Bushnell (DeKalb: Northern Illinois University Press, 1978), 168; Berlin, "Administration of Military Justice," 140.

87. PGW-RW 12:289B91.

88. Harry M. Ward, *George Washington's Enforcers: Policing the Continental Army* (Carbondale: Southern Illinois University Press, 2006) is a point-by-point listing of the kinds of processes, mostly based on European tradition, that Washington put into place to impose discipline on the army. See also Wright's

general discussion of the European influences on Washington. *Continental Army*, 121–52.

89. Ward, *George Washington's Enforcers*, 140–42.

90. This is not to imply that eighteenth-century warfare had not continued to develop with its own logic. But there was also a considerable cultural component to the preference for the linear style of warfare. See Lynn, *Battle*, chap. 4.

91. *Valley Forge Orderly Book*, 19–20.

92. Burke Davis, *The Campaign that Won America: The Story of Yorktown* (New York: Dial, 1970), 185–86.

93. John Shy, "American Society and Its War for Independence," in *Reconsiderations on the Revolutionary War: Selected Essays*, ed. Don Higginbotham (Westport, CT: Greenwood, 1978), 78.

94. Townsend, *Battle of Brandywine*, 29 (emphasis added).

95. Peter H. Wilson, "German Women and War, 1500–1800," *War in History* 3 (1996): 127–60; John A. Lynn II, *Women, Armies, and Warfare in Early Modern Europe* (Cambridge: Cambridge University Press, 2008), 12–14 and passim; Lynn, "Essential Women, Necessary Wives, and Exemplary Soldiers: The Military Reality and Cultural Representation of Women's Military Participation 1600–1815," in *A Companion to Women's Military History*, ed. Barton C. Hacker and Margaret Vining (Leiden, The Netherlands: Brill, forthcoming). Thanks to John Lynn for sharing this manuscript with me.

96. Lee, *Crowds and Soldiers*, 198–99.

97. Mayer, *Belonging to the Army*, 129–30.

98. *Valley Forge Orderly Book*, 62, 81, 173, 187, 212, 276, 306. For a narrative of the formalities, but also the possibilities, of communication, see James Thacher, *Military Journal of the American Revolution, 1775–1783* (Gansevoort, NY: Corner House, 1998), 50–51.

99. PNG 11:128; Cornwallis to Smallwood, November 10, 1780, Gates's Papers. The treatment of prisoners was a major forum for this kind of exchange, including between Washington and Gage (in 1775) and then Washington and Howe (in 1777). In each case, both sides referred not only to the accepted humane practice of war but also to the potential for retaliation in case abuses were not corrected. JCC 5:458; PGW-RW 1:289–90, 301–2, 326–27; 6:76. For the Continental Congress, Washington, and Howe in 1777, see PGW-RW 8:58–61, 91–94, 137–38, 453–54, 522–23; 9:228–30, 496–97; JCC 7:16, 135; abstracts of debates in Continental Congress, February 20, 1777, NCSR 11:381–82; *Pennsylvania Gazette*, January 28, 1778. See also Metzger, *Prisoner in the American Revolution*, 154–58, 160, 162. The Huddy-Asgill and Isaac Hayne controversies near the end of the war also exemplify circumstances in which Continental leadership very nearly resorted to actual retaliation for British actions but in the end held back. Both incidents generated a large body of documents; for short treatments, see Harold E. Selesky, ed. *Encyclopedia of the American Revolution*, 2nd ed. (Detroit: Scribners, 2006), 1:496–97, 529.

100. Cited in Mayer, *Belonging to the Army*, 53.

101. Lt. Uzal Meeker to Edward Hand, March 2, 1779, Edward Hand Papers, New York Public Library.

102. See Thacher, *Military Journal*, 59–60, 68–69. This status sensitivity of Continental officers is now a commonplace among specialists. See Mayer, *Belonging to the Army*, 51–54; Cox, *Proper Sense of Honor*, 23–35, 62–67; Charles Royster, *A Revolutionary People at War: The Continental Army and American Character, 1775–1783* (New York: W.W. Norton, 1979), 87–88, 92–94.

103. Martin, *Ordinary Courage*, 60, 66.

104. Fisher, *Elijah Fisher's Journal*, 11.

105. Benjamin Gilbert, *A Citizen-Soldier in the American Revolution: The Diary of Benjamin Gilbert in Massachusetts and New York*, ed. Rebecca D. Symmes (Cooperstown: New York State Historical Association, 1980), passim, esp. 26, 30, 36–37.

106. Charles Patrick Neimeyer, *America Goes to War: A Social History of the Continental Army* (New York: New York University Press, 1996), 3–4, 8–9. See also John Resch, *Suffering Soldiers: Revolutionary War Veterans, Moral Sentiment, and Political Culture in the Early Republic* (Amherst: University of Massachusetts Press, 2000), 13–46; Martin and Lender, *Respectable Army*, 90–94; Fleming, *Washington's Secret War*, 140–43; Walter Sargent, "The Massachusetts Rank and File of 1777," in *War and Society in the American Revolution: Mobilization and Home Fronts*, ed. John Resch and Walter Sargent (DeKalb: Northern Illinois University Press, 2007), 42–69; John Resch, "The Revolution as a People's War: Mobilization in New Hampshire," in ibid., 70–102; Ward, *George Washington's Enforcers*, 15–20.

107. William Linn, *A Military Discourse Delivered in Carlisle, March the 17th, 1776, to Colonel Irvine's Battalion of Regulars* (Philadelphia: 1776); speech of General Nash, July 21, 1777, An Original Orderly Book of a Portion of the American Army, North Carolina Brigade, 2/7/71–8/13/77, Military Collection, War of the Revolution, Miscellaneous Papers (Box 6), North Carolina Department of Archives, Raleigh, NC; Robert Kirkwood, *The Journal and Order Book of Captain Robert Kirkwood of the Delaware Regiment of the Continental Line*, ed. Joseph Brown Turner (Wilmington: Historical Society of Delaware, 1910), 150–51, 159–60; Lauber, *Orderly Books*, 879–80.

108. The ideological commitment of the soldiers remains debated. In addition to the evidence outlined below, see Royster, *Revolutionary People*, 295–307, 314–18, 373–78; Martin, *Ordinary Courage*, 62–63. Martin and Lender, *Respectable Army*, 53–55, and, contradictorily, 76–77; Jesse Lemisch, "Listening to the 'Inarticulate': William Widger's Dream and the Loyalties of American Revolutionary Seamen in British Prisons," *Journal of Social History* 3 (1969): 1–29; Robert Middlekauff, "Why Men Fought in the American Revolution," *Huntington Library Quarterly* 43 (1980): 135–48. For a review of this literature and more, see Don Higginbotham, "The Early American Way of War: Reconnaissance and Appraisal," *William and Mary Quarterly*, 3rd ser., 44 (1987): 257–61; Wayne E. Lee, "Early American Ways of War: A New Reconnaissance, 1600–1815," *Historical Journal* 44 (2001): 280–84.

109. Pension Record of Jonathan Libby, W24557, Revolutionary War Pension Files, NARA, Washington, DC.

110. Cited in Philip Davidson, *Propaganda and the American Revolution 1763–1783* (Chapel Hill: University of North Carolina Press, 1941), 341.

111. Private Elijah Fisher's diary conveys a remarkable sense of commitment to the army, although his motives are not explicit. He repeatedly reenlisted, even while ill, and despite a significant backlog of pay. Fisher, *Elijah Fisher's Journal*, passim, 3, 5–6, 14.

112. Martin and Lender, *Respectable Army*, 163; JCC 19:83; John A. Nagy, *Rebellion in the Ranks: Mutinies of the American Revolution* (Yardley, PA: Westholme, 2007), 77–166, esp. 127–28, 146, 149, 152.

113. Thomas J. McGuire, *The Philadelphia Campaign*, vol. 2, *Germantown and the Roads to Valley Forge* (Mechanicsburg, PA: Stackpole Books, 2007), 147–49.

114. John Ferling, *Almost a Miracle: The American Victory in the War of Independence* (Oxford: Oxford University Press, 2007), 486; Wayne E. Lee, review of ibid., *Journal of Southern History* 74 (2008): 722–24; Babits and Howard, "Continentals"; Lawrence E. Babits, *A Devil of a Whipping: The Battle of Cowpens* (Chapel Hill: University of North Carolina Press, 1998), 122–23.

115. George W. Troxler, *Pyle's Massacre, February 23, 1781* (Burlington, NC: Alamance County Historical Association, 1973).

116. Starkey, "Paoli to Stony Point."

117. "Sermon against Plundering," in Samuel E. McCorkle Papers, Duke University Special Collections, Durham, NC; Fries, *Records of the Moravians*, 1910.

118. A few include Lee, *Crowds and Soldiers*, chaps. 4–7; Lee, "Restraint and Retaliation"; *An Uncivil War: The Southern Backcountry During the American Revolution*, ed. Ronald Hoffman, Thad W. Tate, and Peter J. Albert (Charlottesville: University Press of Virginia, 1985); Harry M. Ward, *Between the Lines: Banditti of the American Revolution* (Westport, CT: Praeger, 2002); David J. Fowler, "Egregious Villains, Wood Rangers, and London Traders: The Pine Robber Phenomenon in New Jersey During the Revolutionary War" (PhD diss., Rutgers University, 1987); Sung Bok Kim, "The Limits of Politicization in the American Revolution: The Experience of Westchester County, New York," *Journal of American History* 80 (1993): 868–89; Allan Kulikoff, "Revolutionary Violence and the Origins of American Democracy," *Journal of the Historical Society* 11 (2002): 229–60; Michael S. Adelberg, "An Evenly Balanced County: The Scope and Severity of Civil Warfare in Revolutionary Monmouth County, New Jersey," *Journal of Military History* 73 (2009): 9–47. There are many others.

Chapter 8

Epigraph quoted in Albert Hazen Wright, *The Sullivan Expedition of 1779: Contemporary Newspaper Comments and Letters* (Ithaca, NY: Wright, 1943), 2.

1. The Sullivan-Clinton expedition is well served by soldiers' and officers' diaries, some by men apparently moved to describe this entirely new landscape, others by those who incorporated it into longer campaign journals. Most have been published; twenty-six are in Frederick Cook, ed., *Journals of the Military Expedition of Major General John Sullivan Against the Six Nations of Indians in 1779*

(Auburn, NY: Knapp, Peck & Thomson, 1887; repr., Bowie, MD: Heritage Books, 2000). Sullivan's periodic official reports agree in most details with the diarists, although he obviously wrote to highlight his successes. John Sullivan, *Letters and Papers of Major-General John Sullivan, Continental Army*, vol. 3, *1779–1795*, ed. Otis G. Hammond (Concord, NH: New Hampshire Historical Society, 1939), 60–61, 65–66, 75–78, 80–84, 95–100, 107–12, 123–37. The papers of Edward Hand at the New York Public Library provide some supplementary details. Other journals include Daniel Clapp, *Diary and Accounts*, Edward E. Ayer Manuscript Collection, Newberry Library, Chicago; Moses Sproule, "The Western Campaign of 1779: The Diary of Quartermaster Sergeant Moses Sproule," ed. R. W. G. Vail, *New York Historical Society Quarterly* 41 (1957): 35–69; Tjerck Beekman, "Journal of Lieutenant Tjerck Beekman, 1779," *Magazine of American History* 20 (1888): 128–36; Andrew Hunter's diary, Princeton University Library; Obadiah Gore Jr., "Diary of Lieut. Obadiah Gore, Jr., in the Sullivan-Clinton Campaign of 1779," ed. R. W. G. Vail, *New York Public Library Bulletin* 33 (1929): 711–39; Robert Parker, "Journal of Lieutenant Robert Parker, of the Second Continental Artillery, 1779," ed. Thomas R. Bard, *Pennsylvania Magazine of History and Biography* 27 (1903): 404–20, and 28 (1904): 12–25. Orderly books from the Fourth New York Regiment and the campaign diary of John Barr are in Almon W. Lauber, ed. *Orderly Books of the Fourth New York Regiment, 1778–1780, the Second New York Regiment, 1780–1783 by Samuel Tallmadge and Others, with Diaries of Samuel Tallmadge, 1780–1782 and John Barr, 1779–1782* (Albany: State University of New York Press, 1932). Specific details and quotes from the diaries and journals will be referenced by last name and page (where appropriate) only. Officers' letters (often anonymous) written during the actual campaign were published in newspapers and are collected in Wright, *Sullivan Expedition*. Other relevant letters are collected in A. C. Flick, "New Sources on the Sullivan-Clinton Campaign in 1779," *Quarterly Journal of the New York State Historical Association* 10 (1929): 185–24, 265–317.

2. Quotations from Sproule's journal, 50–51; Beekman's journal, 130–31; Barton's journal, in Cook, *Journals*, 6.

3. The best description of the deserters' trials and the details of their punishments is in Hunter's diary (Hunter was also the man reading Burlamaqui). Daniel Clapp's diary described one soldier given fifty lashes for stealing. Clapp's diary, Ayer Manuscript Collection 162, Newberry Library, Chicago. Quotation from Edward Hand's journal, Papers of Edward Hand, New York Public Library.

4. Joseph R. Fischer, *A Well-Executed Failure: The Sullivan Campaign Against the Iroquois, July–September 1779* (Columbia: University of South Carolina Press, 1997), 43.

5. *New York Gazette*, August 30, 1779, in Wright, *Sullivan Expedition*, 34–35. The other raids are enumerated in Wright, *Sullivan Expedition*.

6. Barton's journal, in Cook, *Journals*, 6.

7. Gookin's journal, in Cook, *Journals*, 105–6; Roberts's journal, in Cook, *Journals*, 244; Barton's journal, in Cook, *Journals*, 8, Van Hovenburgh's journal, in Cook, *Journals*, 279; Sullivan, *Letters and Papers*, 110–11; William Smith, *Historical*

Memoirs of William Smith, 1778–1783, ed. W. H. W. Sabine (New York: New York Times, 1971), 230–31. Thanks to Scott Miskimon for this last reference.

8. Van Hovenburgh journal, in Cook, *Journals*, 279.

9. Hand to Yeats, September 25, 1779, Edward Hand Papers, New York Public Library.

10. Max M. Mintz, *Seeds of Empire: The American Revolutionary Conquest of the Iroquois* (New York: New York University Press, 1999), 153–54; Barbara Graymont, *The Iroquois in the American Revolution* (Syracuse, NY: Syracuse University Press, 1972) 229–41.

11. Recent major works on this problem include Richard White, *The Middle Ground: Indians, Empires, and Republics in the Great Lakes Region, 1650–1815* (Cambridge: Cambridge University Press, 1991); Jane T. Merritt, *At the Crossroads: Indians and Empires on a Mid-Atlantic Frontier, 1700–1763* (Chapel Hill: University of North Carolina Press, 2003); Matthew C. Ward, *Breaking the Backcountry: The Seven Years' War in Virginia and Pennsylvania, 1754–1765* (Pittsburgh: University of Pittsburgh Press, 2003); Peter Silver, *Our Savage Neighbors: How Indian War Transformed Early America* (New York: Norton, 2008).

12. Woody Holton, *Forced Founders: Indians, Debtors, Slaves, and the Making of the American Revolution in Virginia* (Chapel Hill: University of North Carolina Press, 1999); Gregory Evans Dowd, *A Spirited Resistance: The North American Indian Struggle for Unity, 1745–1815* (Baltimore: Johns Hopkins University Press, 1992); Dowd, *War Under Heaven: Pontiac, the Indian Nations, and the British Empire* (Baltimore: Johns Hopkins University Press, 2002); Patrick Griffin, *American Leviathan: Empire, Nation, and Revolutionary Frontier* (New York: Hill & Wang, 2007), 22–45.

13. Paul Lawrence Stevens, "His Majesty's 'Savage' Allies: British Policy and the Northern Indians During the Revolutionary War—The Carleton Years, 1774–1778" (PhD diss., State University of New York at Buffalo, 1984), 259.

14. Wayne E. Lee, *Crowds and Soldiers in Revolutionary North Carolina: The Culture of Violence in Riot and War* (Gainesville: University Press of Florida, 2001), 142–43; Rachel N. Klein, "Frontier Planters and the American Revolution: The South Carolina Backcountry, 1775–1782," in *An Uncivil War: The Southern Backcountry During the American Revolution*, ed. Ronald Hoffman, Thad W. Tate, and Peter J. Albert (Charlottesville: University Press of Virginia, 1985), 37–68; Jim Piecuch, "Incompatible Allies: Loyalists, Slaves, and Indians in Revolutionary South Carolina," in *War and Society in the American Revolution: Mobilization and Home Fronts*, ed. John Resch and Walter Sargent (DeKalb: Northern Illinois University Press, 2007), 191–214.

15. Colin G. Calloway, *The American Revolution in Indian Country: Crisis and Diversity in Native American Communities* (Cambridge: Cambridge University Press, 1995), 85–102; James H. Merrell, *The Indians' New World: Catawbas and Their Neighbors from European Contact through the Era of Removal* (Chapel Hill: University of North Carolina Press, 1989), 215–21.

16. Calloway, *American Revolution*, 108; Karim M. Tiro, "A 'Civil' War? Rethinking Iroquois Participation in the American Revolution," *Explorations in*

Early American Culture 4 (2000): 148–65; Tiro, "The Dilemmas of Alliance: The Oneida Indian Nation in the American Revolution," in *War and Society in the American Revolution: Mobilization and Home Fronts*, ed. John Resch and Walter Sargent (DeKalb: Northern Illinois University Press, 2007), 214–34; Joseph T. Glatthaar and James Kirby Martin, *Forgotten Allies: The Oneida Indians and the American Revolution* (New York: Hill & Wang, 2006).

17. Quoted in Dennis P. Ryan, ed., *A Salute to Courage: The American Revolution As Seen Through Wartime Writings of Officers of the Continental Army and Navy* (New York: Columbia University Press, 1979), 161; Tiro, "Dilemmas of Alliance," 226–27; Hand to Yeats, December 24, 1777, Edward Hand Papers, New York Public Library.

18. Calloway, *American Revolution*, 28–29; Troy O. Bickham, *Savages within the Empire: Representations of American Indians in Eighteenth-Century Britain* (Oxford: Clarendon, 2005), chap. 7.

19. Paul B. Moyer, *Wild Yankees: The Struggle for Independence along Pennsylvania's Revolutionary Frontier* (Ithaca, NY: Cornell University Press, 2007), 14–32; Gregory T. Knouff, *The Soldiers' Revolution: Pennsylvanians in Arms and the Forging of Early American Identity* (University Park: Pennsylvania State University Press, 2004), 65–66.

20. Unless otherwise noted, the basic narrative is derived from Graymont, *Iroquois*; Fischer, *Well-Executed Failure*; Mintz, *Seeds of Empire*; Isabel Thompson Kelsay, *Joseph Brant, 1743–1807, Man of Two Worlds* (Syracuse, NY: Syracuse University Press, 1984), 214–71; Glatthaar and Martin, *Forgotten Allies*, 239–56.

21. Graymont, *Iroquois*, 156. Graymont is referring to Cherry Valley, but the same idea applied to the Wyoming Valley, with perhaps an even more personal intensity for many loyalists.

22. Graymont, *Iroquois*, 168–74. The attack is retold in many sources; Graymont's is a balanced account.

23. Luke Swetland, *A Narrative of the Captivity of Luke Swetland in 1778 and 1779, among the Seneca Indians* (Waterville, NY: James J. Guernsey, 1875).

24. Graymont, *Iroquois*, 186–91 (quotation on 190).

25. Fischer, *Well-Executed Failure*, 31, 36; Graymont, *Iroquois*, 192–94.

26. This is an estimate for all six nations of the Iroquois, even though not all of them joined the British cause. The estimate is based on a 1779 count of warriors totaling 1,600. Bruce W. Trigger, *Handbook of North American Indians*, vol. 15, *Northeast* (Washington, DC: Smithsonian Institution, 1978), 421, 527. Multiplying by 4 (a common multiplier for tribal societies) generates 6,400 persons, which may still be a bit low. At the end of the campaign some 5,036 Indian refugees crowded around Niagara for British help (although not all were Iroquois). Graymont, *Iroquois*, 220.

27. Fischer, *Well-Executed Failure*, 43.

28. Campfield's journal, in Cook, *Journals*, 58; Barton's Journal, in Cook, *Journals*, 9, 10.

29. Continental troops failed in their attempts to surprise Chemung, eight towns around Bucktooth (today's Salamanca, New York), Catherine's Town,

Canadesaga, Canandaigua, Unadilla, Oquaga, Tioga, and a Delaware village near Fort Pitt. See Mintz, *Seeds of Empire*, 68–71, 84–85, 116, 129, 136, 138; Graymont, *Iroquois*, 196.

30. Fischer, *Well-Executed Failure*, 57, 104.

31. PNG 3:144–45.

32. For a discussion of the problem of security detachments, see WGW 15:372. For the logistical problems faced by European armies in North America, see Matthew C. Ward, "'The European Method of Warring Is Not Practiced Here': The Failure of British Military Policy in the Ohio Valley, 1755–1759," *War in History* 4 (1997): 255–59.

33. James Clinton to Hand, April [?], 1779, Edward Hand Papers, New York Public Library.

34. Journal of Edward Hand, Edward Hand Papers, New York Public Library.

35. Hand to Yeats, August 26, 1779, Edward Hand Papers, New York Public Library. Afterwards Hand may have felt that some cause for laughter remained. He referred to the campaign as a "peregrination" through Seneca lands and noted that they would soon rejoin Washington's army, "where we expect to turn our Cornstalks into Laurels." Hand to Yeats, October 21, 1779, Edward Hand Papers, New York Public Library. Daniel Clapp also discusses the poor quality of the provisions and the expedients taken to make it edible. Clapp's diary, Ayer Manuscript Collection 162, Newberry Library, Chicago.

36. Sproule's diary, 54, 58; Beekman's diary, 134–35.

37. Fischer, *Well-Executed Failure*, 47–50, 102–28.

38. Fischer, *Well-Executed Failure*, 31, 36, 58.

39. Sullivan, *Letters and Papers*, 48–53.

40. Maryly B. Penrose, comp., *Indian Affairs Papers, American Revolution* (Franklin Park, NJ: Liberty Bell Associates, 1981), 186–90.

41. Penrose, *Indian Affairs Papers*, 190.

42. WGW 14:271.

43. Fogg's journal, in Cook, *Journals*, 98.

44. John Grenier, *The First Way of War: American War Making on the Frontier, 1607–1814* (Cambridge: Cambridge University Press, 2005). See also William L. Shea, *The Virginia Militia in the Seventeenth Century* (Baton Rouge: Louisiana State University Press, 1983), 20, 33; J. Frederick Fausz, "Patterns of Anglo-Indian Aggression and Accommodation along the Mid-Atlantic Coast, 1584–1634," in *Cultures in Contact: The Impact of European Contacts on Native American Cultural Institutions, A.D. 1000–1800*, ed. William W. Fitzhugh (Washington, DC: Smithsonian Institution, 1985), 246; Lee, *Crowds and Soldiers*, 121; Colin G. Calloway, *New Worlds for All: Indians, Europeans, and the Remaking of Early America* (Baltimore: Johns Hopkins University Press, 1997), 109.

45. See, for example, the French use of devastation in John A. Lynn, "A Brutal Necessity? The Devastation of the Palatinate, 1688–1689," in *Civilians in the Path of War*, ed. Mark Grimsley and Clifford J. Rogers (Lincoln: University of Nebraska Press, 2002), 79–110.

46. Emer de Vattel, *The Law of Nations, or, Principles of the Law of Nature, Applied to the Conduct and Affairs of Nations and Sovereigns, with Three Early Essays on the Origin and Nature of Natural Law and on Luxury*, ed. Béla Kapossy and Richard Whatmore; trans. Thomas Nugent (Indianapolis, IN: Liberty Fund, 2008), 544.

47. Bertram Wyatt-Brown, *Southern Honor: Ethics and Behavior in the Old South* (New York: Oxford University Press, 1982), xv, 34. See also Richard Maxwell Brown, *No Duty to Retreat: Violence and Values in American History and Society* (New York: Oxford University Press, 1991); Brown, *Strain of Violence: Historical Studies of American Violence and Vigilantism* (New York: Oxford University Press, 1975); David Hackett Fischer, *Albion's Seed: Four British Folkways in America* (New York: Oxford University Press, 1989), 765–68. Quotation is from "Sermon against Plundering," Samuel E. McCorkle Papers, Duke University Special Collections, Durham, NC.

48. Samuel Von Pufendorf, *De jure naturae et gentium libri octo* (New York: Oceana, 1964), 3:1298; Harry M. Ward, *Between the Lines: Banditti of the American Revolution* (Westport, CT: Praeger, 2002), 78, 126; Lee, *Crowds and Soldiers*, 109, 193–97, 276n27. For the carefully calculated equivalencies discussed by the antagonists at higher political levels, see Charles H. Metzger, *The Prisoner in the American Revolution* (Chicago: Loyola University Press, 1971), 154–58; Betsy Knight, "Prisoner Exchange and Parole in the American Revolution," *William and Mary Quarterly*, 3rd ser., 48 (1991): 201–22; JCC 5:458.

49. Armstrong Starkey, *European and Native American Warfare, 1675–1815* (Norman: University of Oklahoma Press, 1998), 122–23; Graymont, *Iroquois*, 189–90.

50. Gore's diary, 716, 720. James Crawford, whose family had been on the frontiers, also reactivated his Continental commission to accompany the expedition. Crawford to Hand, June 13, 1779, Edward Hand Papers, New York Public Library.

51. Quoted in Mintz, *Seeds of Empire*, 98.

52. Mintz, *Seeds of Empire*, 97; quotation from Clapp's diary, Ayer Manuscript Collection 162, Newberry Library, Chicago.

53. First quotation from Sproule's Diary, 62–63; second quotation from Clapp's diary, Ayer Manuscript Collection 162, Newberry Library, Chicago. Other sample descriptions of Boyd's fate are in the journals of Barton, Beatty, and Van Hovenburgh, all in Cook, *Journals*, 11, 32, 281. These are merely samples. Other Continentals were killed and scalped during some of the small harassing-fire incidents. See Barton's journal, in Cook, *Journals*, 6, 11.

54. For accounts that mention the old woman during the return march, see Journals of Barton and Beatty in Cook, 9, 12, 28, 33.

55. Barton's journal, in Cook, *Journals*, 13; Dearborn's journal, in Cook, *Journals*, 77.

56. Quoted in Calloway, *American Revolution*, 53; Governor Blacksnake, *Chainbreaker's War: A Seneca Chief Remembers the American Revolution*, ed. Jeanne Winson Adler (Hensonville, NY: Black Dome, 2002), 98.

57. Quoted in editor's introduction to William Wood, *New England's Prospect*, ed. Alden T. Vaughan (Amherst: University of Massachusetts Press, 1977), 12.

58. Hand to Jasper Yeats, October 2, 1777; Hand to Jasper Yeats, March 7, 1778, in Edward Hand Papers, New York Public Library.

59. Lee, *Crowds and Soldiers*, 117–29; Adam J. Hirsch, "The Collision of Military Cultures in Seventeenth-Century New England," *Journal of American History* 74 (1988): 1187–1212; Alden T. Vaughan, "'Expulsion of the Salvages': English Policy and the Virginia Massacre of 1622," *William and Mary Quarterly*, 3rd ser., 35 (1978): 57–84; Jill Lepore, *The Name of War: King Philip's War and the Origins of American Identity* (New York: Knopf, 1998), 10, 16, 45; Merritt, *At the Crossroads*, 11–12, passim.

60. Fogg's journal, in Cook, *Journals*, 98.

61. Loyalists were, however, occasionally forced to move, and of course the revolutionary governments confiscated large amounts of loyalist property. Wallace Brown, *The Good Americans: The Loyalists in the American Revolution* (New York: Morrow, 1969), 127.

62. A manuscript manual on war in North America written in the 1760s advocated a plan closely parallel to Washington's advice to Sullivan. The author suggested that the villages be destroyed and the women and children captured to draw out the men. "On War in North America," Amherst Papers, WO 34/102, f. 13, National Archives, Kew, UK. There were exceptions to this exclusivity. Benjamin Franklin and other American leaders were excited by Lafayette's suggestion that a daring naval commander could take hostages who were from the upper ranks of London society by raids on the English coast. John Ferling, *Almost a Miracle: The American Victory in the War of Independence* (Oxford: Oxford University Press, 2007), 359–60, 372.

63. Quoted in White, *Middle Ground*, 391.

64. Vattel, *Law of Nations*, 570; Francis Bland, *The Souldiers March to Salvation* (York: 1647), 26. In contrast, John Barnwell's 1712 march of destruction through the Tuscarora towns of North Carolina deliberately spared the fruit trees. John Barnwell, "Journal of John Barnwell," *Virginia Magazine of History and Biography* 5 (1898): 396.

65. William Gordon, *The History of the Rise, Progress, and Establishment of the Independence of the United States of America* (London: 1788), 3:311.

66. Campfield's journal, in Cook, *Journals*, 54.

67. Mintz, *Seeds of Empire*, 151–52; Graymont, *Iroquois*, 219–20.

68. Nathaniel B. Shurtleff, ed., *Records of the Colony of New Plymouth*, vol. 5 (Boston: William White, 1856), 205.

69. John Barnwell, "Journal of John Barnwell," *Virginia Magazine of History and Biography* 6 (1899): 54.

70. P. J. Marshall, "A Nation Defined by Empire, 1755–1776," in *Uniting the Kingdom? The Making of British History*, ed. Alexander Grant and Keith J. Stringer (London: Routledge, 1995), 209–11, 214–15, 221; Marshall, *The Making and Unmaking of Empires: Britain, India, and America c. 1750–1783* (Oxford: Oxford University Press, 2005), 190–93, 340; Griffin, *American Leviathan*, 19–31, 254–55. This relatively charitable interpretation of British intentions toward the Indians as subjects is

debated. See Dowd, *War Under Heaven*, 178–90. Jenny Hale Pulsipher also outlines the conflict between local authorities and public opinion over the legal rights of Christianized Indians in Massachusetts during King Philip's War. *Subjects Unto the Same King: Indians, English, and the Contest for Authority in Colonial New England* (Philadelphia: University of Pennsylvania Press, 2005), chap. 6. For the September 1760 treaty, see Anthony Pagden, *Lords of All the World: Ideologies of Empire in Spain, Britain and France c. 1500–c. 1800* (New Haven, CT: Yale University Press, 1995), 89.

71. Mintz, *Seeds of Empire*, 114, 186.

72. Knouff, *Soldiers' Revolution*, 65.

73. Hand to Yeats, May 19, 1779; Hand to Yeats July 20, 1779; Hand to Yeats, October 21, 1779, all in Edward Hand Papers, New York Public Library.

74. David Day, *Conquest: How Societies Overwhelm Others* (Oxford: Oxford University Press, 2008), 6 and passim; Ben Kiernan, *Blood and Soil: A World History of Genocide and Extermination from Sparta to Darfur* (New Haven, CT: Yale University Press, 2007), 213–48.

Conclusion

1. Quotations from Sir William Waller, *Vindication of the Character and Conduct of Sir William Waller, Knight; Commander in Chief of the Parliament Forces in the West: Explanatory of His Conduct in Taking Up Arms Against King Charles the First* (London: J. Debrett, 1793), xiii–xiv, 7; John Birch, *Military Memoir of Colonel John Birch, Sometime Governor of Hereford in the Civil War between Charles I. and the Parliament*, ed. T. W. Webb (Westminster, UK: Camden Society, 1873), 217; Mark Grimsley, *The Hard Hand of War: Union Military Policy Toward Southern Civilians, 1861–1865* (Cambridge: Cambridge University Press, 1995), 62–63; Joseph T. Glatthaar, *Partners in Command: The Relationships between Leaders in the Civil War* (New York: Free Press, 1994), 154; PNG 2:283

2. Geoffrey Best, *Humanity in Warfare* (New York: Columbia University Press, 1980), 148–51; Leon Friedman, ed., *The Law of War: A Documentary History* (New York: Random House, 1972), 1:156–57; Ian Brownlie, *International Law and the Use of Force by States* (Oxford: Clarendon, 1963), 19–25.

3. Richard Shelly Hartigan, *Lieber's Code and the Law of War* (Chicago: Precedent, 1983), 15–16, 48–50; Burrus M. Carnahan, "Lincoln, Leiber and the Laws of War: The Origins and Limits of the Principle of Military Necessity," *The American Journal of International Law* 92 (1998): 213–31. Halleck also wrote late in the war on the necessity of retaliation, but also on the limits thereon. Henry Wager Halleck, "Retaliation in War," *American Journal of International Law* 6 (1912): 107–18.

4. Edward M. Coffman, *The Old Army: A Portrait of the American Army in Peacetime, 1784–1898* (New York: Oxford University Press, 1986).

5. Timothy D. Johnson, *Winfield Scott: The Quest for Military Glory* (Lawrence: University Press of Kansas, 1998), 168–69, 189–94, 207; Brian McAllister Linn, *The Echo of Battle: The Army's Way of War* (Cambridge, MA: Harvard University Press, 2007), 75.

6. Earl J. Hess, *The Union Soldier in Battle: Enduring the Ordeal of Combat* (Lawrence: University Press of Kansas, 1997), 56–58. For the continued significance of drill and training, see Mark A. Weitz, "Drill, Training, and the Combat Performance of the Civil War Soldier: Dispelling the Myth of the Poor Soldier, Great Fighter," *Journal of Military History* 62 (1998): 263–90. In contrast to these expectations, the Civil War soldier proved intensely resistant to the culture of discipline outside the immediate battlefield environment. Reid Mitchell, *The Vacant Chair: The Northern Soldier Leaves Home* (New York: Oxford University Press, 1993), 21–2; Gerald F. Linderman, *Embattled Courage: The Experience of Combat in the American Civil War* (New York: Free Press, 1987), 43; Joseph T. Glatthaar, *General Lee's Army: From Victory to Collapse* (New York: Free Press, 2008), 176–79, 271–72.

7. The actual decisiveness of these battles has been challenged by Irving W. Levinson, "A New Paradigm for an Old Conflict: The Mexico-United States War," *Journal of Military History* 73 (2009): 393–416.

8. Peter Paret, ed., *Makers of Modern Strategy: From Machiavelli to the Nuclear Age* (Princeton, NJ: Princeton University Press, 1986), 130–31, 138, 154–55, 180, 414–15, 424–26; Johnson, *Winfield Scott*, 234–35. For early war conciliatory strategies, and the focus on Richmond, see Glatthaar, *Partners in Command*, 59, 81–82; George C. Bradley and Richard L. Dahlen, *From Conciliation to Conquest: The Sack of Athens and the Court-Martial of Colonel John B. Turchin* (Tuscaloosa: University of Alabama Press, 2006); Ethan S. Rafuse, *McClellan's War: The Failure of Moderation in the Struggle for the Union* (Bloomington: Indiana University Press, 2005); Russell F. Weigley, *A Great Civil War: A Military and Political History, 1861–1865* (Bloomington: Indiana University Press, 2000), 29–32.

9. Gary W. Gallagher reviews the older literature and the interpretation followed here in *Lee and His Army in Confederate History* (Chapel Hill: University of North Carolina Press, 2001), 151–90, esp. 159–60, 171–78. See also Emory N. Thomas, "Rebellion and Conventional Warfare: Confederate Strategy and Military Policy," in *Writing the Civil War: The Quest to Understand*, ed. James M. McPherson and William J. Cooper Jr. (Columbia: University of South Carolina Press, 1998); Glatthaar, *Partners in Command*, 21, 41; Donald Stoker and Joseph G. Dawson III, "Forum: Confederate Military Strategy in the U.S. Civil War," *Journal of Military History* 73 (2009): 571–613 (esp. 587–88). Thanks to Mark Grimsley for discussions on this issue.

10. Stephen Conway, *The War of American Independence 1775–1783* (London: Edward Arnold, 1995).

11. David A. Bell, *The First Total War: Napoleon's Europe and the Birth of Warfare as We Know It* (Boston: Houghton Mifflin, 2007).

12. Many of these numbers are presented in earlier chapters. The figure for the American population in 1780 comes from U.S. Bureau of the Census, *Historical Statistics of the United States: Colonial Times to 1970* (White Plains, NY: Kraus, 1989), 2:1–19, table Z.

13. That potential had been frightfully realized during the American Revolution and had been noted across Europe during the Spanish guerilla war against

Napoleon's rule. See note 118 in chapter 7; Charles J. Esdaile, *Fighting Napoleon: Guerrillas, Bandits, and Adventurers in Spain, 1808–1814* (New Haven, CT: Yale University Press, 2004).

14. Examples include Robert R. Mackey, *The Uncivil War: Irregular Warfare in the Upper South, 1861–1865* (Norman: University of Oklahoma Press, 2004); Noel C. Fisher, *War at Every Door: Partisan Politics and Guerrilla Violence in East Tennessee, 1860–1869* (Chapel Hill: University of North Carolina Press, 1997); Clay Mountcastle, *Punitive War: Confederate Guerrillas and Union Reprisals* (Lawrence: University Press of Kansas, 2009).

15. Hartigan, *Lieber's Code*, 2, 31–44, 92; Scott Reynolds Nelson and Carol Sheriff, *A People at War: Civilians and Soldiers in America's Civil War, 1854–1877* (New York: Oxford University Press, 2007), 148–59.

16. Joseph T. Glatthaar, *Forged in Battle: The Civil War Alliance of Black Soldiers and White Officers* (New York: Meridian Books, 1991), 155–57, 203; Glatthaar, *The March to the Sea and Beyond: Sherman's Troops in the Savannah and Carolinas Campaigns* (Baton Rouge: Louisiana State University Press, 1995), 40–80; John Gauss, *Black Flag! Black Flag!: The Battle at Fort Pillow* (Lanham, MD: University Press of America, 2003); James M. McPherson, *For Cause and Comrades: Why Men Fought in the Civil War* (New York: Oxford University Press, 1997), 152–54.

17. Quotation and analysis are from Glatthaar, *Partners in Command*, 144 (quotation), 154, 204–5, 208; and Benjamin Franklin Cooling, *Fort Donelson's Legacy: War and Society in Kentucky and Tennessee, 1862–1863* (Knoxville: University of Tennessee Press, 1997).

18. Sherman to Halleck, December 24, 1864, in *The War of the Rebellion: A Compilation of the Official Records of the Union and Confederate Armies* (Washington, DC: GPO, 1880–1901), ser. 1, 44:799.

19. Grimsley, *Hard Hand of War*.

20. For works that emphasize vindictiveness and violence, see Charles Royster, *The Destructive War: William Tecumseh Sherman, Stonewall Jackson, and the Americans* (New York: Knopf, 1991); Harry S. Stout, *Upon the Altar of the Nation: A Moral History of the American Civil War* (New York: Viking, 2006). Grimsley traces the southern myth of Sherman's march and Union violence in general. *Hard Hand of War*, 219–20.

21. Grimsley's *Hard Hand of War* is supported by Glatthaar, *March to the Sea*, 66–80, 119–55; Mitchell, *Vacant Chair*, 97–113; Mark E. Neely Jr., *The Civil War and the Limits of Destruction* (Cambridge, MA: Harvard University Press, 2007); Neely, "Was the Civil War a Total War?" *Civil War History* 37 (1991): 5–28; Noah Andre Trudeau, *Southern Storm: Sherman's March to the Sea* (New York: Harper, 2008).

22. In addition to the sources listed in notes 14 and 20, see George S. Burkhardt, *Confederate Rage, Yankee Wrath: No Quarter in the Civil War* (Carbondale: Southern Illinois University Press, 2007); Walter Brian Cisco, *War Crimes against Southern Civilians* (Gretna, LA: Pelican, 2007); Linderman, *Embattled Courage*, 180–201; Lonnie R. Speer, *War of Vengeance: Acts of Retaliation against Civil War POWs* (Mechanicsburg, PA: Stackpole Books, 2002).

23. Glatthaar, *General Lee's Army*, 174–85; Mark A. Weitz, "Shoot Them All: Chivalry, Honour and the Confederate Army Officer Corps," in *The Chivalric Ethos and the Development of Military Professionalism*, ed. D. J. B. Trim (Leiden, The Netherlands: Brill, 2003), 321–47; McPherson, *For Cause and Comrades*, 74, 150.

24. Holly Brewer has pointed out to me that Christianity played a key role in this conceptualization: the oath of allegiance required of a subject was one made to God. Inability to swear a Christian oath undermined their ability to be seen as subjects. For the Irish, the problem was not so much the Irishmen's Christianity as their presumed primary loyalty to the pope instead of the queen.

25. The colonial "imperial" wars with France could be considered normal international wars, but the Americans experienced them primarily as wars against Indians, not against France as such.

26. Mark A. Grimsley, "'Rebels' and 'Redskins': U.S. Military Conduct toward White Southerners and Native Americans in Comparative Perspective," in *Civilians in the Path of War*, ed. Mark Grimsley and Clifford J. Rogers (Lincoln: University of Nebraska Press, 2002); Lance Janda, "Shutting the Gates of Mercy: The American Origins of Total War, 1860–1880," *Journal of Military History* 59 (January 1995), 7–26; Robert M. Utley, "Cultural Clash on the Western North American Frontier: Military Implications," in *The Military and Conflict Between Cultures: Soldiers at the Interface*, ed. James C. Bradford (College Station: Texas A&M University Press, 1997), 91–108; Brian McAllister Linn, *The Philippine War, 1899–1902* (Lawrence: University Press of Kansas, 2000), esp. 219–24, 322–28. John W. Dower, *War Without Mercy: Race and Power in the Pacific War* (New York: Pantheon Books, 1986); Peter Schrijvers, *The GI War Against Japan: American Soldiers in Asia and the Pacific During World War II* (Basingstoke, UK: Plagrave, 2002); Craig M. Cameron, *American Samurai: Myth, Imagination, and the Conduct of Battle in the First Marine Division, 1941–1951* (Cambridge: Cambridge University Press, 1994); John A. Lynn, *Battle: A History of Combat and Culture* (Boulder, CO: Westview, 2003), 219–80.

27. Stephen C. Neff, *War and the Law of Nations: A General History* (Cambridge: Cambridge University Press, 2005).

28. Jonathan B. A. Bailey, "Military History and the Pathology of Lessons Learned: The Russo-Japanese War, a Case Study," in *The Past as Prologue: The Importance of History to the Military Profession*, ed. Williamson Murray and Richard Hart Sinnreich (Cambridge: Cambridge University Press, 2006), 181–83; Sharon Korman, *The Right of Conquest: The Acquisition of Territory by Force in International Law and Practice* (Oxford: Clarendon, 1996), 29–40; Karma Nabulsi, *Traditions of War: Occupation, Resistance and the Law* (Oxford: Oxford University Press, 2000), passim, esp. 184–85, 199–204, 213, 220–27; Bell, *First Total War*, 77, 82, 115; David Kaiser, *Politics and War: European Conflict from Philip II to Hitler* (Cambridge, MA: Harvard University Press, 1990), 272–73.

29. Quoted in Rudolf Stadelmann, *Moltke und der Staat* (Krefeld, West Germany: Sherpe-Verlag, 1950), 244, 258 (translation with assistance from Karen Hagemann).

30. Quoted in Hew Strachan, *The First World War* (New York: Viking, 2004), 26.

31. Strachan, *First World War*, 215; Roger Chickering, "Militärgeschichte als Totalgeschichte im Zeitalter des totalen Krieges," in *Was ist Militärgeschichte?*, ed. Thomas Kühne and Benjamin Ziemann (Paderborn, Germany: Schüningh, 2000), 306; Chickering, "Total War: The Use and Abuse of a Concept," in *Anticipating Total War: The German and American Experiences, 1871–1914*, ed. Manfred F. Boemeke, Roger Chickering, and Stig Förster (Washington, DC: German Historical Institute, 1999), 26–27; Alexander B. Downes, *Targeting Civilians in War* (Ithaca, NY: Cornell University Press, 2008).

Index

Made in the USA
Coppell, TX
31 January 2022

72655694R00208